Geometry of Submanifolds

CONTEMPORARY MATHEMATICS

756

Geometry of Submanifolds

AMS Special Session on Geometry of Submanifolds
in Honor of Bang-Yen Chen's 75th Birthday
October 20–21, 2018
University of Michigan, Ann Arbor, Michigan

Joeri Van der Veken
Alfonso Carriazo
Ivko Dimitrić
Yun Myung Oh
Bogdan D. Suceavă
Luc Vrancken
Editors

AMS AMERICAN MATHEMATICAL SOCIETY

The photographs in this book are reprinted with permission.

2010 *Mathematics Subject Classification*. Primary 53A10, 53A55, 53B20, 53B21, 53B25, 53B35, 53C25, 53C40. 53C42, 53C43.

Library of Congress Cataloging-in-Publication Data

Names: Veken, Joeri Van der, editor.
Title: Geometry of submanifolds : AMS special session in honor of Bang-Yen Chen's 75th birthday, October 20–21, 2018, University of Michigan, Ann Arbor, Michigan / Joeri Van der Veken, Alfonso Carriazo, Ivko Dimitrić, Yung Myung Oh, Bogdan D. Suceavă, Luc Vrancken, editors.
Description: Providence, Rhode Island : American Mathematical Society, 2020. | Series: Contemporary mathematics, 0271-4132 ; volume 756 | Includes bibliographical references. |
Identifiers: LCCN 2019059344 | ISBN 9781470450922 (paperback) | ISBN 9781470456665 (ebook)
Subjects: LCSH: Submanifolds. | Geometry, Differential. | AMS: Differential geometry – Classical differential geometry – Minimal surfaces, surfaces with ... | Classical differential geometry – Differential invariants – Local differential geometry – Local Riemannian geometry. | Differential geometry – Local differential geometry – Methods of Riemannian geometry. | Differential geometry – Local differential geometry – Local submanifolds. | Differential geometry – Local differential geometry – Hermitian and Kahlerian... . | Differential geometry – Global differential geometry.
Classification: LCC QA649 .G48 2020 | DDC 516.3/6–dc23
LC record available at https://lccn.loc.gov/2019059344
Contemporary Mathematics ISSN: 0271-4132 (print); ISSN: 1098-3627 (online)
DOI: https://doi.org/10.1090/conm/756

---

10 9 8 7 6 5 4 3 2 1 25 24 23 22 21 20

Dedicated to Professor Bang-Yen Chen, in honor of his 75th birthday

Contents

Foreword by the Editors

About a century ago, the geometry of submanifolds gained a lot of momentum through the study of the Schläfli's conjecture, which stated that a real analytic Riemannian manifold of dimension n can be locally isometrically embedded into any real analytic Riemannian manifold of dimension $\frac{1}{2}n(n+1)$. M. Janet (1926), É. Cartan (1927) and C. Burstin (1931) made essential contributions to the understanding of the importance of the immersion problems and to a result that today bears their names.

On behalf of the organizers of the AMS Special Session in Bang-Yen Chen's honor, Luc Vrancken presents Bang-Yen Chen with a plaque. The text on the plaque reads: *A man is to be judged not just by his own accomplishments, but by those he has inspired in others.*

A major development for the theory was the much celebrated Embedding Theorem, proved by John Forbes Nash, Jr. (in a series of three papers published in 1954, 1956, and 1966). Nash's work received outstanding recognition from the mathematical community, including his election as a fellow of the American Mathematical Society in 2012, and the Abel Prize in 2015. After the important moment represented by the Nash's Embedding Theorem, many outstanding mathematicians, as e.g. Shiing-Shen Chern (2011–2004), Manfredo Perdigão do Carmo (1928–2018), or Shoshichi Kobayashi (1932–2012) directed some of their best efforts towards the geometry of submanifolds. To better shed a light on the geometry of submanifolds and its historical development, the editors of the present volume invited Leopold Verstraelen to write an essay describing from a broad perspective his views on this important chapter of mathematics.

The development of geometry of submanifolds benefited greatly from Bang-Yen Chen's contributions. Bang-Yen Chen is the author of the first monograph titled *Geometry of Submanifolds*, published in 1973 by Marcel Dekker, a work cited by many mathematicians since its publication. In 2019, the Dover paperback edition of this fundamental volume was published and is available today. The editors of the present volume share a common experience, as they have been attracted by this domain of differential geometry by reading and studying Bang-Yen Chen's *Geometry of Submanifolds.*

After 1973, many authors became interested by the geometry of submanifolds. Richard S. Palais and Chuu-Lian Terng wrote a monograph titled *Critical Point Theory and Submanifold Geometry*, published by Springer-Verlag in 1988, in their prestigious Lecture Notes in Mathematics series. In their *Introduction* to their volume titled *Submanifolds and Holonomy* (published in its first edition in 2003), Jürgen Berndt, Sergio Console, and Carlos Enrique Olmos wrote that "the study of submanifolds in Euclidean space has a long tradition, and many beautiful results and theories emerge from it".

The present volume is, therefore, a celebration of the geometry of submanifolds and its connections with other areas of mathematics. It was a fortunate coincidence that in the month when Bang-Yen Chen turned 75, the American Mathematical Society scheduled their Fall Central Sectional Meeting with number 1143 in Ann Arbor, at the University of Michigan, in the proximity of East Lansing, where Bang-Yen Chen spent most of his academic career, with Michigan State University. Motivated by their gratitude towards a researcher who inspired their work, the editors of the present volume proposed an *AMS Special Session on Geometry of Submanifolds, in Honor of Bang-Yen Chen's 75th Birthday.* The event took place on October 20-21, 2018.

For a more complete perspective of the academic context in which this volume was prepared, it is useful to remind here a few biographical pieces of information. Bang-Yen Chen was born on October 3, 1943, in Toucheng Township, Yilan County, Taiwan. He received his B.S. from Tamkang University in 1965 and his M.S. from National Tsing Hua University in 1967. He obtained his Ph.D. degree from University of Notre Dame in 1970 under the supervision of Tadashi Nagano. Before his graduate years, Bang-Yen Chen taught at Tamkang University between 1966 and 1968, and at the National Tsing Hua University during the academic year 1967-1968. After his doctoral years (1968-1970) at University of Notre Dame, Bang-Yen Chen joined the faculty at Michigan State University as a research associate from

David E. Blair (left), Bang-Yen Chen (center), and Yun Myung Oh (right). Yun Myung Oh defended her doctoral dissertation under Bang-Yen Chen's supervision at Michigan State University in 2000. The title of her thesis was *Explicit Construction of Lagrangian Isometric Immersion of a Real Space Form* $M^n(c)$ *into a Complex Space Form* $M^n(4c)$.

1970-1972, then became associate professor in 1972, and full professor in 1976. He was presented with the title of University Distinguished Professor in 1990. After 2012, he became University Distinguished Professor Emeritus. Bang-Yen Chen is the author of over 500 works including 12 books, mainly in differential geometry and related subjects. His works have been cited over 25,000 times.

Three of the editors of the present volume, Ivko Dimitrić, Yun Myung Oh, and Bogdan Suceavă are his former doctoral students at Michigan State University. Bang-Yen Chen also directed Susumu Ishikawa's thesis at Kyushu University, as well as several other dissertations at Michigan State University, for Young Ho Kim, Yhuji Shibuya, Yoshihiko Tazawa, and Jie Yang.

Oscar Garay presenting his contribution titled *Binormal evolution surfaces swept out by elastica-like extremals.* During his intervention, he recalled his first academic visit at Michigan State University, at the beginning of the 1980s, when the geometry of submanifolds of finite type received a lot of attention from many geometers.

Knowing very well how valuable Dr. Bang-Yen Chen's personal and academic experience is, the editors extended him the invitation to contribute to this volume with an essay describing his most valuable experiences. Bang-Yen Chen wrote the essay *My Education in Differential Geometry and My Indebtedness*, hereby included, where he described his academic interactions with his mathematical mentors Tadashi Nagano, Shiing-Shen Chern, Tominosuke Otsuki, and Kentaro Yano.

Bang-Yen Chen is today known for several fundamental ideas in differential geometry, e.g., Chen inequalities, Chen invariants, Chen's conjectures, Chen surface, Chen-Ricci inequality, Chen submanifolds, Chen equality, submanifolds of finite type, the (M_+, M_-) method for compact symmetric spaces and the 2-numbers of Riemannian manifolds (this last concept based on a joint contribution with Tadashi

Bang-Yen Chen discussing with Ye-Lin Ou. In the background: Patrick J. Ryan (left) and Tim Buttsworth (right).

Nagano), and slant submanifolds, to mention just a few. In 1991, Bang-Yen Chen formulated the biharmonic conjecture, which claims that minimal submanifolds are the only biharmonic submanifolds in Euclidean spaces. Additionally, in 1996 he conjectured that every finite type spherical hypersurface is either of 1-type or of 2-type. At the time when the AMS Special Session took place, Bang-Yen Chen was working with Ye-Lin Ou on a volume titled *Biharmonic submanifolds and biharmonic maps in Riemannian geometry*. This work was published by World Scientific in 2020.

The editors would like to extend their thanks to all the scholars who participated in the AMS Special Session. Their expertise and their interactions have been particularly valuable and interesting. While their papers are not included in the present volume, the contributions of David E. Blair (Michigan State University), Alfonso Carriazo (University of Sevilla, Spain), Oscar J. Garay (University of the Basque Country, Spain), Elsa Ghandour (Université Polytechnique Hauts-de-France, France), Martin Magid (Wellesley College), Timothy James Buttsworth (The University of Queensland, Australia), have been extremely valuable to and tremendously appreciated by the editors. Also, many thanks to the co-authors of the contributors to the special sessions: Paul Baird (University of Brest, France), Martha Patricia Dussan Angulo and A. P. Franco Filho (both from Universidade de São Paulo), and Leonard M. Giugiuc (Colegiul National Traian, Drobeta-Turnu Severin, Romania).

Yusuf Doğru (left), Bang-Yen Chen (center), and Cihan Özgür (right), at the 6th Geometry Symposium in Uludağ University, in Bursa, Turkey, in July 2008

The editors of the present volume express their thanks to Georgia Benkart, who served as AMS Secretaries in the academic year 2018–2019, when the AMS Special Session from Ann Arbor was organized.

While the editors prepared the present volume, their work benefited from the outstanding support and expert consultations of several referees. Without their expertise the quality of the present volume would not be the same. Last, but not least, many thanks to Sergei Gelfand and Christine Thivierge for their editorial guidance and support during the preparation of the present volume.

The participants to the AMS Special Session on Geometry of Submanifolds, in Honor of Bang-Yen Chen's 75th Birthday. Photo taken at the end of the morning session on October 20, 2018

Oscar Garay (left), Bang-Yen Chen (center), and Ivko Dimitrić (right). Ivko Dimitrić defended his doctoral dissertation in 1989 at Michigan State University, under Bang-Yen Chen's coordination. The title of his work is *Quadric Representation and Submanifolds of Finite Type*. Photo taken on October 20, 2018 in Ann Arbor, MI.

Contemporary Mathematics
Volume **756**, 2020
https://doi.org/10.1090/conm/756/15195

Reflections on some research work of Bang-Yen Chen

Joeri Van der Veken, Alfonso Carriazo, Ivko Dimitrić, Yun Myung Oh, Bogdan D. Suceavă, and Luc Vrancken

ABSTRACT. This essay provides a brief sketch of selected mathematical research work of Bang-Yen Chen done during the last fifty years.

1. δ-invariants and basic inequalities

According to Nash's embedding theorem, every Riemannian manifold can be isometrically embedded in some Euclidean space with sufficiently high codimension. Nash's theorem was aimed for in the hope that if a Riemannian manifold could always be regarded as a Euclidean submanifold, this would then yield the opportunity to use extrinsic help. However, this hope had not been materialized for many years. There were several reasons why it is so difficult to apply Nash's theorem. One reason is that it requires very large codimension for a Riemannian manifold to admit an isometric embedding in a Euclidean space. On the other hand, submanifolds of higher codimension are very difficult to understand. Another reason is that at that time there did not exist general optimal relations between known intrinsic invariants and the main extrinsic invariants for arbitrary submanifolds except the three fundamental equations of Gauss, Codazzi and Ricci.

In order to overcome the difficulties mentioned above, Chen introduced in the early 1990s new types of intrinsic invariants, $\delta(n_1, \ldots, n_k)$, different in nature from the "classical" Ricci and scalar curvatures. At the same time he was able to establish universal optimal relations between his new invariants and the main extrinsic invariants for Riemannian submanifolds. The corresponding optimal inequalities for Lagrangian submanifolds in complex space forms involving δ-invariants were established and studied in his joint articles with F. Dillen, J. Van der Veken, L. Verstraelen and L. Vrancken in [**57**–**59**] among others. Submanifolds satisfying the equality case of the above inequalities are called *ideal*. Similar type inequalitites for submanifolds in various ambient spaces have also been studied by many authors.

The δ-invariants and their corresponding inequalities have many applications to several areas in mathematics (cf. e.g., [**16, 19, 20, 31, 39**]). For example, it was

---

2010 *Mathematics Subject Classification.* Primary 53-03, 53B20, 53B25 ; Secondary 53A15, 53A30,53A55, 53B35, 53C15, 53C25, 53C35,53C40, 53C42, 53C43, 53C55, 53C80, 53C99, 91B99.

Key words and phrases. δ-invariants and basic inequalities, compact symmetric spaces, Chen and Nagano's two-numbers, finite type theory for maps and immersions, Kaehler manifolds, spectral geometry, Riemannian submersions, minimal submanifolds, affine hypersurfaces.

Bogdan Suceavă (left) and Bang-Yen Chen (right), in Bucharest, Romania, in 2008. Bogdan Suceavă defended his doctoral dissertation *New Riemannian and Kählerian Curvature Invariants and Strongly Minimal Submanifolds* in 2002 at Michigan State University, under Bang-Yen Chen's coordination.

applied in [**49**] to prove that if $\pi : M \to N$ is a covering map between two compact irreducible homogeneous spaces and if the first eigenvalues of the Laplacian satisfy $\lambda_1(M) \neq \lambda_1(N)$, then N doesn't admit an ideal embedding in any Euclidean space regardless of codimension, although M may.

2. A new approach to compact symmetric spaces

A new approach to compact symmetric spaces M, known as the (M_+, M_-)-theory, was introduced by Chen and T. Nagano in a series of their papers [**65**–**68**] during the 1970–1980s. Their approach was based on antipodal points and fixed point sets, which is different from other approaches done by É. Cartan and others. Their approach works as follows: A pair of points $\{o, p\}$ in M is called *antipodal* if there exists a closed geodesic γ in M such that p is the midpoint of γ from o. For each antipodal pair $\{o, p\}$ in a compact symmetric space M, they introduced a pair of orthogonal totally geodesic submanifolds $M_+^o(p), M_-^o(p)$ through p such that $\dim M_+^o(p) + \dim M_-^o(p) = \dim M$ and $rk(M_-^o(p)) = rk(M)$.

The totally geodesic submanifolds M_+'s and M_-'s are called *polars* and *meridians*, respectively. One important property of polars and meridians states that

Bang-Yen Chen (left) and Shihshu Walter Wei (right).

David Blair introducing Bang-Yen Chen, at the AMS Special Session on Recent Advances in the Geometry of Submanifolds: Dedicated to the Memory of Franki Dillen (1963–2013), held from March 14–15, 2015, at Michigan State University, East Lansing, Ml, U.S.A. For this event, Bang-Yen Chen presented his work *A survey on Ricci solitons on Riemannian submanifolds.* Many of the contributions presented in this AMS Special Session are included in the volume 674 of the *Contemporary Mathematics* series.

M is determined (globally) by any pair $(M_+(p), M_-(p))$. If B is a complete totally geodesic submanifold of a compact symmetric space M, then for each pair $(B_+(p), B_-(p))$ of B there exists a pair $(M_+(q), M_-(q))$ of M such that $B_+(p)$ and $B_-(p)$ are totally geodesic submanifolds of $M_+(q)$ and $M_-(q)$, respectively. The same argument applies to the totally geodesic submanifold $B_+(p) \subset M_+(q)$ and to $B_-(p) \subset M_-(q)$ and so on. Thus, begining with $B \subset M$, one obtains strings of conditions.

A polar of o consisting of a single point is called a *pole*. For a pole $\{p\}$ of o, the set $C(o,p)$ consisting of the midpoints of all geodesics joining o and p is called a *centrosome* of $\{o,p\}$. Each connected component of $C(o,p)$ is again a totally geodesic submanifold of M. Polars, meridians and centrosomes play important roles in Chen and Nagano's approach. For instance, centrosomes were used by J. M. Burns in 1992 (Glasgow Math. J) to compute homotopy of compact symmetric spaces. For some applications of their new approach, see, e.g., [**9, 13, 53, 66**].

3. Two-numbers and applications

The notion of two-numbers was introduced in 1982 by Chen and Nagano in [**67**]. A subset S of a Riemannian manifold M is called *antipodal* if any two points in S are antipodal on some closed geodesic connecting them. An antipodal set A_2M in M is called a *maximal antipodal set* if it doesn't lie in a antipodal set of M as a proper subset. The supremum of the cardinality of all maximal antipodal set of M

is called the *two-number* of M, denoted by $\#_2 M$. Clearly, one has $\#_2 M \geq 2$ for every compact Riemannian manifold M, since every compact Riemannian manifold contains a closed geodesic. The two-number $\#_2 M$ is an obstruction to the existence of a totally geodesic embedding $f : B \to M$, since the existence of f implies $\#_2 B \leq \#_2 M$.

The invariant $\#_2 M$ has certain bearings on the topology of M; e.g., Chen and Nagano proved in [**68**] that $\#_2 M$ is equal to the Euler number $\mathcal{X}(M)$ of M, if M is a semisimple Hermitian symmetric space. In general they proved that $\#_2 M \geq \mathcal{X}(M)$ for every compact symmetric space M. Further, M. Takeuchi proved in 1989 (Nagoya Math. J.) that $\#_2 M = \dim H(M; \mathbb{Z}_2)$ holds for any symmetric R-space. The two-numbers also relate closely with group theory. For a group manifold G, Chen and Nagano proved in [**67**] that $\#_2 G = 2^{r_2 G}$, where $r_2 G$ is the 2-rank of G in the sense of A. Borel and J.-P. Serre (1953, Comm. Math. Helv.). As an immediate application, 2-rank of compact connected Lie groups were completely determined in [**68**] via the 2-numbers of Lie groups.

For a detailed survey on two numbers and applications, see [**13, 38, 53**].

4. Finite type theory for maps and immersions

The main objects in algebraic geometry are algebraic varieties. One has the notion of degree for algebraic varieties via their algebraic structures. Although each Riemannian manifold can be realized as a Riemannian submanifold in a Euclidean space according to Nash's embedding theorem, one lacks the notion of degree for Euclidean submanifolds. Inspired by this observation, Chen introduced the notions of order and finite type submanifolds in the late 1970s (cf. e.g., [**7, 11, 12**]).

Let $x : M \to \mathbb{E}^m$ be an isometric immersion of a compact Riemannian manifold M into the Euclidean m-space $\mathbb{E}^m$. Since the eigenvalues of the Laplacian Δ of M form a discrete sequence $0 = \lambda_0 < \lambda_1 < \lambda_2 < \ldots \nearrow \infty$, one has the assoiated spectral decomposition of the immersion given by

$$x = x_0 + \sum_{t=p}^{q} x_t, \quad (\text{in } L^2\text{-sense}), \quad \Delta x_t = \lambda_t x_t, \quad x_t \neq 0, \tag{4.1}$$

where q is either ∞ or an integer $\geq p$. The set $T(x) = \{t > 0 : x_t \neq 0\}$ is called the *order* of x. The smallest p and the supremum q of $T(x)$ are called the *lower* and the *upper order* of x, respectively. And x is said to be of *finite type* if $T(x)$ is a finite set. Just like minimal submanifolds, maps and submanifolds of finite type are characterized by a spectral variation principle (see [**56**]). During the last four decades, many results on maps and submanifolds of finite type have been obtained by many geometers.

A Euclidean submanifold is called *biharmonic* if its immersion x is biharmonic, i.e., $\Delta^2 x = 0$. The study of biharmonic submanifolds was initiated in the middle of the 1980s in Chen's program to understand finite type submanifolds. The long standing Chen's conjecture states that minimal submanifolds are the only biharmonic submanifolds in Euclidean spaces. The study of biharmonic submanifolds is nowadays a very active research subject. In particular many results on biharmonic submanifolds were obtained by many authors in this century. The first results on finite type theory were summarized in [**12**]. For the most recent survey, see [**44**] and [**89**].

Bang-Yen Chen is the guest of Koichi Ogiue's family in Tokyo, in 2001.

5. Kaehler manifolds

5.1. Bochner-Kaehler metrics. A Kaehler metric g on a complex manifold is called *Bochner-Kaehler* if its Bochner curvature tensor vanishes. Bochner-Kaehler metrics (BK metrics) on complex surfaces are exactly self-dual Kaehler metrics. Using algebraic geometric method, Chen classified compact complex surfaces admitting BK metrics in 1978, see [**5**]. The same result in terms of self-dual Kaehler surfaces was obtained independently by J.-P. Bourguignon (1981), A. Derdzinski (1983), and M. Itoh (1984) via different methods. Some results on BK metrics for higher-dimensional complex manifolds were obtained in [**64**, **69**].

5.2. Chern numbers. Chen and Ogiue [**71**] proved in 1975 the following sharp inequality for a compact Einstein-Kaehler n-manifold M:

$$(nc_1(M)^2 - 2(n+1)c_2(M)) \wedge \omega^{n-2} \leq 0,$$

with equality holding if and only if M is a complex space form, where ω is the cohomology class of the Kaehler 2-form of M. This was extended by M. Lübke (1982, Math. Ann.) to Einstein-Hermitian vector bundles.

5.3. Stablility. In 1981, Chen, P.F. Leung and T. Nagano established the second variational formula of compact Lagrangian submanifolds in Kaehler manifolds. They applied it to prove the following stablility results (cf. [**9**]).

Let $f : M \to \tilde{M}$ be a minimal Lagrangian immersion of a compact manifold M into a Kaehler manifold $\tilde{M}$. Then we have:

(i) *If $\tilde{M}$ has positive Ricci curvature, the index of f satisfies $i(f) \geq \beta_1(M)$, where $\beta_1(M)$ is the first Betti number of M. In particular, if $H^1(M;\mathbb{R}) \neq 0$ holds, then M is unstable.*

(ii) *If $\tilde{M}$ has nonpositive Ricci curvature, then M is stable.*

5.4. Spectral geometry. Let M^n be a compact Kaehler n-manifold and CP^n denote the complex projective n-space with the Fubini-Study metric of constant holomorphic sectional curvature. One interesting problem in spectral geometry is that "Can CP^n be characterized by the spectrum for p-forms for a fixed p among Kaehler manifolds?" S. Tanno proved that this is true if $p = 0$ and $n \leq 6$, or $p = 1$ and $8 \leq n \leq 51$. Chen and Vanhecke proved in [**79**] that this is also true for $p = 2$ and $n \neq 8$. For the complex Laplacian $\square^{1,1}$ (acting on $(1,1)$-form on M^n), Chen and K. Ogiue proved in [**72**] that the same property is true for $n \geq 2$.

6. Riemannian submersions

6.1. A link between submersions and minimality. Let $\pi : M \to B$ be a Riemannian submersion. By applying O'Neill's tensor π, Chen introduced in [**26**] a submersion δ-invariant, denoted by $\breve{A}_\pi$. He has shown that if π has totally geodesic fibers, then for every isometric immersion ψ of M into a Riemannian manifold $\tilde{M}$ one has

$$\breve{A}_\pi \leq \frac{n^2}{4}|H|^2 + b(n-b)\max \tilde{K}, \quad n = \dim M, \quad b = \dim B, \tag{6.1}$$

where $\max \tilde{K}(p)$ is the maximum value of the sectional curvature function of $\tilde{M}$ restricted to M. By applying the inequality (6.1), he discovered the following link between submersions and minimality: *If a Riemannian manifold admits a non-trivial Riemannian submersion with totally geodesic fibers, then it doesn't admit any minimal immersion into any Riemannian manifold of non-positive curvature.*

6.2. Submersions with minimal fibers. For Riemannian submersions $\pi : M \to B$ with minimal fibers, Chen proved in [**27**] the following.

(i) *If M is positively curved, then the horizontal distribution of π is always a non-totally geodesic distribution.*

(ii) *If M is negatively curved (respectively, non-positively curved), then fibers have negative (respectively, non-positive) scalar curvature.*

6.3. A link between submersions and affine hypersurfaces. In [**27**] Chen also proved the following result: *If a Riemannian manifold M can be realized as an improper hypersphere (or as an elliptic proper centroaffine hypersphere centered at the origin) in an affine space, then every Riemannian submersion $\pi : M \to B$ with minimal fibers has non-totally geodesic horizontal distribution.*

7. Production functions and constant elasticity of substitution property in economics

The notion of production functions is one of the key concepts of mainstream neoclassical theories in economics, being used in modeling of the relationship between the output of a production process and the inputs that have been used in obtaining it. The simplest production model is the famous 2-input Cobb-Douglas (CD) production function introduced in 1928 (Amer. Econ. Rev.) in order to describe the distribution of the national income of the United States. The CD production model was generalized by K. J. Arrow et. al. in (1961, Rev. Econ. Stat.) to Constant Elasticity of Substitution (CES) production functions; later extended by H. Uzawa (1962, Rev. Econ. Stud.) and D. McFadden (1963, Rev. Econ. Stud) to generalized CES or ACMS production functions. The ACMS production functions include as special cases many other famous production models in economics, like multinomial or Leontief production functions. The German economist J. von Thünen was the first to formulate the relationship between output and input as a map $f : \mathbb{R}^n_+ \to \mathbb{R}_+$, where f is the quantity of output, n the number of the inputs such as labor, capital, land, raw materials, etc. In 2011, A.D. Vilcu and G.E. Vilcu used differential geometry to study production functions via the graphs in $\mathbb{R}^{n+1}$.

In [**33**–**35**], Chen classified several classes of product functions satisfying the constant elasticity of substitution property. For example, he proved that if an n-input production homogeneous production function satisfies the constant elasticity of substitution property, it is either the generalized Cobb-Douglas production function or the generalized ACMS production. Further, he studied the relationship between production functions and geometry in [**32, 36, 37, 41, 85**]; e,g., Chen and G.E. Vilcu proved in [**85**] that the graph of an n-input homogeneous production function is a flat hypersurface if and only if it has constant return to scale or it is a multinomial production function. For more detailed in this respect, see [A survey on the geometry of production models in economics, Arab J. Math. Sci. **23** (2017), 18–31] by A.D. Vilcu and G.E. Vilcu.

8. Some other topics studied by Bang-Yen Chen

Besides the topics mentioned above, Chen, many joint with his coworkers, also studied the following topics, among others.

1. Total mean curvature and total absolute curvatures.
2. Conformally flat manifolds and submanifolds.
3. Parallel surfaces and surfaces with parallel mean curvature vectors in real and complex space forms.
4. Lagrangian, total real and purely real submanifolds.
5. Slant submanifolds.
6. CR-submanifolds.
7. Real hypersurfaces of complex space forms.
8. Warped product submanifolds and warped product CR-submanifolds.
9. Lorentzian surfaces in pseudo-Riemannian (Kaehlerian) manifolds.
10. Exact solutions of (ordinary or partial) differential equations.
11. Robertson-Walker spacetimes and marginally trapped surfaces.
12. Relations between λ_1 on surfaces and conformal mappings.
13. Ricci and Yamabe solitons.
14. Geometry of position vector fields of Euclidean submanifolds.
15. Geometric inequalities.
16. Differential geometry of geodesic spheres.
17. Symmetries and reflections on Riemannian and Hermitian manifolds.
18. Contact numbers of Euclidean submanifolds.
19. Curves in Euclidean 3-spaces.
20. Realizations of Riemannian manifolds as affine hypersurfaces.

For a selection of papers related to these topics, see [**1–4, 8, 10, 14, 22, 25, 29, 42, 45, 54, 55, 60–63, 70, 73–77, 80–83, 86–88**].

The references below list some works of Chen which are closely related to all the topics mentioned above. In particular, see his books [**6, 9, 12, 15, 31, 44, 52, 84**] and his survey articles [**13, 17, 18, 21, 23, 24, 28, 30, 40, 43, 46–48, 50, 51**] for an overview of his research work.

References

[1] D. E. Blair and B.-Y. Chen, *On CR-submanifolds of Hermitian manifolds*, Israel J. Math. **34** (1979), no. 4, 353–363 (1980), DOI 10.1007/BF02760614. MR570892

[2] B.-Y. Chen, *On the total curvature of immersed manifolds. I. An inequality of Fenchel-Borsuk-Willmore*, Amer. J. Math. **93** (1971), 148–162, DOI 10.2307/2373454. MR278240

[3] B.-Y. Chen, *On the total curvature of immersed manifolds. II. Mean curvature and length of second fundamental form*, Amer. J. Math. **94** (1972), 799–809, DOI 10.2307/2373759. MR319114

[4] B.-Y. Chen, *On the total curvature of immersed manifolds. III. Surfaces in euclidean 4-space*, Amer. J. Math. **95** (1973), 636–642, DOI 10.2307/2373732. MR334119

[5] B.-Y. Chen, *Some topological obstructions to Bochner-Kaehler metrics and their applications*, J. Differential Geometry **13** (1978), no. 4, 547–558 (1979). MR570217

[6] B.-Y. Chen, *Geometry of submanifolds*, Marcel Dekker, Inc., New York, 1973. Pure and Applied Mathematics, No. 22. MR0353212

[7] B.-Y. Chen, *On total curvature of immersed manifolds. IV. Spectrum and total mean curvature*, Bull. Inst. Math. Acad. Sinica **7** (1979), no. 3, 301–311. MR543046

[8] B.-Y. Chen, *Conformal mappings and first eigenvalue of Laplacian on surfaces*, Bull. Inst. Math. Acad. Sinica **7** (1979), no. 4, 395–400. MR558248

[9] B.-Y. Chen, *Geometry of submanifolds and its applications*, Science University of Tokyo, Tokyo, 1981. MR627323

[10] B.-Y. Chen, *On the total curvature of immersed manifolds. V. C-surfaces in Euclidean m-space*, Bull. Inst. Math. Acad. Sinica **9** (1981), no. 4, 509–516. MR657022

[11] B.-Y. Chen, *On the total curvature of immersed manifolds. VI. Submanifolds of finite type and their applications*, Bull. Inst. Math. Acad. Sinica **11** (1983), no. 3, 309–328. MR726979

[12] B.-Y. Chen, *Total mean curvature and submanifolds of finite type*, Series in Pure Mathematics, vol. 1, World Scientific Publishing Co., Singapore, 1984. MR749575

[13] B.-Y. Chen, *A new approach to compact symmetric spaces and applications: A report on joint work with Professor T. Nagano*, Katholieke Universiteit Leuven, Louvain, 1987. MR952556

[14] B.-Y. Chen, *Slant immersions*, Bull. Austral. Math. Soc. **41** (1990), no. 1, 135–147, DOI 10.1017/S0004972700017925. MR1043974

[15] B.-Y. Chen, *Geometry of slant submanifolds*, Katholieke Universiteit Leuven, Belgium, 1990. MR1099374

[16] B.-Y. Chen, *Some pinching and classification theorems for minimal submanifolds*, Arch. Math. (Basel) **60** (1993), no. 6, 568–578, DOI 10.1007/BF01236084. MR1216703

[17] B.-Y. Chen, *A Riemannian invariant and its applications to submanifold theory*, Results Math. **27** (1995), no. 1-2, 17–26, DOI 10.1007/BF03322265. Festschrift dedicated to Katsumi Nomizu on his 70th birthday (Leuven, 1994; Brussels, 1994). MR1317820

[18] B.-Y. Chen, *A report on submanifolds of finite type*, Soochow J. Math. **22** (1996), no. 2, 117–337. MR1391469

[19] B.-Y. Chen, *Strings of Riemannian invariants, inequalities, ideal immersions and their applications*, The Third Pacific Rim Geometry Conference (Seoul, 1996), Monogr. Geom. Topology, vol. 25, Int. Press, Cambridge, MA, 1998, pp. 7–60. MR1751063

[20] B.-Y. Chen, *Some new obstructions to minimal and Lagrangian isometric immersions*, Japan. J. Math. (N.S.) **26** (2000), no. 1, 105–127, DOI 10.4099/math1924.26.105. MR1771434

[21] B.-Y. Chen, *Riemannian submanifolds*, Handbook of differential geometry, Vol. I, North-Holland, Amsterdam, 2000, pp. 187–418, DOI 10.1016/S1874-5741(00)80006-0. MR1736854

[22] B.-Y. Chen, *Exact solutions of a class of differential equations of Lamé's type and its applications to contact geometry*, Rocky Mountain J. Math. **30** (2000), no. 2, 497–506, DOI 10.1216/rmjm/1022009277. MR1786994
[23] B.-Y. Chen, *Segre embedding and related maps and immersions in differential geometry*, Arab J. Math. Sci. **8** (2002), no. 1, 1–39. MR1938982
[24] B.-Y. Chen, *Geometry of warped products as Riemannian submanifolds and related problems*, Soochow J. Math. **28** (2002), no. 2, 125–156. MR1897183
[25] B.-Y. Chen, *When does the position vector of a space curve always lie in its rectifying plane?*, Amer. Math. Monthly **110** (2003), no. 2, 147–152, DOI 10.2307/3647775. MR1952443
[26] B.-Y. Chen, *Riemannian submersions, minimal immersions and cohomology class*, Proc. Japan Acad. Ser. A Math. Sci. **81** (2005), no. 10, 162–167 (2006). MR2196721
[27] B.-Y. Chen, *Riemannian submersions with minimal fibers and affine hypersurfaces*, Monatsh. Math. **151** (2007), no. 2, 143–152, DOI 10.1007/s00605-007-0451-y. MR2322940
[28] B.-Y. Chen, *Realization of Robertson-Walker spacetimes as affine hypersurfaces*, J. Phys. A **40** (2007), no. 15, 4241–4250, DOI 10.1088/1751-8113/40/15/014. MR2320066
[29] B.-Y. Chen, *δ-invariants, inequalities of submanifolds and their applications*, Topics in differential geometry, Ed. Acad. Române, Bucharest, 2008, pp. 29–155. MR2484670
[30] B.-Y. Chen, *Submanifolds with parallel mean curvature vector in Riemannian and indefinite space forms*, Arab J. Math. Sci. **16** (2010), no. 1, 1–46. MR2663981
[31] B.-Y. Chen, *Pseudo-Riemannian geometry, δ-invariants and applications*, World Scientific Publishing Co. Pte. Ltd., Hackensack, NJ, 2011. With a foreword by Leopold Verstraelen. MR2799371
[32] B.-Y. Chen, *On some geometric properties of h-homogeneous production functions in microeconomics*, Kragujevac J. Math. **35** (2011), no. 3, 343–357. MR2881132
[33] B.-Y. Chen, *Classification of h-homogeneous production functions with constant elasticity of substitution*, Tamkang J. Math. **43** (2012), no. 2, 321–328, DOI 10.5556/j.tkjm.43.2012.321-328. MR2948428
[34] B.-Y. Chen, *Classification of homothetic functions with constant elasticity of substitution and its geometric applications*, Int. Electron. J. Geom. **5** (2012), no. 2, 67–78. MR2990592
[35] B.-Y. Chen, *Geometry of quasi-sum production functions with constant elasticity of substitution property*, J. Adv. Math. Stud. **5** (2012), no. 2, 90–97. MR3024745
[36] B.-Y. Chen, *On some geometric properties of quasi-sum production models*, J. Math. Anal. Appl. **392** (2012), no. 2, 192–199, DOI 10.1016/j.jmaa.2012.03.011. MR2917297
[37] B.-Y. Chen, *A note on homogeneous production models*, Kragujevac J. Math. **36** (2012), no. 1, 41–43. MR2990634
[38] B.-Y. Chen, *The 2-ranks of connected compact Lie groups*, Taiwanese J. Math. **17** (2013), no. 3, 815–831, DOI 10.11650/tjm.17.2013.2606. MR3072263
[39] B.-Y. Chen, *A tour through δ-invariants: from Nash's embedding theorem to ideal immersions, best ways of living and beyond*, Publ. Inst. Math. (Beograd) (N.S.) **94(108)** (2013), 67–80, DOI 10.2298/PIM1308067C. MR3137491
[40] B.-Y. Chen, *Recent developments of biharmonic conjectures and modified biharmonic conjectures*, Pure and Applied Differential Geometry – PADGE 2012, 81–90, Ber. Math., Shaker Verlag, Aachen, 2013.
[41] B.-Y. Chen, *Solutions to homogeneous Monge-Ampère equations of homothetic functions and their applications to production models in economics*, J. Math. Anal. Appl. **411** (2014), no. 1, 223–229, DOI 10.1016/j.jmaa.2013.09.029. MR3118479
[42] B.-Y. Chen, *Ricci solitons on Riemannian submanifolds*, Riemannian Geometry and Applications—Proceedings RIGA 2014, Editura Univ. Bucur., Bucharest, 2014, pp. 30–45. MR3330253
[43] B.-Y. Chen, *Some open problems and conjectures on submanifolds of finite type: recent development*, Tamkang J. Math. **45** (2014), no. 1, 87–108, DOI 10.5556/j.tkjm.45.2014.1564. MR3188077
[44] B.-Y. Chen, *Total mean curvature and submanifolds of finite type*, 2nd ed., Series in Pure Mathematics, vol. 27, World Scientific Publishing Co. Pte. Ltd., Hackensack, NJ, 2015. With a foreword by Leopold Verstraelen. MR3362186
[45] B.-Y. Chen, *Einstein manifolds as affine hypersurfaces*, Int. Electron. J. Geom. **8** (2015), no. 1, 33–44. MR3340280

[46] B.-Y. Chen, *A survey on Ricci solitons on Riemannian submanifolds*, Recent advances in the geometry of submanifolds—dedicated to the memory of Franki Dillen (1963–2013), Contemp. Math., vol. 674, Amer. Math. Soc., Providence, RI, 2016, pp. 27–39. MR3565242

[47] B.-Y. Chen, *CR-warped submanifolds in Kaehler manifolds*, Geometry of Cauchy-Riemann submanifolds, Springer, Singapore, 2016, pp. 1–25. MR3525850

[48] B.-Y. Chen, *CR-submanifolds and δ-invariants*, Geometry of Cauchy-Riemann submanifolds, Springer, Singapore, 2016, pp. 27–55. MR3525851

[49] B.-Y. Chen, *Covering maps and ideal embeddings of compact homogeneous spaces*, J. Geom. Symmetry Phys. **44** (2017), 13–20, DOI 10.7546/jgsp-44-2017-13-20. MR3700454

[50] B.-Y. Chen, *Topics in differential geometry associated with position vector fields on Euclidean submanifolds*, Arab J. Math. Sci. **23** (2017), no. 1, 1–17, DOI 10.1016/j.ajmsc.2016.08.001. MR3589496

[51] B.-Y, Chen, *Euclidean submanifolds via tangential components of their position vector fields*, Mathematics **5** (2017), Art. 51, pp. 17.

[52] B.-Y. Chen, *Differential geometry of warped product manifolds and submanifolds*, World Scientific Publishing Co. Pte. Ltd., Hackensack, NJ, 2017. With a foreword by Leopold Verstraelen. MR3699316

[53] B.-Y. Chen, *Two-numbers and their applications—a survey*, Bull. Belg. Math. Soc. Simon Stevin **25** (2018), no. 4, 565–596. MR3896273

[54] B.-Y. Chen and S. Deshmukh, *Classification of Ricci solitons on Euclidean hypersurfaces*, Internat. J. Math. **25** (2014), no. 11, 1450104, 22 pp., DOI 10.1142/S0129167X14501043. MR3285304

[55] B.-Y. Chen and F. Dillen, *Rectifying curves as centrodes and extremal curves*, Bull. Inst. Math. Acad. Sinica **33** (2005), no. 2, 77–90. MR2151806

[56] B.-Y. Chen, F. Dillen, L. Verstraelen, and L. Vrancken, *A variational minimal principle characterizes submanifolds of finite type* (English, with English and French summaries), C. R. Acad. Sci. Paris Sér. I Math. **317** (1993), no. 10, 961–965. MR1249369

[57] B.-Y. Chen, F. Dillen, J. Van der Veken, and L. Vrancken, *Curvature inequalities for Lagrangian submanifolds: the final solution*, Differential Geom. Appl. **31** (2013), no. 6, 808–819, DOI 10.1016/j.difgeo.2013.09.006. MR3130570

[58] B.-Y. Chen, F. Dillen, J. Van der Veken, and L. Vrancken, *Classification of $\delta(2, n-2)$-ideal Lagrangian submanifolds in n-dimensional complex space forms*, J. Math. Anal. Appl. **458** (2018), no. 2, 1456–1485, DOI 10.1016/j.jmaa.2017.10.044. MR3724740

[59] B.-Y. Chen, F. Dillen, L. Verstraelen, and L. Vrancken, *An exotic totally real minimal immersion of S^3 in $\mathbf{C}P^3$ and its characterisation*, Proc. Roy. Soc. Edinburgh Sect. A **126** (1996), no. 1, 153–165, DOI 10.1017/S0308210500030651. MR1378838

[60] B.-Y. Chen, F. Dillen, and L. Vrancken, *Lagrangian submanifolds in complex space forms attaining equality in a basic inequality*, J. Math. Anal. Appl. **387** (2012), no. 1, 139–152, DOI 10.1016/j.jmaa.2011.08.066. MR2845740

[61] B.-Y. Chen and O. J. Garay, *Maslovian Lagrangian immersions of real space forms into complex space forms*, Japan. J. Math. (N.S.) **30** (2004), no. 2, 227–281, DOI 10.4099/math1924.30.227. MR2111552

[62] B.-Y. Chen, P.-F. Leung and T. Nagano, *Totally geodesic submanifolds of symmetric spaces III*, (arXiv:1307.7325v2 [math.DG], 06.30.2013).

[63] B.-Y. Chen and S.-J. Li, *The contact number of a Euclidean submanifold*, Proc. Edinb. Math. Soc. (2) **47** (2004), no. 1, 69–100, DOI 10.1017/S0013091503000038. MR2064737

[64] B.-Y. Chen and H.-S. Lue, *Chern classes and Bochner-Kaehler metrics*, Colloq. Math. **44** (1981), no. 2, 203–208 (1982), DOI 10.4064/cm-44-2-203-208. MR652579

[65] B.-Y. Chen and T. Nagano, *Totally geodesic submanifolds of symmetric spaces. I*, Duke Math. J. **44** (1977), no. 4, 745–755. MR458340

[66] B.-Y. Chen and T. Nagano, *Totally geodesic submanifolds of symmetric spaces. II*, Duke Math. J. **45** (1978), no. 2, 405–425. MR487902

[67] B.-Y. Chen and T. Nagano, *Un invariant géométrique riemannien* (French, with English summary), C. R. Acad. Sci. Paris Sér. I Math. **295** (1982), no. 5, 389–391. MR684733

[68] B.-Y. Chen and T. Nagano, *A Riemannian geometric invariant and its applications to a problem of Borel and Serre*, Trans. Amer. Math. Soc. **308** (1988), no. 1, 273–297, DOI 10.2307/2000963. MR946443

[69] B.-Y. Chen and K. Ogiue, *A characterization of complex space forms in terms of Bochner curvature tensor and Chern classes and its application*, Indiana Univ. Math. J. **24** (1974/75), no. 11, 1087–1091, DOI 10.1512/iumj.1975.24.24090. MR380688

[70] B.-Y. Chen and K. Ogiue, *On totally real submanifolds*, Trans. Amer. Math. Soc. **193** (1974), 257–266, DOI 10.2307/1996914. MR346708

[71] B.-Y. Chen and K. Ogiue, *Some characterizations of complex space forms in terms of Chern classes*, Quart. J. Math. Oxford Ser. (2) **26** (1975), no. 104, 459–464, DOI 10.1093/qmath/26.1.459. MR405303

[72] B.-Y. Chen and K. Ogiue, *On the spectrum of* $cm^{1,1}$ *on Kaehler manifolds*, Arch. Math. (Basel) **40** (1983), no. 4, 367–371, DOI 10.1007/BF01192798. MR708600

[73] B.-Y. Chen and B. D. Suceavă, *Classification theorems for space-like surfaces in 4-dimensional indefinite space forms with index 2*, Taiwanese J. Math. **15** (2011), no. 2, 523–541, DOI 10.11650/twjm/1500406219. MR2810166

[74] B.-Y. Chen and Y. Tazawa, *Slant submanifolds of complex projective and complex hyperbolic spaces*, Glasg. Math. J. **42** (2000), no. 3, 439–454, DOI 10.1017/S0017089500030111. MR1793811

[75] B.-Y. Chen and J. Van der Veken, *Marginally trapped surfaces in Lorentzian space forms with positive relative nullity*, Classical Quantum Gravity **24** (2007), no. 3, 551–563, DOI 10.1088/0264-9381/24/3/003. MR2294231

[76] B.-Y. Chen and J. Van der Veken, *Spatial and Lorentzian surfaces in Robertson-Walker space times*, J. Math. Phys. **48** (2007), no. 7, 073509, 12, DOI 10.1063/1.2748616. MR2337690

[77] B.-Y. Chen and J. Van der Veken, *Complete classification of parallel surfaces in 4-dimensional Lorentzian space forms*, Tohoku Math. J. (2) **61** (2009), no. 1, 1–40, DOI 10.2748/tmj/1238764545. MR2501861

[78] B.-Y. Chen and J. Van der Veken, *Classification of marginally trapped surfaces with parallel mean curvature vector in Lorentzian space forms*, Houston J. Math. **36** (2010), no. 2, 421–449. MR2661254

[79] B.-Y. Chen and L. Vanhecke, *The spectrum of the Laplacian of Kähler manifolds*, Proc. Amer. Math. Soc. **79** (1980), no. 1, 82–86, DOI 10.2307/2042392. MR560589

[80] B.-Y. Chen and L. Vanhecke, *Differential geometry of geodesic spheres*, J. Reine Angew. Math. **325** (1981), 28–67, DOI 10.1515/crll.1981.325.28. MR618545

[81] B.-Y. Chen and L. Vanhecke, *Reflections and symmetries in compact symmetric spaces*, Bull. Austral. Math. Soc. **38** (1988), no. 3, 377–386, DOI 10.1017/S000497270002774X. MR966712

[82] B. Y. Chen and L. Vanhecke, *Isometric, holomorphic and symplectic reflections*, Geom. Dedicata **29** (1989), no. 3, 259–277, DOI 10.1007/BF00572443. MR995302

[83] B.-Y. Chen and L. Vanhecke, *Symplectic reflections and complex space forms*, Ann. Global Anal. Geom. **9** (1991), no. 3, 205–210, DOI 10.1007/BF00136812. MR1143403

[84] B.-Y. Chen and L. Verstraelen, *Laplace transformations of submanifolds*, Centre for Pure and Applied Differential Geometry (PADGE), vol. 1, Katholieke Universiteit Brussel, Group of Exact Sciences, Brussels; Katholieke Universiteit Leuven, Department of Mathematics, Leuven, 1995. MR1347686

[85] B.-Y. Chen and G. E. Vîlcu, *Geometric classifications of homogeneous production functions*, Appl. Math. Comput. **225** (2013), 345–351, DOI 10.1016/j.amc.2013.09.052. MR3129653

[86] B.-Y. Chen and S. W. Wei, *Submanifolds of warped product manifolds* $I \times_f S^{m-1}(k)$ *from a p-harmonic viewpoint*, Bull. Transilv. Univ. Braşov Ser. III **1(50)** (2008), 59–77. MR2478006

[87] B.-Y. Chen and S. W. Wei, *Growth estimates for warping functions and their geometric applications*, Glasg. Math. J. **51** (2009), no. 3, 579–592, DOI 10.1017/S0017089509990012. MR2534009

[88] B. Y. Chen and K. Yano, *Special conformally flat spaces and canal hypersurfaces*, Tohoku Math. J. (2) **25** (1973), 177–184, DOI 10.2748/tmj/1178241376. MR334032

[89] Y.-L. Ou and B.-Y. *Chen, Biharmonic submanifolds and biharmonic maps in Riemannian geometry*, World Scientific Publishing Co. Pte. Ltd., Hackensack, NJ, 2020.

Contemporary Mathematics
Volume **756**, 2020
https://doi.org/10.1090/conm/756/15205

My education in differential geometry and my indebtedness

Bang-Yen Chen

1. My education in differential geometry in Taiwan

I was born on October 3, 1943 in Toucheng Township, Yilan County, Taiwan. I enrolled in 1961 at Department of Mathematics, Tamkang University. My teacher of differential geometry at Tamkang University is Professor Tsing Houa Teng (cf. [**3**] for his brief biography). The textbook he used is "Lectures on Classical Differential Geometry" by D. J. Struik.

After I received my B.S. degree in June 1965 from Tamkang University, I enrolled at the Institute of Mathematics, National Tsing Hua University in Hsinchu, Taiwan for my M.S. program. Graduate schools of mathematics in Taiwan at that time had only master programs. Although Tsing Hua University had very few graduate students, the University had enough resources to invite famous professors from Japan and United States to teach graduate courses for one or two semesters. My first teacher in differential geometry at Tsing Hua University is Professor Tominosuke Otsuki from Tokyo Institute of Technology in the Fall semester of 1965. I recall that, among other topics, he lectured on "Total absolute curvature of Chern and Lashof" as well as on "Otsuki's lemma" (named by Professor Chern) which solved a conjecture of S. S. Chern and N. H. Kuiper posed in [**30**] (see [**31**]). The other teacher of differential geometry at Tsing Hua University is Professor Tadashi Nagano from the University of Tokyo. While Professor Nagano stayed at Tsing Hua University for the Fall semester of 1966 and taught one course on "Geometry of G-structures", I wrote my master thesis [**1**] under his guidance. Before he left Taiwan he told me that he would be teaching at University of Notre Dame starting from Fall semester of 1967.

During that time, every physical able male college student needed to serve one year in military after graduation. Luckily, I was borrowed from the Ministry of Defense of Taiwan by National Tsing Hua University to teach "Differential Geometry" to senior students in the 1967–1968 academic year instead. During 1966–1968, I also taught courses to senior students at Tamkang University for several semesters and taught "Differential Geometry" at National Taiwan Normal University in Taipei for one semester during 1966–1967.

2010 *Mathematics Subject Classification.* Primary 01A60, 01A61, 53-03, 53B20, 53B25; Secondary 53A15, 53A30,53A55, 53B35, 53C15, 53C25, 53C35, 53C40, 53C42, 53C43, 53C55, 53C80, 53C99.

From left to right: Tadashi Nagano, Bang-Yen Chen, Yoshihiro Ohnita, and Yoshihiko Tazawa, at Sophia University, in Tokyo, in 1998.

2. My study at University of Notre Dame and my indebtedness to Professor Tadashi Nagano

In September 1968, I enrolled at the University of Notre Dame for my Ph.D. program with two-years research fellowship. At Notre Dame, I took the course "Kaehler Manifolds" from Professor Yozo Matsushima and the course "Several Complex Variables" from Professor Yum-Tong Siu".

After I arrived at University of Notre Dame, I went to see Professor Nagano to seek his advice. He told me that I could choose research problems by myself under the condition that I shall see him at least once a week to let him know my progress. Each week I reported to him, Professor Nagano always gave me his best suggestions to improve my results. It turned out that his advice provided the **best research training for me**. Under his advice, I had complete freedom to seek my research problems including my doctoral thesis [**2**]. On the other hand, I had to work hard in order to report to him every week. For these reasons, Professor Nagano had made extremely important life-long influence on my research. Although I did not write any joint paper with Professor Nagano during my two years at Notre Dame, we did write joint papers [**9**–**14**] on several subjects starting about five years after my graduation (see also [**5**–**7**]).

3. My indebtedness to Professor Shiing-Shen Chern

I met Professor Shiing-Shen Chern for the first time when I was a student at Tamkang University while I attended his inspiring talk on "Mathematical Contributions of É. Cartan and H. Weyl" in Taipei. Later, I attended a series of Chern's lectures on "Integral Geometry" and also a series of Wu-Yi Hsiang's lectures on "Lie Groups and their Representations" at the Summer School on Mathematics sponsored by Academia Sinica, National Taiwan University and National Tsing Hua University.

In early 1971, about eight months after I received my doctoral degree from the University of Notre Dame, I went back to Notre Dame to attend Professor Chern's talk on "Characteristic forms and geometric invariants". After his talk, Professor Chern invited me to visit University of California at Berkeley for one month. During my visit at Berkeley in the Fall semester of 1971, we had many discussions, mostly during our lunch hour with only two of us. From those personal discussions, I learned much from him about his philosophy of research. His main suggestion to me is that I shall always choose problems with very clear meaning; the results I obtained shall be very easy to understand and easy to explain. Professor Chern's important suggestions planted the seeds of a profound influence on my research later on. At the end of my visit, Professor Chern suggested me to write a book to include results from my earlier works. Soon after, I received an invitation letter from Professor Shoshichi Kobayashi to write a book on "Submanifold Theory" to be included in the Marcel Dekker series "Pure and Applied Mathematics'. As a result, my first book "Geometry of Submanifolds" [**3**] was published in early 1973 as volume 22 of this book series.

4. My indebtedness to Professor Tominosuke Otsuki

After his retirement from Tokyo Institute of Technology, Professor Otsuki moved to Science University of Tokyo. As chairman of the Department of Mathematics, he invited me as a visiting professor at Science University of Tokyo for three months in 1980–1981. During my visit at SUT, I was invited to give talks at University of Tokyo, Kyushu University, Nagoya University, Osaka University, Tsukuba University, Tokyo Institute of Technology, Ochanomizu University and Tokyo Metropolitan University. During my visit, Professor Otsuki went with me to visit Professor Shigeo Sasaki at his home as well as to meet Professor Shun-ichi Tachibana at Ochanomizu University. He also traveled with me from Tokyo to Kyushu, Nagoya and Osaka either by airplane or by bullet train (*shinkansen*). In those trips he told me a lots about his research philosophy and his mathematical life which inspired me a lot and made important influence on my research work. At the end of my visit, Professor Otsuki asked me to write a book to summarize my lectures delivered in Japan. My book "Geometry of Submanifolds and Its Applications" [**4**] was then published by Science University of Tokyo in 1981.

Professor and Mrs. Otsuki together with Bang-Yen Chen and wife Pi-mei Chen in Yokohama, Japan, on December 26, 1997.

5. My indebtedness to Professor Kentaro Yano

I met Professor Kentaro Yano for the first time in late 1970 at Michigan State University while he was a visiting professor under the invitation of Professors D. E. Blair and G. D. Ludden. Since then he had visited MSU as visiting professor for a few times until 1973; about one month for each visit. During his visits at MSU, we had many discussions on geometry of conformally flat manifolds and submanifolds, among others. Usually, we discussed after lunch until 5 pm almost every week day. Hence many of our joint articles done during that period are related to conformally flat manifolds and submanifolds [**19**–**22**, **32**]. From his visits at MSU we also published several articles on integral formulas and pseudo-umbilical or quasi-umbilical submanifolds (see [**15**–**18**, **24**–**27**]). Later, we published two joint articles [**28**, **29**] on different subjects via communications by air mails.

From his visits at MSU, I learned very much from him. Most importantly, I tried to learn from his skill as well as from his philosophy on research. Consequently, Professor Yano had very important impact on my research as well.

Bang-Yen Chen (left) and Sir Michael Francis Atiyah (right), in East Lansing, Michigan, in the spring of 2002.

6. In conclusion

First, I thank the organizers and speakers of the special session of the 1143rd Meeting of the American Mathematical Society in honor of my 75th birthday. I also would like to express my many thanks to all contributors of this special volume of Contemporary Mathematics for their excellent works.

During my over 50 years research career, I have written joint articles with more than 60 colleagues; mostly while they visited me at Michigan State University or I visited them at their universities. I would like to use this unique opportunity to express my many thanks to all of them for the friendship and the cooperations.

Finally, I like to thank Pi-Mei, my wife of over fifty years, for her unselfish and untiring support, without which my works would have been impossible to accomplish.

Bang-Yen Chen (left) and Katsumi Nomizu (right).

References

[1] B.-y. Chen, *On the total absolute curvature of manifolds immersed in Riemannian manifold*, Kōdai Math. Sem. Rep. **19** (1967), 299–311. MR221436

[2] B.-Y. Chen, *On the G-total curvature and topology of immersed manifolds*, ProQuest LLC, Ann Arbor, MI, 1971. Thesis (Ph.D.)–University of Notre Dame. MR2940079

[3] B.-y. Chen, *Geometry of submanifolds*, Marcel Dekker, Inc., New York, 1973. Pure and Applied Mathematics, No. 22. MR0353212

[4] B.-y. Chen, *Geometry of submanifolds and its applications*, Science University of Tokyo, Tokyo, 1981. MR627323

[5] B.-Y. Chen, *A new approach to compact symmetric spaces and applications*, Katholieke Universiteit Leuven, Louvain, 1987. A report on joint work with Professor T. Nagano. MR952556

[6] B.-Y. Chen, *The 2-ranks of connected compact Lie groups*, Taiwanese J. Math. **17** (2013), no. 3, 815–831, DOI 10.11650/tjm.17.2013.2606. MR3072263

[7] B.-Y. Chen, *Two-numbers and their applications—a survey*, Bull. Belg. Math. Soc. Simon Stevin **25** (2018), no. 4, 565–596. MR3896273

[8] B.-y. Chen, M. L. Chu, H. C. Lai, H. W. Pu, H. J. Shyr, G. S. Yang, and N. C. Yeh, *Foreword*, Tamkang J. Math. **4** (1973), no. 2, i-ii. MR342331

[9] B.-y. Chen and T. Nagano, *Totally geodesic submanifolds of symmetric spaces. I*, Duke Math. J. **44** (1977), no. 4, 745–755. MR458340

[10] B.-y. Chen and T. Nagano, *Totally geodesic submanifolds of symmetric spaces. II*, Duke Math. J. **45** (1978), no. 2, 405–425. MR487902

[11] B.-y. Chen and T. Nagano, *Un invariant géométrique riemannien* (French, with English summary), C. R. Acad. Sci. Paris Sér. I Math. **295** (1982), no. 5, 389–391. MR684733

[12] B.-Y. Chen and T. Nagano, *Harmonic metrics, harmonic tensors, and Gauss maps*, J. Math. Soc. Japan **36** (1984), no. 2, 295–313, DOI 10.2969/jmsj/03620295. MR740319

[13] B.-Y. Chen and T. Nagano, *A Riemannian geometric invariant and its applications to a problem of Borel and Serre*, Trans. Amer. Math. Soc. **308** (1988), no. 1, 273–297, DOI 10.2307/2000963. MR946443

[14] B.-Y. Chen, P.-F. Leung and T. Nagano, *Totally geodesic submanifolds of symmetric spaces III*, Technical Report (1980) (arXiv:1307.7325v2 [math.DG]).

[15] B.-y. Chen and K. Yano, *On submanifolds of submanifolds of a Riemannian manifold*, J. Math. Soc. Japan **23** (1971), 548–554, DOI 10.2969/jmsj/02330548. MR281121

[16] K. Yano and B.-y. Chen, *Minimal submanifolds of a higher dimensional sphere*, Tensor (N.S.) **22** (1971), 369–373. MR0287495

[17] B.-y. Chen and K. Yano, *Integral formulas for submanifolds and their applications*, J. Differential Geometry **5** (1971), 467–477. MR288699

[18] B. Y. Chen and K. Yano, *Pseudo-umbilical submanifolds in a Riemannian manifold of constant curvature*, Differential geometry (in honor of Kentaro Yano), Kinokuniya, Tokyo, 1972, pp. 61–71. MR0328829

[19] B.-y. Chen and K. Yano, *Hypersurfaces of a conformally flat space*, Tensor (N.S.) **26** (1972), 318–322. Commemoration volumes for Prof. Dr. Akitsugu Kawaguchi's seventieth birthday, Vol. III. MR0331283

[20] B.-y. Chen and K. Yano, *Special quasi-umbilical hypersurfaces and locus of spheres* (English, with Italian summary), Atti Accad. Naz. Lincei Rend. Cl. Sci. Fis. Mat. Nat. (8) **53** (1972), 255–260 (1973). MR346667

[21] B.-y. Chen and K. Yano, *Sous-variétés localement conformes à un espace euclidien* (French), C. R. Acad. Sci. Paris Sér. A-B **275** (1972), A123–A125. MR388284

[22] B. Y. Chen and K. Yano, *Conformally flat spaces of codimension* 2 *in a Euclidean space*, Canadian J. Math. **25** (1973), 1170–1173, DOI 10.4153/CJM-1973-126-6. MR326602

[23] B. Y. Chen and K. Yano, *Special conformally flat spaces and canal hypersurfaces*, Tohoku Math. J. (2) **25** (1973), 177–184, DOI 10.2748/tmj/1178241376. MR334032

[24] B.-y. Chen and K. Yano, *Umbilical submanifolds with respect to a nonparallel normal direction*, J. Differential Geometry **8** (1973), 589–597. MR341347

[25] B. Y. Chen and K. Yano, *Submanifolds umbilical with respect to a quasi-parallel normal direction*, Tensor (N.S.) **27** (1973), 41–44. MR0334040

[26] B.-y. Chen and K. Yano, *Submanifolds umbilical with respect to a non-parallel normal subbundle*, Kōdai Math. Sem. Rep. **25** (1973), 289–296. MR336661

[27] B.-y. Chen and K. Yano, *Pseudo-umbilical submanifolds of codimension* 3 *with constant mean curvature*, Kōdai Math. Sem. Rep. **25** (1973), 490–501. MR336662

[28] B.-y. Chen and K. Yano, *Manifolds with vanishing Weyl or Bochner curvature tensor*, J. Math. Soc. Japan **27** (1975), 106–112, DOI 10.2969/jmsj/02710106. MR355895

[29] B.-y. Chen and K. Yano, *On the theory of normal variations*, J. Differential Geometry **13** (1978), no. 1, 1–10. MR520597

[30] S.-s. Chern and N. H. Kuiper, *Some theorems on the isometric imbedding of compact Riemann manifolds in euclidean space*, Ann. of Math. (2) **56** (1952), 422–430, DOI 10.2307/1969650. MR50962

[31] T. Otsuki, *On the existence of solutions of a system of quadratic equations and its geometrical application*, Proc. Japan Acad. **29** (1953), 99–100. MR60281

[32] K. Yano, C. S. Houh, and B. Y. Chen, *Intrinsic characterization of certain conformally flat spaces*, Kōdai Math. Sem. Rep. **25** (1973), 357–361. MR331286

2231 Tamarack Drive, Okemos, Michigan 48864, U.S.A.
Email address: chenb@msu.edu

Contemporary Mathematics
Volume **756**, 2020
https://doi.org/10.1090/conm/756/15214

Submanifold theory—A contemplation of submanifolds

Leopold Verstraelen

1. It is a great honour to have the opportunity to write an opening article for these proceedings of the 2018 AMS Special Session at Ann Arbor at the occasion of MSU Distinguished Professor Bang-Yen Chen's 75th anniversary. According to the Editor's proposal, hereafter follows *an attempt to describe what I see when looking well at submanifolds, in a way that better be clear to a general audience of mathematicians (i) and that includes some historical notes (ii).*

With respect to *(i)*, luckily, as *Gauss* sometimes said: "*Die Mathematik ist mehr eine Wissenschaft das Auge als eine für das Ohr*". Actually, as shown by the neuroscientists *S. Dehaene and M. Amalric, the regions in our brains that are involved when doing mathematics are the regions in our brains that are involved when making visual and temporal perceptions*, whereas it are the regions linked to language that are solicited when dealing with matters about history or geography, for instance; and, as formulated by Amalric: "*La pensée mathématique serait donc indépendante du langage, (...). Elle préexisterait même à l'acquisition du langage et à l'apprentissage des maths à l'école*" [**19**].

The present article freely makes use of *the Forewords* that I had the pleasure to write for Professor Chen's books [**8**–**10**], (*'Pseudo-Riemannian Geometry, δ-Invariants and Applications'* -2011-, *'Total Mean Curvature and Submanifolds of Finite Type. Second Edition'* -2015- and *'Differential Geometry of Warped Product Manifolds and Submanifolds'* -2017-), as well as of *the papers* [**72**–**75**] that were written herewith in connection. And, by these texts, and maybe, in addition, some of their references, one can get some impression of the manifold and fundamental contributions that Professor Chen has been making to *the differential geometry of submanifolds*, that is, to *submanifold theory*, year after year, since the 1960s; (and, eventually, one moreover may have some fun when looking at some of the drawings in [**72**, **75**]). In the course of preparing to write these Forewords and these papers and, like now, the present article, I am so much confronted with the severe limitations of my knowledge and understanding. Certainly, we should all be looking forward to publications of deeper and broader expositions of *the development of some main topics of submanifold theory*, in a near future. Also, I would like to express the hope that some colleagues would write *on the origins and developments*

2010 *Mathematics Subject Classification.* Primary 01A60, 01A61, 53-03, 53B20, 53B25; Secondary 53A15, 53A30,53A55, 53B35, 53C15, 53C25, 53C35,53C40, 53C42, 53C43, 53C55, 53C80, 53C99.

Bang-Yen Chen (left) and Leopold Verstraelen (right), in Antwerpen, Belgium, in 1987.

within or closely related with submanifold theory of some main fields of wide interest in mathematics today, and, *on the crucial rôle that submanifold theory has played in the solutions of some serious problems in other fields of mathematics*, and, *on some of the beautiful applications of submanifold theory in the natural and in the medical and in the human sciences and in engineering.* Here and now, for further readings, it seems well to draw attention to the two volumes so far of the '*Handbook of Differential Geometry*' [**40**], (eds. F.J.E. Dillen e.a., -2000-2005-), and to the six volumes on geometry so far of the '*Encyclopaedia of Mathematical Sciences*' [**30**], (ed. R.V. Gramkrelidze, -1991-2001-).

And, with respect to *(ii)*, such historical notes hereafter are mostly intended to serve in the construction of kind of a "décor" wherein is put *my contemplation of submanifolds at present, the basic nature of which is psychological.* At this stage, it may suffice to suggest the following for further readings *on the history of submanifolds*: [**63**] by *D. J. Struik*, [**78**] by *P. Vincensini*, [**34**] by *L. Godeaux*, [**58**] by

B. Rouxel, [**26–28**] by *P. Dombrowski*, [**57**] by *K. Reich*, [**48**] by *D. Laugwitz*, [**51**] edited by *A. N. Kolmogorov and A. P. Yushkevich* and [**64**] edited by *T. Gowers and J. Barrow-Green and I. Leader*.

Many geometers may very well have somewhat different up to diametrically opposed opinions about various aspects and even most basic aspects of their beloved geometry. But, I do suppose that, *anno 2019*, almost all are able to rather agree with the following *rudimentary division of the life of geometry in three phases*: after a long pregnancy between Tigris and Eufrat and in Egypt, geometry was born and passed its childhood in Ancient Greece as *axiomatic Euclidean geometry* (α); during its adolescence in Europe, the event of utmost importance was *the extension of its synthetic powers by the algebraic-analytic forces* (β); still there, geometry then grew into full manhood as *differential geometry*, and as such has been flourishing all over the world (γ).

René Descartes and *Jan Koenderink* did formulate the following assertions: $\sim$ "*Of most importance in the study of geometry is the way in which it cultivates the mind*" ($\sim$ "=" maybe not 100% literally exact quote) and, respectively, "*The mind is a natural, that is a biological product of the constant interaction of the embodied brain and the world*". And, concerning what came above and what will come below, me thinks that, *for specimens of the human kind, 'nature' essentially stands for their organised thoughts about sensations and perceptions of 'their worlds outside and inside' and 'doing mathematics' basically stands for their thoughtful living in 'the universe' of their idealisations and abstractions of these sensations and perceptions"*.

2. Now, at once, I would like to add the following further references for literature on the history of mathematics: concerning (α) and (β), see *K. Ueno's* [**68**] *on the development of algebraic geometry*; concerning (α) and also giving good general advice for reading on history, see *B. L. Van der Waerden's* [**70**] and *O. Neugebauer's* [**52**] *on Egyptian, Babylonian and Greek mathematics*; and, concerning the times before (α), see *B. L. Van der Waerden's* [**69**] on the beginnings of quantitative studies of events in "our" space-time, that is, *on the prehistory of the mathematical sciences*.

Next, it might be not too amiss to take over here a few lines from the last chapter of [**70**] about the "between (α) and (β)", namely, *about the decadence of mathematical activity in Old Greece*, for which is given as a capital reason, *the interruption of the personal, living, oral teachings on the works of the great mathematicians by true mathematics teachers*: $\sim$ "In an oral presentation, the teacher can draw pictures and make on them indications of what he is talking about, he can put special attention to the more important matters of the story, he can explain how the proofs of the theorems showed up. All this is lacking in the written formulations in strict, classic style: the proofs are logically sound, but not suggestive. It is like being locked up in a logical trap for mouses; the arguments are correct and complete, but their lines of thought cannot be seen. As long as each generation could pass on its scientific knowledge and understanding to the next generation, science kept on blossoming. But as soon as –by one or other external cause– the oral transfers of science were interrupted and only classic books were left over, it became extremely difficult to really understand the work of the great predecessors, and virtually impossible to go beyond". Not unrelated herewith are the following two quotes from *H. Weyl's* 1953 text on '*Universities and Science in*

Germany': "... *the idea of a university implies an object and two conditions*: the object is *Wissenschaft*, knowledge in the most exalted sense of that term, namely, the ardent, methodological, independent search after truth in any and all its forms but wholly irrespective of utilitarian applications. The conditions are *Lehrfreiheit and Lehrfreiheit.* Lehrfreiheit means that the one who teaches is free to teach what he chooses as he chooses. Lehrfreiheit denotes the emancipation of the student from all compulsion and compulsory drill, recitations, quizzes, tests, etc.", and: "*The teacher's greatest asset is a real scholarly education*". In any case, when, at present, reflecting a bit more on the situation of the teaching of mathematics, and this does apply worldwide, one should be strongly aware of the fact that, *even of much more importance than the teaching at a human level of mathematics and in general in the institutions for higher educations, are the teachings at a human level of mathematics and in general in the kindergartens and in the primary- and secondary schools.*

Actually, in his '*Secret Knowledge*', *David Hockney* did compactly summarise this "between (α) and (β)" when writing that "*If science does not pass on its knowledge to the next generations, then, within the shortest time possible, we'll enter again in dark ages*"; and, the '*science*' hereby mentioned stands for '*art and science*'. The above '*human level*' and the below '*joy of teaching*' are proposed to be interpreted in the light of the following citations from *Huntley*'s '*The divine proportion –a study in mathematical beauty*' and from *Bronowski*'s '*Science and human values*': $\sim$ "... one of the most intense joys that the soul of man can experience is that of creative activity. Ask the artist. Ask the scientist. They all know the deep spiritual satisfaction associated with the orgasm of creation", and, "The discoveries of science, the works of art are explorations, more, are explosions, of a hidden likeness. The discoverer or the artist presents in them two aspects of nature and fuses them into one. This is the act of creation in which an original thought is born, and it is the same act in original science and original art. This view alone gives a meaning to the act of appreciation; for the appreciator must see the movement, wake to the echo which was started in the creation of the work. In the moment of appreciation we live again the moment when the creator saw and held the hidden likeness. We re-enact the creative act, and we ourselves make the discovery again. The great poem and the deep theorem are new to every reader, and yet are his own experiences, because he himself re-creates them. They are the marks of unity in variety, and in the instant when the mind seizes this for itself, the heart misses a beat", hereby suggesting as a hint for such interpretation *the basic principle of mathematical instruction of Hilton and Holton and Pederson: "Mathematics should be taught so that students have a chance to comprehend how and why mathematics is done by those who do it successfully".*

To close this aside on "between (α) and (β)" in as positive a way as possible, here are statements by *Erasmus* on education in general and by *Gaspard Bosteels* on the teaching of mathematics (G.B. was a post-doc of Van der Waerden who became Head-Master of the Atheneum of Berchem at Antwerp where he organised all teaching activities according to an optimal secondary school adaptation of the norms given in the former citation from a text by Weyl), respectively: "*A human being is not born but educated, formed*" and "*The greatest joy a person can have is knowingly to arouse people's love for mathematics*".

3. Among others, *Leonardo da Vinci* and *Albrecht Dürer* played important rôles in the awakening of the young geometry (α) out of its deep sleep during a long period of darkness. And, somewhat before *Simon Stevin*'s 1585 '*De Thiende*' and 1586 '*De Beghinselen der Weeghconst*' which were of basic value in making possible the beginnings of (β) and (γ) –cfr. [**34, 73**]–, in the *Introduction* to his 1543 '*De humani corporis fabrica*', Vesalius wrote the following about teaching at the universities in his time (still mostly "in the dark") and about teaching in general: "Thus all is taught wrongly and the days pass by with the treatment of ridiculous matters" and "*There is no one who has not personally experienced either in geometry proper or in other fields of mathematics, how much drawings –far better than the longest verbal descriptions– can help to understand matters of importance*". And, in his *Introduction* to the '*W.P. Atlas van het heelal*' (by Ernst and De Vries, with co-operation of De Jaeger), also for more general communications, *Marcel Minnaert* too stressed this latter point, (M.M. was a pioneer of solar physics, and an extremely successful teacher to pupils or students from the kindergartens up to the universities): "*To explain something that matters, certainly when about scientifical phenomena, naturally and automatically we go for paper and pencil* (in a classroom, we take some chalk and go to the blackboard; L.V.); *like by magic, misunderstanding and imprecision disappear, all of a sudden one understands; and what we afterwards recall of some explanation usually is the schematic drawing itself which summarises all essential traits of the phenomena*".

From '*The Origins of Knowledge and Imagination*' by *Jacob Bronowski* are taken the following two quotes: "*The place of sight in human evolution is cardinal*" and "*The world of science is wholly dominated by the sense of sight*". To make sure, here and anywhere, it seems not so bad to agree with *Einstein*'s formulation according to which "*Science is the attempt to make the chaotic diversity of sense-experiences correspond to a logically coherent uniform system of thought*"; (and, "*Wass man von der Naturwissenschaften wissen sollte*" can be well read in *E. P. Fischer*'s '*Die andere Bildung*' [**31**]). And, now follows part of the letter of *Einstein* that was published in *Hadamard*'s book on '*The Psychologogy of Invention in the Mathematical Field*': "The words or the language, as they are written or spoken, do not seem to play any role in my mechanism of thought. The physical entities which seem to serve as elements in thought are certain signs and more or less clear images which can be voluntary reproduced and combined. There is, of course, a certain connection between these elements and relevant logical concepts. It is also clear that the desire to arrive finally at logically connected concepts is the emotional basis of this rather vague play with the above mentioned elements. But taken from a psychological viewpoint, this combinatory play seems to be the essential feature in productive thought –before there is any connection with logical constructions in words or other kinds of signs which can be communicated to others. The above mentioned elements are, in my case, of visual and some of muscular type. Conventional words or other signs have to be sought for laboriously only in secondary stage, when the mentioned associative play is sufficiently established and can be reproduced at will", and, finally, his quote "*Das eigentlich schöpferische Prinzip aber liegt in der Mathematik*", taken from [**56**].

Like by writing a lot one may learn well to read much better, by drawing and painting a lot one may learn well to see much better. Connected with the latter, in his '*Praise of Our Eyes*', *da Vinci* argued that every true science begins with sense-experiences of "our reality", and, that most important among our senses in this respect is our sight, while in his '*Treatise on Painting*', he argued that no human inquiries, on whatever, that do not seriously base on mathematics, can be called scientific, and, that *painting is the science of human vision*, this latter statement concerning the art of making correct 2D representations of our visual perceptions of 3D natural scenes. And *Dürer* came far in the art of well making such representations in *2D drawings of 3D geometrical scenes.* Certainly in the 16th and 17th centuries, there were Jesuit colleges offering better scientific educations than many universities, and, for instance, almost a century after Dürer's stay at Antwerp, rector *F. d'Aiguillon* of the local such college published his *lectures on optics* making use of the *orthogonal projection* (later called *Monge projection*), the *stereographic projection* (named as such by him and which had been introduced by *Hipparchus* around –150 or so and the *conformal character* of which was shown by *Ptolemaus* around 150), and the *scenographic- or central projection.* And, in *Monge*'s 1798 '*Géométrie descriptive*', the goals of this field were stated as follows: "*... représenter avec exactitude, sur des dessins qui n'ont que deux dimensions, les objets qui en ont trois, et qui sont capables de définitions rigoureuses. Déduir de la description exacte des corps tout ce qui suit nécessairement de leurs formes et de leurs positions respectives*". By the way, *Kommerell*'s [**46**] contains next comment on Monge's sort of "*haptic mental sight*": "...das phänomenale räumliche Anschauungsvermögen, das ≪den Vater der darstellende Geometrie≫ zu Gebote stand; man hat den Eindruck, das die Geraden und Ebenen, die Kurven und Flächen, um die es sich handelt, mit gerade zu greifbarer Deutlichkeit vor Monge's geistlichen Auge standen". And, here are *some excerpts of Gauss' review of this course on descriptive geometry*, cfr. [**26**]: "It is not to be denied that the advantages of an analytical treatment over a geometric treatment, its conciseness, simplicity, uniformity, and especially its generality, usually become more and more decisive as the investigations become more difficult and more complicated. However, *it is always very important to continue to cultivate the geometrical method.* (...). In particular we must praise the work under consideration for its great clarity ..., and therefore recommend its study as nourishing intellectual substance by which undoubtedly much can be contributed to the revival and conservation of the genuine geometric spirit, sometimes missing in the mathematics of these times. (...). ... *(the geometric method will) remain indispensable in the early study of young people*, to prevent one-sideness, ... and to give to the understanding a lineliness and directness, which are much less developed and –occasionally– rather jeopardized by the analytic method". Clearly, the elementary geometric 2D descriptions of the idealisations of some simple natural objects as they are perceived in 3D are not able to well describe the idealisations of more complicated natural objects as they are perceived in 3D. Hereto, necessarilly *the algebraic and the analytic descriptions then came to the rescue*, thus yielding the evolutions from (α) to (β) and (γ).

This seems a good place for some quotations from *Klein*'s '*Elementary mathematics from an advanced standpoint* –geometry' [**42**], notes of lectures given to teachers of mathematics in 1908: "*As a matter of fact, the source of all fundamental geometric concepts and axioms is our naive geometric perception*"; "*The axioms of (Euclidean; L.V.) geometry are –according to my way of thinking– not arbitrary, but sensible, statements, which are, in general, introduced by space perception and* *are determined as to their precise content by expediency*"; "*... we consider it desirable ... to develop space perception as a whole as early as possible, and consequently to accustom the pupil from the beginning to three-dimensional figures, rather than to restrict artificially his first instruction to the plane*"; "*I am urging you to make a thorough study of descriptive geometry, which is, I think, indispensable for every teacher of mathematics*".

Eventually to be taken into account when willing *to psychologically reflect* a bit more *about the visual arts and about the sciences*, now are mentioned some facts taken from [**4, 55**]. Already *since about 40 000 years ago, figurative 3D and 2D representations of men and animals are being made*, among the first such sculptures being 'the Lion-man of Holenstein-Stadel' and 'the Venus of Hohle Fels' and among the first such drawings and paintings being the well known ones in caves in Spain and France. Already long before, *abstract geometrical 2D "drawings"* were being made. Indeed, *since about 100 000 years ago*, in South-Africa, on shells of ostrich eggs, images were made of "*horizontal and vertical lines*" (cfr. the "*gravitational cross*" as discussed e.g. in [**5**]), and, in particular, of "*square tilings*" –i.e., of *two sets of more or less equidistant more or less parallel lines*, which are *symmetric* in that the lines of distinct sets cut each other more or less in a same angle, of about 90°– , while, *since about 75 000 years ago*, also in South-Africa, on red ochre coloured stones, images were made of "*equilateral triangular tiling*" –i.e., of *three sets of more or less equidistant more or less parallel lines*, which are *symmetric* in that the lines of distinct sets cut each other more or less in a same angle, of about 60°– .

And, this may be a good place *to mathematically reflect* somewhat more about *the rôle played by parallelism from the prehistory of geometry till now*; cfr. a.o. [**6, 17, 27, 44**]. In this respect, in particular, one might hereby take into consideration the following quote from *Coolidge*: $\sim$ "*... it is to the doubts about Euclid's parallel postulate, and efforts of many thinkers to settle these doubts, that we do owe the whole modern abstract conception of mathematical science*", whereby, in a kind of ultimate state, notably by the contributions of *Hilbert* and *Gödel*, one arrived at the present day understanding of *the logical basis of mathematics* as summarised in the concluding sentence, "*It is better to be aware of our limitations than to live in a fool's paradise*", of the Chapter '*The Shape of Logic*' in Stewart's '*Taming the Infinite*' [**62**].

4. As far as I know, the Greek "*theorema*" is derived from the Greek "*theorein*" which means "*to seriously look at things in a way so as to really see what there is to be seen*", or, still, "*to look at things with the purpose of understanding*", and which is also at the origin of the Greek "*theoria*", of which "*contemplation*" is the Latin equivalent. And while "contemplation" seems more to refer to a meaning including *an aesthetical appreciation* –say, in a way, refers to "*early vision*"– , "theoria" seems more to refer to a meaning of *an intellectual appreciation* –say, in a way, refers to "*cognitive vision*"– .

The following is a citation from *R. Dontot*'s contribution on geometry in the '*general 1937 Larousse*' [**29**]: "*L'objet de la géométrie est, en principe, l'étude des formes et des propriétés des corps naturels.* Ceux-ci sont beaucoup trop variés pour qu'une pareille étude soit possible: ainsi, le géomètre substitue-t-il à ces corps des figures, appelées *figures géométriques*, qui en sont des images schématisées, figures qu'il est possible de définir rigoureusement et par conséquent d'étudier avec précision. La géométrie n'est donc pas une science expérimentale, puisque son objet est, non pas d'étudier certains aspects de la nature, mais une reproduction nécessairement arbitraire de celle-ci: c'est pourquoi il est d'usage de dire que *la géométrie est une science abstraite*, bien qu'à la vérité elle puise ses inspirations dans l'étude des phénomènes expérimentaux et qu'elle contrôle ses résultats, en principe théorique, par des mesures applicables à des examples concrets. La notion expérimentale la plus simple est celle de volume. On dit qu'un corps occupe un certain volume lorsqu'il occupe dans l'espace une certaine place. Nous admettrons comme une chose évidente qu'un volume est limité par une surface; mais si l'existence du volume est physiquement contrôlable et physiquement mesurable, *la surface est une création de l'esprit*: c'est quelque chose d'analogue à une baudruche, par exemple, enveloppant un volume quelconque, mais d'analogue seulement. *C'est un être géométrique*". Accordingly, in its main notions and in its main statements about these notions, *classical geometry –Euclidean and conformal and projective geometry–* essentially deals with abstractions and generalisations of some primitive objects and their properties in the way that these objects and their properties are perceived by *human vision* in "our" surrounding world, *statically* as well as *dynamically* and *directly* as well as *through natural processes like various sorts of projections and of sections.* And, from *Freudenthal*'s '*Initiation to Geometry*', (written before the above mentioned findings in South-Africa were made), comes the following: "Ornaments from the dawn of mankind seem to support the thesis that *symmetry has been the first geometrical idea that arose in the human mind. Symmetry in its broad sense probably was the first mathematical idea that caught the attention of human beings, in particular before quantifications and countings of any kind, human beings appear to have been interested in similarities of forms, of shapes, in its wide sense.* Through an infinity of instances from nature and technics, symmetry may be approached by the childish spirit (...). One may argue that geometry teaching should start with stereometry. Solid bodies are less abstract than plane shapes. The child cap grasp them in a literal sense. Stereometry meets the children's creative wishes. Figures are drawn, solids are made. *Geometry, as a logical system, is a means –and even the most powerful means– to make children feel the strength of the human spirit, that is: of their own spirit*".

In the same '*general 1937 Larousse*' [**18**], *A. Cuvellier* wrote the following: "*Esthétique* signifie étymologiquement: *théorie de la sensibilité.* Mais, depuis le 18ième siècle, le mot s'applique spécialement à *la sensibilité relative au beau. Le beau est le résultat* d'une action réciproque de l'homme et de la nature, ou, si l'on veut, *d'une action de l'homme* sur la nature, mais *sur une nature donnée, à l'aide de procédés techniques donnés, et c'est cette action qui s'appelle l'art. ... certaines impressions élémentaires de symétrie ou d'harmonie, liées aux conditions d'exercise de nos sens, entrent comme composantes dans le plaisir esthétique*; des recherches expérimentales poursuivies notamment par *Fechner*, ont établi qu'*il existe par example une proportion déterminée (la section d'or) entre les côtes d'un rectangle, qui nous le rend agréable à l'œil*". *The "golden" rectangles*, that is, having as proportion between their width and their length this *"golden" number* φ, are characterised by the property that *their gnomonic growth* –adding a square on one of their lengths– *produces a conformal rectangle*, or, still, *they are the rectangles of which the shape, the form, is preserved under gnomonic growth.* And, as pointed out by *M. Borisavljević* in his 1958 '*The golden number and the scientific aesthetics* *of architecture*': "*The visual field for both eyes represents an oval shape inscribed in a (horizontal) golden rectangle*". Actually, the *vertical* –V– and the *horizontal* –H– spans of normal human sight are about 130° and 200°, respectively, of which most people are not aware though. This lies at the basis of the so-called *HV effect*, whereby any given line segment when presented in a vertical position is perceived to be significantly longer than when presented in a horizontal position. This *HV effect* was –in a most unsubtle way– taken into account in the *geometrical description of human visual sensation*, and the –not so unrealistic– corresponding *geometrical description of human visual perception*, was illustrated for *the irradiation illusion of Helmholtz* and for *the bulge illusion of Kitayoka* in [**53**].

From *C. Vandersleyen*'s French edition of '*Das Alte Ägypten*' comes the following: "La science égyptienne se fondait sur l'idée que le monde était connaissable. Ne fut donc représenté ou décrit que ce qui avait été reconnu. La question des causes, des lois physiques, d'après la représentation égyptienne aboutissait à *Maat*, cet *ordre ≪correct≫* institué par les dieux, *organisation parfaite du macrocosme et du microcosme.* (...). *Les Égyptiens*, en fin de compte, *pratiquaient la science pour reconnaitre Maat et s'y rendre toujours plus conformes.* Lorsqu'ils arrivaient aux frontières des possibilités du savoir, le mythe entrait en jeu. (...). *Dans la construction d'édifices, on utilisait le triangle de Pythagore pour la construction de l'angle droit; la section d'or ainsi que les nombres de Fibonacci servaient à donner une structure harmonieuse aux édifices*", while, in the '*Glossaire*' one finds the following presentation of *Maat*: "Mot égyptien féminin désignant une notion philosophique fondamentale de la pensée égyptienne. C'est *l'équilibre* qui maintient le monde dans l'ordre prévu dès l'origine par le créateur. *Maat symbolise donc aussi la vérité, la justice, l'ordre.* Elle est figuré sous la forme d'une déesse portant sur la tête la plume d'autruche écrivant son nom. Elle est l'essence même de la vie des dieux".

The "Nederduitse" name for "*geometry*" is "*meetkunde*", which actually expresses much better than the term geometry itself what geometry is really about. "*To measure the earth*" was and remains a fascinating endeavour, and, for instance, *the geometer and cartographer Mercator's conformal planar representation of the earth whereby spherical loxodromes are mapped onto straight lines*, for always is a great example of "meetkunde", both from the pure and from the applied points of view. But the name "meetkunde" literally, stands for "*the art to achieve measure, balance, what is just* (*meaning here correct as well as precise*)", and the Dutch "*meten*" means "*to measure*", i.e. "*to take the measure*", measure in Dutch being "maat". Kepler tried in vain to introduce the term "Masskunst" in stead of "Geometrie" also in "Hoogduits". And, for instance, "*symmetry*" means "*same measure*"; (in this respect, see also [**61**]). Next follows one more citation from 'Das Alte Ägypte's *Livre des Morts*': "... le défunt est conduit par un dieu devant le tribunal. Là, il est d'abord soumis à la pesée. *Son coeur, en tant que siège de la compréhension et de la mémoire, est placé sur le plateau de balance opposée à celui de Maat.* (...). Maat est représentée sur un plateau de balance par son symbol, une plume ou une déesse accroupie portant une plume. Le but recherché dans cet épisode de la pesée est d'amener les deux plateaux de la balance à s'équilibrer. (...). A côté de la balance, *le scribe et dieu du calcul, Thot, observe et note le resultat.* Ensuite, le défunt est conduit devant le juge de l'au-delà qui doit annoncer la justification".

Finally, as put by *Rik Verhulst* (one of the most distinguished mathematics teachers of the past half century in the Low Countries), in his 2016 philosophy book '*De zes hamerslagen van de westerse rationaliteit –Van speculatief denken tot baanbrekende wetenschap–*' :∼"*It was in Ancient Greece that for the first time appeared a philosophical inclination, gradually turning into a general attitude, to try to understand and to explain our reality and existence within a collection of rational, theoretically connected models* instead of by usual mythological or religious stories".

5. Here is a quote from the *Introduction* to the 1986 '*Kleine Enzyclopaedie Mathematik*' [**41**]: "In ihre historische Entwicklung ist die Mathematik zunächtst einmal in ganz naïver Weise vorgegangen. *Sie hat angefangen bei der Zahlen* 1,2,3 usw. *und bei den anschaulich ganz selbstveständlichen Figuren der Geometrie*, den Punkten, Strecken, Geraden, Ebenen in Raum, Winkeln, Dreiecken, Kreisen, und ist zu komplizierteren Gebilden aufgestiegen, *wobei das Reich der Zahlen und der Figuren nicht getrennt voneinander entwicklen, sondern durch ein Begriff des Messens miteinander verbunden sind.* In diese vom anschaulich Einfachen und Selbstverständlichen zu komplizierteren Problemen fortschreitenden Entwicklung vollzog sich der Aufbau der Mathematik zum Beispiel bei der Babyloniern und der Ägyptern, und dabei wurden schon erstaunliche Leistungen erzielt wie etwa in der Astronomie die Vorausberechnung von Mondfinsternissen. Auf eine völlig neue Stufe der Entwicklung wurde die Mathematik durch *die alten Griechen* gehoben, die sich veranlasst sahen, nicht nur immer weiter vorwärts zu stürmen, sondern sich Rechenschaft darüber abzulegen, was man eigentlich tut, wenn man Mathematik treibt. *Der Erfolg war, dass durch sie die Mathematik zu einer Wissenschaft in heutigen Sinne wurde*".

As far as I know, the golden proportion φ first showed up some 3500 years ago, in Egypt. *However easy to construct by ruler and compass regular* 3*-gons or equilateral triangles, and thus also regular* 6*-gons and* 12*-gons, etc., and however easy to construct by ruler and compass regular* 4*-gons or squares and thus also regular* 8*-gons, etc., the construction by ruler and compass of regular* 5*-gons or pentagons and of regular* 10*-gons or decagons etc. is possible, but, is something else*; (and, it was Gauss who settled the matter of which regular n-gons actually are possible to be constructed like that and which are not). Likely in Babylonia, where equilateral triangles did carry a divine status, *the measure of angles* in degrees resulted from giving as value for their three equal angles the basic number of their mostly used numerical system, namely 60. And, correspondingly, the full angle all around measures 360°, a right angle measures 90° and the sum of the interior angles of any triangle in a Euclidean plane is 180°.

For regular inscribed decagons in a circle with center C, the corresponding "central point" triangles (= the isosceles triangles ABC with C as top and a side AB of a decagon as basis) must have a top angle of 36°. So, one studied *isosceles triangles ABC with top angle $\angle C = 36°$ and consequently with base angles $\angle A = \angle B = 72°$*, with as purpose to hopefully find a way to construct them by ruler and compass. Now, when looking at such a triangle ABC, me thinks, it is difficult to resist to draw just like that *the bissectrice of the angle at A* (or at B), *hereby dividing the angle $\angle A = 72°$* (or $\angle B = 72°$) *into two angles of* 36° *which* then in particular *are equal to the top angle C of the triangle ABC*. And, when calling $\tilde{B}$ the intersection point of this bisectrice of the angle at A in the triangle ABC *with the side BC*, and by giving the points A and B as alternative names sometimes also $\tilde{C}$ and $\tilde{A}$, respectively, one readily observes that the triangle $\tilde{A}\tilde{B}\tilde{C}$ is just like the triangle ABC, an isosceles triangle with top angle $\angle\tilde{C} = 36°$ and base angles $\angle\tilde{A} = \angle\tilde{B} = 72°$. Hence, *the initial triangle ABC and the smaller triangle $\tilde{A}\tilde{B}\tilde{C}$ are similar*, they *are conformal*. When calling their bases $AB = c$ and $\tilde{A}\tilde{B} = \tilde{c}$, respectively, and their two equal sides $CA = CB = a$ and $\tilde{C}\tilde{A} = \tilde{C}\tilde{B} = \tilde{a}$, respectively, at once one can see that $\tilde{a} = c$ and that $\tilde{c} = a - c$. And, since the conformality of the two triangles implies that $\tilde{c}/\tilde{a} = c/a$, it then follows that $\tilde{c}/\tilde{a} = c/a = \varphi = (\sqrt{5} - 1)/2$ $(\star)$, or, still, *that the triangles ABC and $\tilde{A}\tilde{B}\tilde{C}$ are "golden" triangles* (i.e. *by "adding" to the isosceles triangle $\tilde{A}\tilde{B}\tilde{C}$*, by prolonging its base $\tilde{A}\tilde{B}$ with the segment $\tilde{B}C = \tilde{B}\tilde{C}$, *the associated isosceles triangle $A\tilde{B}C$* (with top angle 108°), *one obtains a larger isosceles triangle ABC which is conformal with the isosceles triangle $\tilde{A}\tilde{B}\tilde{C}$*, and *the isosceles triangles with top angles of* 36° *are characterized by this property*). And, with $(\star)$, it is not difficult to construct golden triangles by ruler and compass indeed, in particular also making use of the *Theorem of Pythagoras*. It was the interest in regular pentagons as next most elementary most symmetric polygons (all sides have the same measure and all angles have the same measure) after the equilateral triangles and the squares, that led to the golden proportion.

In his history of mathematics, *Proclus* wrote that *it was in the school of Pythagoras that first appeared pairs of mutually incommensurable line segments.* And, here and there, one may find three versions of *logo's of this school, namely, the regular pentagon and the regular pentagram* (formed by the diagonals of a regular pentagon) and "*the three triangles*" (the figure obtained by adding to a golden triangle ABC, say, standing on its base AB, at each of its sides CA and CB, left and right, an associated golden triangle $AC\tilde{B}$ of the golden triangle $\tilde{A}\tilde{B}\tilde{C}$, putting the side AC of such a triangle $AC\tilde{B}$ first on the side CA of the triangle ABC and then on the side CB of the triangle ABC.) That *the sides AC and AB of a golden triangle ABC are not commensurable line segments*, or, equivalently, *that the sides and the diagonals of a regular pentagon are not commensurable*, can be seen as follows. *Assuming*, to the contrary, *that AB and AC would be commensurable, there would exist a "common measure"*, i.e. *a line segment, say DE, fitting exactly as well into AB as into AC*. But, *then*, since for the smaller golden triangle $\tilde{A}\tilde{B}\tilde{C}$ one has $\tilde{A}\tilde{B} = BC - \tilde{B}C = CB - AB$ and $\tilde{A}\tilde{C} = AB$, *the segment DE would also be a common measure for $\tilde{A}\tilde{B}$ and $\tilde{A}\tilde{C}$*, (as one can visualise the side CB as being measured up by DE, which is a common measure of CB and $C\tilde{B} = AB$). Now, going from the golden triangle $\tilde{A}\tilde{B}\tilde{C}$ to a smaller golden triangle, say, $\bar{A}\bar{B}\bar{C}$, in the same manner as going from the triangle ABC to $\tilde{A}\tilde{B}\tilde{C}$, then likewise, the same line segment DE would also be a common measure for the sides $\bar{A}\bar{B}$ and $\bar{A}\bar{C}$. And so on. *In this process, DE remains fixed,* while *the golden triangles* A(something)B(something)C(something) *step by step get smaller and smaller*, and, with them, also their sides A(something)B(something) and A(something)C(something) get smaller and smaller too. Hence, whatever the size of DE, sooner or later, but in any case, this process produces some golden triangle A(something)B(something)C(something) of which the sides would be smaller than DE, so that DE cannot at all anymore be a measure of those sides. *This contradiction proves that the assumption made above is false.* Thus, *AB and AC are not commensurable line segments*, or, in other words, *the golden proportion φ is not a fraction of two natural numbers*, or, still, *φ is not a rational number.* –On the side, the manifold uses of, say, pentagon-like logo's in our times for what not, are not at all so rational as this logo was for the school of Pythagoras–. And, subsequently, *Eudoxos developed a theory which properly dealt with all proportions, the rational and the irrational ones, thus giving a geometrical treatment of the real number system.* *Kepler*'s well known assertion "*Geometrie besitzt zwei grosse Schätze: einer is das Theorem des Pythagoras, das andere ist der Goldene Schnit*" may well be seen in this context: by the knowledge and understanding both of mutually commensurable and incommensurable line segments, essentially *one could deal with the distances of all pairs of points on a straight line.*

For a line segment S, its "golden part" is given by $\varphi \cdot S$, and, in a way in $2D$ analogy, the "*golden angle*" was defined as $\varphi^2 \cdot 360° \approx 137,5°$. In particular, this value roughly agrees with *the angles* occuring for instance *between two consecutive twigs along the stems of many plants*, like rose roses, which finding is attributed to *da Vinci*. And, *in the "10-stairs units" of human DNA molecules, the proportion between the diameter and the height of the right circular cylinder through the two circular helices that support the stairs with the bases, roughly equals φ.* Also, *the shift between these two helices in the direction of the DNA axis basically equals φ.* In its perfect geometrical model, using, for such a minuscule "unit", say, a circular cylinder in vertical position, this corresponds to the property that in the circular horizontal sections of this cylinder, –one for each of the ten "steps"– , *from the center, the step-chords are seen under a golden angle.* When taking into account the width of the stairs in reality, when looking into such a unit top-down from a point higher on the axis, *the* 10 *stairs form two pentagrams*, and around the rim of the cylinder, *their tips form a regular decagon.*

There are several proofs of the fact that the golden ratio is irrational. And, on the way to the proof in '*Mathematics. A Very Short Introduction*' [**35**], *Gowers* writes: "Since a large proportion of the human brain is devoted to the processing of visual data, it is not surprising that many arguments exploit the powers of visualization. To illustrate this, I shall give another proof of irrationality, this time of the *so-called golden ratio.* This is a number that has fascinated non-mathematicians (and to a lesser extent mathematicians) for centuries. *It is the ratio of the side lengths of a rectangle with the following property: if you cut a square off it then you are left with a smaller, rotated rectangle of exactly the same shape as the original one*". In my opinion, there are good reasons for which mathematicians should be not less fascinated by the golden ratio than non-mathematicians could be. And, related to the *Preface* of [**35**], certainly also "*Hilbert space theory*" is one of the "*some main fields*" mentioned in Section 1. Besides, *Stewart*'s history book [**62**] beginning with "*Mathematics began with numbers*", one question might be: "*Really*?"; (if only in accordance with Van der Waerden's above mentioned advice when reading books on history).

The following essentially comes from *Fuchs* [**32**]: ∼"*Proclus* in the 5th century *commented on the parallel postulate like this*: ≪This statement has to be completely omitted from the postulates. For it is a theorem, and it involves many difficulties, which Ptolemaus tried to solve in his book. To prove this theorem, one needs many definitions as well as further theorems≫. But, as it did not succeed to prove 'this theorem', *attempts were made to replace it by some equivalent, more transparent postulate.* One of them was ≪In any triangle the sum of the interior angles is equal to two right angles≫. And, *Wallis* proposed the following: ≪*For any arbitrary* *figure, there is always a similar figure of arbitrary size≫. This postulate means that it is always possible to diminish or enlarge any arbitrary figure by a process which changes its magnitude, but preserves its shape.* Based on this postulate, Wallis succeeded in making Euclid's parallel postulate a provable theorem. What this really comes to in the end is that *Wallis replaced a statement of an only moderate degree of transparency by one which is much more accessible to intuition*", since our visual recognition of conformal figures is so very basically natural, indeed.

6. The following is a citation from *Hermann Helmholtz*'s 1870 '*Ueber den Ursprung und die Bedeutung der geometrischen Axioma*': "*Die Thatsache, das eine Wissenschaft von der Art bestehen und in der Weise aufgebaut werden kann, wie es bei der Geometrie der Fall ist, hat von jener die Aufmerksamkeit aller derer, welche für die principiellen Fragen der Erkenntnisstheorie Interesse fühlten, im höchsten Grade in Anspruch nehmen müssen. Unter allen Zweigen menschlicher Wissenschaft giebt es keine zweite, die gleich ihr fertig, wie eine erzgerüstete Minerva aus dem Haupte des Zeus, hervorgesprungen erscheint* ... Dabei fällt ihr in keiner Weise die mühsame und langwierige Aufgabe zu, Erfahrungsthatsachen sammeln zu müssen, wie es die Naturwisschenschaften im engeren Sinne zu thun haben, *sondern die ausschliessliche Form ihres wisschenschaftlichen Verfahrens ist die Deduction.* Schluss wird aus Schluss entwickelt, und doch zweifelt schliesslich Niemand von gesundene Sinnen daran, dass diese geometrischen Sätze ihr sehr praktische Anwendung auf die uns umgebende Wirklichkeit finden müssen. (...). Die grösste Schwierigkeit in diesen Untersuchungen bestand und besteht immer darin, dass sich mit den logischen Begriffsentwicklungen gar zu leicht Ergebnisse der alltäglichen Erfahrung als scheinbare Denknothwendigkeiten vermischten, so lange die einzige Methode der Geometrie die von Euklides gelehrte Methode der Anschauung war. (...). Denken wir uns –darin liegt keine logische Unmöglichkeit– verstandbegabten Wesen von nur zwei Dimensionen, die an der Oberfläche irgend eines unserer festen Körper leben und sich bewegen. Wir nehmen an, dass sie nicht die Fähigkeit haben, irgend etwas ausserhalb dieser Oberfläche wahrzunehmen, wohl aber Wahrnehmungen, ähnlich den unsrigen, innerhalb der Ausdehnung der Fläche, in der sie sich bewegen, zu machen... . *Anschauungen, die man hat, sich wegdenken ist leicht; aber Anschauungen, für die man nie ein Analogon gehabt hat, sich sinnlich vorstellen, ist sehr schwer. When sie deshalb zum Raume von drei Dimensionen übergehn, so sind wir in unserem Vorstellungsvermögen gehemmt durch den Bau unserer Organe und die damit gewonnenen Erfahrungen, welche nur zu dem Raum passen, in dem wir leben.... Diese letzteren Verhältnisse hat Herr Beltrami dadurch der Anschauung zugänglich gemacht, dass er zeigt, wie man die Punkte, Linien und Fläche eines pseudosphärischen Raumes von drei Dimensionen im Innern einer Kugel des Euklides'schen Raum so abbilden kan, das jede geradeste Linie des pseudosphärischen Raume in der Kugel durch eine gerade vertreten wird, jede ebenste* Fläche der ersteren durch eine Ebene in der letzteren. Die Kugeloberfläche selbst entspricht dabei der unendlich entfernten Punkte des pseudospärischen Raumes. (...). *Während Riemann von der allgemeinsten Grundfragen der analytische Geometrie her dieses neue Gebiet betrat, war ich selbst theils durch Untersuchungen über die räumliche Darstellung des System der Farben*, also durch Vergleichung einer dreifach ausgedehnten Mannigfaltigkeit mit einer anderen, *theils durch Untersuchungen über den Ursprung unseres Augenmasses für Abmessungen des Gesichtsfeldes*, zu ähnlichen Betrachtungen, wie Riemann, gekommen.

Während dieser von dem oben erwähnten algebraischen Ausdrucke, welche die Entfernung zweier einander unendlich naher Punkte in allgemeinster Form darstellt, als seiner Grundannahme ausgeht, ($ds^2 = g_{ij}dx^i dx^j$; L.V), binn ich anderseits von der Thatsache der Betrachtung ausgegangen, dass in unserem Raum die Bewegung fester Raumgebilde mit demjeniger Grade von Freiheit möglich ist, den wir kennen, und habe aus dieses Thatsache die Nothwendigkeit jenes algebraischen Ausdrucks hergeleitet, den Riemann als Axiom hinstelt.... *Es musste vielmehr eine Reihe alltäghliger Erfahrungen, namentlich die Anschauung von der Änlichkeit grosser und kleiner Körper*, welche nur in ebenen Raum möglich ist, *darauf führen jede geometrische Anschauung, die dieser Thatsache wiedersprach, als unmöglich zu verwerfen ... aber es ist dies eine empirische dur Häufung und Verstärkung gleichartig wiederkehrender Eindrücke in unserem Gedächtnis gewonnene Kenntniss*, keine transcendentale und vor aller Erfahrung gegebene Anschauungsform". Thus, in short, part of the above says that *Riemann* came to the geometry of his spaces $(M^n, g = ds^2)$ essentially *by mathematical abstraction and generalisation of Gauss' intrinsic geometry of surfaces* M^2 *in 3D Euclidean spaces* $\mathbb{E}^3$, and *Helmholtz* came to the geometry of Riemannian spaces $(M^3, g = ds^2)$ *basing on his studies of our visual sense.*

The "*research proposal*" of Gauss to go from his "inner" geometry of surfaces M^2 in $\mathbb{E}^3$ to "*abstract*" intrinsic geometry, seems very much inspired by the opening sentence of *Descartes*' 1637 '*Géométrie*', in which he presented his programme *to base the whole Euclidean geometry on the determination of the distances between all pairs of points by means of the Theorem of Pythagoras properly expressed in Cartesian co-ordinates.* In this analytical geometry, the axiomatical foundation of the former synthetic geometry basically is replaced by a geometrical foundation of the real number system, and this programme works equally well in all dimensions. *In this co-ordinate setting of Euclidean geometry, the infinitesimal calculus could well be developed geometrically.* And, in turn, in particular, this made possible to *Newton* ($\sim$1670) to obtain the general analytic formula's for *the curvature of the curves* M^1 *in Euclidean planes* $\mathbb{E}^2$ *at any of their points* and to *Euler* ($\sim$1760) to describe *the curvature behaviour of the surfaces* M^2 *in Euclidean spaces* $\mathbb{E}^3$ *at any of their points*, in terms of the curvatures there of the planar normal sections in all tangent directions of M^2.

In connection with Helmholtz's reference to "*Erkenntnisstheorie*", next follows *Einstein*'s "*erkenntnisstheoretische Credo*" taken from his '*Autobiographisches*': "Ich sehe auf der einen Seite die Gesamtheit der Begriffe und Sätze, die in Bücher niedergelegt sind. Die Beziehungen zwischen den Begriffen und Sätze unter einander sind logischer Art, und das Geschäft des logischen Denkens ist strikte beschränkt auf die Herstellung der Verbindung zwischen Begriffen und Sätzen unter einander nach festgesetzten Regeln, mit denen sich die Logic beschäftigt. Die Begriffe und Sätze erhalten ≪Sinn≫ bezw. ≪Inhalt≫ nur durch ihre Beziehung zu Sinnen-Erlebnisse. Die Verbindung der letzteren mit den erstenen ist rein intuitiv, nicht selbst von logischer Natur. Der Grund der Sicherheit, mit der diese Beziehung bezw. intuitive Verknüpfung vorgenommen werden kan, und nichts anderes, underscheidet die leere Phantasterei von der wissenschaftlichen ≪Wahrheit≫. Das Begriffssystem ist eine Schöpfung des Menschen samt den syntaktischen Regeln, welche in die Struktur der Begriffssysteme ausmachen. Die Begriffssysteme sind zwar an sich logisch gänzlich willkürlich, aber gebunden durch das Ziel, eine möglichts sichere (intuitive) und vollständige Zuordnung zu der Gesamtheit der Sinnen-Erlebnisse zuzulassen; zweitens entstreben sie möglichtste Sparsamkeit inbezug auf ihr logisch unabhängingen Elemente (Grundbegriffe und Axiome) d.h. nicht definierte Begriffe und nicht erschlossene Sätze. Ein Satz ist richtig, wenn er innerhalb eines logischen Systems nach den acceptierten logischen Regeln abgeleitet ist. Ein System hat Wahrheitsgehalt, entsprechend der Sicherheit und Vollständigkeit seinen Zuordnungs-Möglichkeit zu der Erlebniss-Gesamtheit. Ein richtiger Satz erbörgt seine ≪Wahrheit≫ von dem Wahrheit-Gehalt des System, dem er angehört".

7. *Vincensinni began his 1972 article on the history of differential geometry in 'La Revista Internationale di Sintesi Scientifica'* [**78**] *with* "Il serait téméraire de vouloir essayer de donner une idée tant soit peu précise de l'interêt qu'il convient d'attacher aux différents sujets dont l'ensemble constitue la géométrie différentielle. (...) . Quant à la notion même de l'interêt que peut offrir un thème scientifique, lorsqu'on n'en perçoit pas les rapports immédiats ou lointains, avec les réalités inhéritantes au monde concret dans lequel nous vivons, elle est elle aussi, dans une large mesure, affaire de foi, et particulièrement en géométrie, où, plus peut-être que dans les autres branches de la Science, l'intérêt d'un sujet déterminé est susceptible d'affecter des formes très diverses. Indépendamment des qualités intrinsèques du sujet, considéré dans ses relations avec l'ensemble de l'édifice mathématique, des facteurs tenant plus spécialement à la nature de l'esprit humain ou à la sensibilité même de l'âme humaine, peuvent intervenir dans son appréciation. Et il n'est pas jusqu'à la simple mode, qui ne puisse influencer la marche de la pensée géométrique, et le jugement que l'on peut être amené à porter sur la valeur de son évolution. Cette évolution est d'ailleurs intimement liée au courant de conscience qui, depuis le fond des âges, anime l'humanité" (cfr. "*Avant-propos. Encyclopédie française*", 1937, par *Montel*). Depuis les temps les plus reculés en effet, par instinct de conservation d'abord, puis pour améliorer progressivement leurs conditions de vie, les hommes se sont retrouvés dans la nécessité d'inventer et de construire des instruments de plus en plus perfectionnés, de procéder à des comparaisons quantitatives et qualitatives mettant en jeu les notions de nombre ou d'étendu, bref, de se livrer aux différentes opérations qui constituent l'essence même de la pensée mathématique. La part prise par les différents aspects de la pensée mathématique dans l'évolution du courant de conscience auquel il vient d'être fait allusion a naturellement varié, non seulement avec le temps, mais aussi avec les influences que chaque époque, avec ses besoins propres et ses exigences particulières, spirituelles ou matérielles, naturelles ou fortuites, a exercée sur cette même pensée. Mais il semble bien qu'aucune doctrine mathématique n'ait été plus intimement liée que la géométrie à cette évolution. La géométrie représente incontestablement l'une des expressions les plus parfaites de ce besoin de spéculation qui est le propre et la raison d'être de l'esprit humain, et c'est à cette forme particulière de la pensée mathématique, et plus spécialement à la géométrie différentielle qui en est à la fois la quintoscence et le remarquable aboutissement, qu'est consacrée l'étude qui va suivre", *and ended it by referring to Chern's 1970 ICM lecture 'Differential Geometry; its Past and its Future'* [**16**]. Herefrom comes the following: "*Differential Geometry is the study of geometry by the methods of infinitesimal calculus or analysis. Among mathematical disciplines it is probably the least understood* ($\bar{\star}$). (...). The advent into higher and even infinitely many dimensions does make the intuition unreliable and the dependence on algebra and analysis mandatory. The basis of algebra is the algebraic operations and the basis of analysis is the topological structure. I would like to surmise that *the core of differential geometry is the Riemannian structure (in its broad sense)"*. To make sure, in many studies of geometry even in dimensions "only" up to 3, dependence on algebra and analysis is mandatory too, and, maybe "some main fields" of mathematics alluded to in Section 1 did originate in the developments of geometry that were indicated there by (β) and (γ). In any case, the following citation from *Bergson*, (also taken from "*Tome 1. Encyclopédie française*"), very much indeed applies to mathematics in particular: "*L'humanité ne comprend pas bien le nouveau que s'il prend la place de l'ancien*".

The above ($\tilde{\star}$) points to a footnote in which Chern quoted *Birkhoff*'s "... *disturbing secret fear that geometry may ultimately turn out to be no more than the glittering intuitional trappings of analysis*". This might be a good place then to illustrate the possibly distinct firm opinions that mathematicians eventually can have about geometry, and, about mathematics in general, for that matter, by mentioning now the following two passages from *Geometry* I by *Alekseevskÿ and Lychagin and Vinogradov* [**30**]: "*Mathematics is the science that deals with pure thought" (taken from Chapter 1. Introduction: A Metamathematical View of Differential Geometry), and, (taken from Chapter 3. The Field Approach of Riemann),* "The history of the discovery of the main concepts of Riemannian geometry is essentially the process of transferring by degrees to the language of differential calculus the regularities to which our intuitive ideas about the surrounding world are subject. (...). ..., *we explain by a ≪picture≫ which aspects of differential calculus is Riemannian geometry*".

From *Oswald Veblen and Henry Whitehead's 'Foundations of Differential Geometry'* [**71**] comes the following: "During the nineteenth century mathematicians arrived at the notion that there are not one, but many geometries. This idea, which originated in attempts to prove Euclid's axiom of parallelism, reduced in the end to the following: *≪Any mathematical science is a body of theorems deduced from a set of axioms. A geometry is a mathematical science≫. The question then arises why the name geometry is given to some mathematical sciences and not to others.* It is likely that there is no definite answer to this question, (any objective definition of geometry would probably include the whole of mathematics, $\bar{\star}$), but that *a branch of mathematics is called a geometry because the name seems good, on emotional and traditional grounds, to a sufficiently large number of competent people"*. With respect to the above "*any mathematical science is a body of theorems deduced from a set of axioms*", I vividly recall Professor *Shing-Shen Chern* insisting to be so happy not precisely to know what differential geometry really is. And, *Elie Cartan* enthusiastically supported the above "*a branch of mathematics is called geometry because the name seems good, on emotional and traditional grounds, to a sufficiently large number of competent people*", as several of his former geometry students liked to remember, and, from them, Professor *Radu Roşca*, told me that PapaCartan told him that this sufficiently large number of competent people could very well be 1. And, finally, the above $\bar{\star}$ likely is not unrelated to some hope expressed in Section 1.

8. From *M. Berger*'s "*Peut-on définir la géométrie aujourd'hui?*" referring to conversations with *E. Calabi* and *R. Bott* respectively, come the following statements: "*Geometry is any branch of mathematics in which you can trace back your* *primary source of information to intuition in your sensorial experience*" and "*Geometry is the abstraction of our visual perception*". And, in his "*The Foundations of Arithmetic*", Frege wrote the following: "Empirical propositions hold good of what is physically and psychologically existent, the truths of (Euclidean) geometry govern all that is spatially intuitable, whether existent or product of our fancy. The wildest visions of delirium, the boldest inventions of legend or poetry... remain, so long as they remain intuitable, still subject to the axioms of geometry. Conceptual thought can after fashion shake off their yoke, when it postulates, say, a space of four dimensions or of positive curvature. To study such conceptions is not useless by any means; but it is to leave the ground of intuition entirely behind. If we do make use of intuition even here, as an aid, it is still the same old intuition of Euclidean space, the only space of which we can have any picture", what the visual artist *Escher* most compactly expressed by saying that "*Our 3D space is the only reality that we know*".

In any case, *geometry started off as the natural science of human vision, static as well as dynamic*. And, in his "*Introduction to Mathematics*", *A.N. Whitehead* wrote that "The aim of scientific thought is to see what is general in what is particular and what is permanent in what is transitory," and, in this sense overseeing the histories of the natural sciences, he observed that "*All science as it grows to perfection becomes mathematical in its ideas*". Thus, *mathematics developed in the process of the development of the science of our vision, starting from about* −600, *in Old Greece, resulting in the first theories on the Euclidean and conformal and projective geometries*. This way of looking at the period (α) in the life story of geometry may indicate why *originally 'mathematics' and "geometry" were very close to being synonyms*.

In general, *when we "open our eyes" we notice the sorts of things* (basically the things that in geometry are curves, surfaces and solids) *that actually became of interest for us to see in the context of the natural evolution of our human kind* (and, the above mentioned HV visual effect for instance, should also be considered in this way). Of course, all over in these kinds of thoughts, essentially one is wondering about one of the most basic questions in real psychology: *scientifically speaking, what are human perceptions and sensations?* At this stage, here are some statements in this respect, from *Helmholtz* and from *Koenderink-Van Doorn*, respectively taken from '*On the facts of perception*' and from '*On the nature of observations*': "*The impressions of the senses are only signs for the constitution of the external world, the interpretation of which must be learned by experience. ... we are compelled in every explanation of natural phenomena to leave the sphere of sense, and to pass to things which are not objects of sense, and are defined by abstract conceptions.* (...). Perhaps the relation between our senses and the external world may be best enunciated as follows: our sensations are for us only symbols of the objects of the external world, and correspond to them only in some such way as written characters or articulate words to the things they denote. They give us, it is true, information respecting the properties of things without us, but no better than information that we can give a blind man about color by verbal description", and : " Despite the deep cleft between the realm of smooth functions and observed scalar fields, scientists have not been hesitant to apply the methods of differential calculus to actual data. That this actually works fine is to be considered a tribute to common sense. However, a more principled approach is desirable. (...). Consider a single image... In summary, scalar densities can only be defined relative to a fiducial resolution, and the one-parameter family of resolution forms a central atlas and avoids spurious resolution if and only if the observations take place with a Gaussian weighing function, –which in [**53**] was rudimentary amended taking into account the HV effect; L.V.– . This structure is well known today and generally designated as *scale space.* (...). The scale space setting immediately allows us to define differentiation even for scalar fields that are not differentiable at all. (...). *The point is that we know nothing about the "real image", except for the observation...* . In particular, we have no knowledge whatsoever concerning its smoothness or differentiability. However, we can rest assured that the observed density is smooth by design. For all we know, there might well exist infinitely many "real" images that would yield the identical observation. We shouldn't care, since it is irrelevant. They can't be observed! It is entirely a matter of personal taste whether one wants to believe in the existence of non-observable entities or not". For the time being, in these respects, one may be referred to "*Color for the sciences*" [**45**] and to "*De Clootcrans Press*" [**20**]. And, finally, it is well to be aware that, at every moment, human vision is always restricted to only some limited part of whatever sceneries being looked at.

In the above "ordre d'idées" and "en gros", *the geometry of Ancient Greece, by a most natural mathematical evolution, became the submanifold theory of our time. Expressed in an infinitesimal analytical way, first of all, the geometry in 3D Euclidean spaces* $\mathbb{E}^3$, that is, *in the spaces* $\mathbb{R}^3$ *of triply composed real numbers, in which the distance ds between two 'nearby' points* (x, y, z) *and* $(x+dx, y+dy, z+dz)$ *is determined by the Theorem of Pythagoras* –$ds^2 = dx^2 + dy^2 + dz^2$; whereby "the s" likely comes from "*de schuine zijde*", Dutch for "hypothenusa" in a rectangular triangle with HV sides at its right angle– , *and in which accordingly are determined the measures and forms of the curves and surfaces that are situated in these spaces, do well agree with our kind's most primitive visual sensations and perceptions*, and thus could well be called "*the immediate geometry of the human kind*". *Next*, for instance along the lines of the above quote of Frege, *one could after all rather well make the natural generalisation from the geometry of the curves* M^1 *and the surfaces* M^2 *in* $\mathbb{E}^3$ *to the geometry of the submanifolds* M^n *in Euclidean spaces* $\mathbb{E}^{n+m}$, *for arbitrary dimensions n and co-dimensions m*; (on this matter will be devoted some more attention later on). Geometrical perspectives, sections and projections of all kinds, submanifolds of submanifolds, etc., in this way are all notions and procedures that, for specimens of our human kind, are "*really common*" for $m + n = 3$ and "*pretty common*" for $n + m > 3$. In this state of mind, somewhat or maybe too romantically, *the general theory of submanifolds in Euclidean spaces by some has been called "the geometry of the human kind"*.

9. Actually, the essence of most what came up in the preceeding sections, in his way, Gauss was able to write down in one (1) sentence in an 1830 letter to Bessel: "*Nach meiner innigsten Ueberzeugung hat die Raumlehre zu unserm Wissen a priori eine gans andere Stellung, wie die reine Grössenlehre*; *es geht unserer Kentniss von jener durchauss diejenige vollständige Ueberzeugung von ihrer Nothwendigkeit (also auch von ihrer absoluten Wahrheit) ab, die der letzteren eigen ist*; *wir müssen in Demuth zugeben, dass, wenn die Zahl bloss unseres Geistes Product ist, der Raum auch ausser unseres Geiste ein Realität hat, der wir a priori ihre Gesetze nicht vollständig vorschreiben können*". Or, as put by *Kronecker*: "Die principielle

Unterschied zwischen der *Geometrie* und die *Mechanik* einerzeits und zwischen der übrigen hier unter der Bezeichung "Arithmetik" zusammengefassten mathematische Disciplinen anderzeits, besteht nach Gauss darin, das der Gegenstand der letzteren, die *Zahl*, bloss unseres Geistes Product is, während der *Raum* ebens wie die *Zeit* auch aussen unsere Geiste eine Realität hat, die wir a priori ihr Gesetze nicht vollständig vorschreiben können", from which formulation moreover one may pretty well understand *the rough division of the mathematical sciences* into geometry and mechanics, or, say, –static and dynamic– *geometry, on the one hand, and, on the other hand*, the other mathematical disciplines, which Kronecker grouped together as "*arithmetic*" and which later on, alternatively, were rather grouped together as "*algebra and analysis*". And, in his way, *Chern* wrote this down in two (2) sentences, together the opening paragraph of his *Introduction* to the 2000 *Volume I* of '*Handbook of Differential Geometry*': "*While algebra and analysis provide the foundations of mathematics, geometry is at the core. This was already recognized by Euclid, whose book contains a geometrical treatment of the number system*".

'*Raum und Zeit*' is the title of the 1908 talk by Minkowski in which *indefinite Euclidean geometry* was started by *the fusion of our kind's perceptions of locations (x, y, z) in space and of our kind's perceptions of time t into natural measures of events $(x, y, z; t)$ by the "Theorem of Pythagoras in space-times"*, which is well adapted to the different natures of our experiences of space and time, or, put in his own words: "*Man kan danach das Wesen dieses* (Welt-; L.V.) *Postulates mathematisch sehr prägnant in die magische Formel kleiden: 300 000 km = i sek*", –wereby the three letter combination *ict* likely showed up for the first and best time–. The opening lines of *Minkowski*'s lecture were the following: "Die Anschauungen über Raum und Zeit die ich Ihnen entwickeln möchte, sind auf experimentell-physikalischen Boden erwachsen. Darin liegt ihre Stärke. Ihre Tendenz ist eine Radikale. *Von Stund an sollen Raum für sich und Zeit für sich völlig zu Schatten herabsinken, und nur noch eine Art Union der beiden soll Selbstandigkeit bewahren*". A sort of still further stage of what thus began in 1908, was formulated as follows by *Einstein*, when, around 1930, he was asked for a most compact possible summary of his work: "*Früher hat man geglaubt wenn alle Dinge aus der Welt verschwinden, so bleiben noch Raum und Zeit übrig; nach der Relativitätstheorie verschwinden aber Zeit und Raum mit der Dingen*". Next follows a final quote in this respect, from *Thom*'s '*Paraboles et Catastrophes*': "Je pense que les seuls concepts que l'on peut considérer comme raisonablement définis de manière intrinsèque sont ceux susceptible d'être finalement définis (...) uniquement au moyen de considérations liées à la localisation spatio-temporelle. De ce point de vue, *les seuls concepts vraiment ≪scientifiques≫ sont ceux liés à la géométrie de l'espace-temps*". And, based on these 4D Minkowski space-times, the *pseudo Euclidean spaces of arbitrary dimensions and with any signatures*, and, moreover, the *pseudo Riemannian spaces*, could be defined in a natural manner by mathematical generalisation, since, *by that time, arbitrary dimensional Euclidean spaces and Riemannian spaces already were around for a while.*

Putting the two former citations of *Chern* together, in short: "*At the core of mathematics is geometry and at the core of differential geometry is the Riemannian structure*", the *importance of Riemannian geometry cannot much be in doubt, really.* And, this being so, hereafter will be recalled that *Riemannian geometry is part of the geometry of submanifolds*, and, *within submanifold theory, main notions of*

Riemannian geometry can mentally pretty well be seen, like main notions of the intrinsic geometry of surfaces in Euclidean 3D spaces can mentally very well be seen. For some references of what will follow, a.o., see [**40**], (with in particular, *B.-Y. Chen*'s '*Chapter 3. Riemannian submanifolds*' and *U. Limiste*'s '*Chapter 7. Submanifolds with parallel fundamental form*' in *Volume I*, and, *K. Fukaya*'s '*Chapter 4. Metric Riemannian geometry*' and *F. J. Palomo and A. Romero*'s '*Chapter 8. Certain actual topics in modern Lorentzian geometry*' in *Volume II*); *B. L. Laptev and B. A. Rozenfel'd*'s '*Chapter 1. Geometry*' of [**51**], and, to a lesser degree, the Forewords in [**8**–**10**] and the papers [**74, 75**], and, also, [**3, 44, 54, 59, 79**], while *W. Kühnel*'s [**47**] offers a solid neat introduction to it all.

Following several instances in geometry proper and in mechanics, and, also, in mathematics in general and in other sciences, *3D Euclidean geometry has been extended to nD Euclidean geometry, for all natural dimensions n.* Of course, historically and psychologically, this rather was quite a different matter than, say, the consideration of n-gons for any n after having studied triangles and quadrangles in some detail, at least, before one had become well acquainted with "*co-ordinate geometry*". But, then, for instance, the extension of the complex number system in its capacity of algebraically expressing the translations and rotations in Euclidean 2D planes to the quaternion number system having the capacity to algebraically express the translations and rotations in Euclidean 3D spaces, led to a "*mathematically real*" $\mathbb{R}^4$. So did also, as other such for instance, the sort of graph of a complex function of one complex variable, given by the system of the straight lines through the arguments and their corresponding values assigned by these functions, when these are situated in two parallel planes in $\mathbb{E}^3$, while these lines form a subset of the "*mathematically real*" $\mathbb{R}^4$ of all straight lines in $\mathbb{E}^3$. The above "*mathematically real*" 's intend to indicate that, as far as I can see, foremost it were professional, but maybe not too mathematically inclined philosophers who for some time opposed the introduction of spaces of dimensions greater than 3, probably because such spaces could not fit into " their realities"; in his 2015 '*Ontology of the Mirror World*', *Jan Koenderink* wrote the following: "*Reality* (Wirklichkeit) *is an achievement of the mind, although science and analytical philosophy consider it as an ≪input≫ –very strange, I would even say embarrassing*". For explicit descriptions on the first publications of multi dimensional geometry, i.p. see *Laptev and Rosenfel'd*'s [**51**], from which are taken the following quotes: "The terminology of multi-dimensional geometry first appeared in a mathematical work in a small memoir of *Cayley* entitled ≪*Chapters in the analytical geometry of* (n) *dimensions*≫, (1843)", (and the year after did appear Cayley's first article on his *octaves*; L.V.); "The first fundamental investigation in multi-dimensional geometry was *Grassmann*'s ≪*Lineale Ausdehnungslehre*≫, (1844)"; "Another approach to the concept of multi-dimensional geometry was proposed by *Plücker* in his ≪*System der Geometrie des Raumes*≫, (1846)"; "In 1851 *Ludwig Schläfli* (...) carried out wide-ranging research on multi-dimensional geometry"; " In 1875 *Camille Jordan* published his ≪*Essai sur la géométrie à n dimensions*≫ (...). In contrast to the majority of papers on multi-dimensional geometry of the time, Jordan made use (of since then; L.V.) everyday terminology", (and, in particular, in this work he determined the *orthogonal transformations* and, besides, the *Euclidean angles between any two linear subspaces in arbitrary dimensional spaces* $\mathbb{E}^n$; L.V.).

After thus having gone from 3 dimensional Euclidean spaces to Euclidean spaces with an arbitrary natural number of dimensions, then to go *from curves M^1 and surfaces M^2 in $\mathbb{E}^3$ to submanifolds M^n of any dimension n and with any co-dimension m in Euclidean spaces* $\mathbb{E}^{n+m}$ was a natural mathematical matter. As such, in 1874, *Jordan treated the general differential geometry of curves M^1 in $\mathbb{E}^{1+m}$ and carried over Euler-Meusnier's work on the curvature of surfaces M^2 in $\mathbb{E}^3$ to general submanifolds M^n in* $\mathbb{E}^{n+m}$, thereby, in particular, determining *a canonical, distinguished, orthogonal tangent frame on all such submanifolds*, namely, determining their *extrinsic principal tangent directions*, (what *Kronecker* had done in 1869 for the case of *hypersurfaces*). And, recently, *Kostadin Trenčevski* succeeded in determining, in addition, *a canonical, distinguished, orthogonal normal frame on all submanifolds M^n in* $\mathbb{E}^{n+m}$, (or, in any Riemannian ambient space, for that matter), namely, determining their *principal normal directions*; cfr., a.o., [**66, 67**] and, also [**38, 72**].

Following Euler, Gauss basically studied surfaces M^2 in $\mathbb{E}^3$ using "padges", i.e., by expressing the *position vector* $X = (x, y, z)$ of M^2 in terms of *two parameters*: $X(p, q) = (x(p, q), y(p, q), z(p, q))$. From *Peter Dombrowski*'s [**26**] on *Gauss 'Disquisitiones generales circa superficies curvas'* is taken what follows: (out of a letter from Gauss to Schumacher, in 1816), "I had thought of an interesting problem, namely: in the general case, *to project (map) a given surface onto another (given) one in such a way, that the image and the original become infinitesimally similar.* A special case arises when the first surface is a sphere and the second a plane. Here *the stereographic projection* and *Mercator's projection* are particular solutions. However, one wants the general solution for all types of surfaces, containing these as particular cases"; to the title of his 1822 essay solving this problem, Gauss added in Latin: "*Ab his via sternitur ad maiora*", hereby *announcing the "inner differential geometry" based on the "theorema egregium"* of which his *first analytical proof* was *in isothermal co-ordinates*; (and, from Gauss' 1827 *abstract* of this ≪disquisitiones≫), "These theorems lead to the consideration of *the theory of curved surfaces from a new point of view*, where a wide and still wholly uncultivated field is open to investigation. *If we consider surfaces not as boundaries of bodies, but as bodies of which one dimension vanishes*, and if at the same time we conceive them as *flexible but not extendible*, we see that *two essentially different relations must be distinguished*, namely, *on the one hand, those that presuppose a definite form of the surface in space*; *on the other hand, those that are independent of the various forms which the surface may assume.* The discussion is concerned with the latter. In accordance with what has been said, the measure of (the Gauss; L.V.) curvature belongs to this case. But it is easily seen that the consideration of figures constructed upon the surface, their angles, their areas and their integral curvatures, the joining of points by means of shortest lines, and the like, also belong to this case. *All such investigations must start from this, that the very nature of the curved surface is given by means of an arbitrary linear element* (ds; L.V.) *in the form* $(Edp^2 + 2Fdpdq + Gdq^2)^{1/2}$", the latter giving the expression of the *first* or *metrical fundamental form $g = ds^2$ induced on a surface M^2 in $\mathbb{E}^3$ from the Euclidean metric $dS^2 = dx^2 + dy^2 + dz^2$ on* $\mathbb{E}^3$. Between brackets, the *second fundamental form h* or the *shape operator* of a surface M^2 in $\mathbb{E}^3$ basically goes back to Meusnier. *Thus arose the distinction between the extrinsic and the intrinsic geometries of surfaces M^2 in $\mathbb{E}^3$, and of submanifolds in general*; (in this respect, see also [**24**]).

It always remains a great joy to reread *Riemann*'s 1854 *creation of Riemannian* geometry (in its broad sense), in his '*Ueber die Hypothesen, welche der Geometrie zu Grunde liegen*', (with as contents: I. *Begriff einer n fach ausgedehnte Grösse*, II. *Massverhältnisse* deren eine Mannigfaltigkeit von n Dimensionen fähig ist, unter die Voraussetzung, das die Linien unabhänging von der Lage eine Länge bezitzen, also jede Linie durch jede messbar ist –that is, basically, *giving a line element ds as positive definite homogeneous expression of the first degree in the infinitesimals dx of the co-ordinates x that is invariant under changes of the signs of dx and of which the coefficients are functions of x*– , III. *Anwendung auf den Raum*). *Proper Riemannian geometry* corresponds *to definite metrics of the form* $g = ds^2 = g_{ij}dx^i dy^j$, while *Riemann-Finsler geometry* corresponds to *Riemannian metrics without the "quadratic restriction"* [**2**]; and, *pseudo Riemannian geometry* corresponds to *indefinite non-degenerate metrics of the form* $g = ds^2 = g_{ij}dx^i dx^j$. Maybe it is not without interest to close this paragraph with the last sentences of part I. *Begriff einer n fach ausgedehnte Grösse*: "Es giebt indess auch *Mannigfaltigkeiten, in welchen die Ortsbestimmung* nicht eine endliche Zahl, sondern entweder *eine unendliche Reihe oder eine stetige Mannigfaltigkeit von Grössenbestimmungen erfordert*. Solche Mannigfaltigkeiten bilden z.B. die möglichen Bestimmungen einer Funktion für ein gegebenes Gebiet, die mögliche Gestalten einer raumlichen Figur u.s.w.".

By the 1956 "*Isometric Embedding Theorem of Nash*", *every (proper) Riemannian manifold* (M^n, g) *can be identified with submanifolds* M^n *in Euclidean spaces* $\mathbb{E}^{n+m}$, (having sufficiently high co-dimensions). So, *Riemannian geometry essentially is the intrinsic geometry of the submanifolds of Euclidean spaces.* Likewise, *pseudo Riemannian geometry essentially is the intrinsic geometry of the non-degenerate submanifolds of pseudo Euclidean spaces.* And, it seems difficult to be surprised by the relevance of the study of the relations between the intrinsic and the extrinsic geometries of submanifolds, ever since 1827.

The extrinsic geometry of a submanifold M^n *in an ambient space* $\widetilde{M}^{n+m}$, i.e., *the geometrical properties of a submanifold* M^n *which are determined by the form or shape that* M^n *assumes in* $\widetilde{M}^{n+m}$, is basically determined by its *second fundamental form h* and by the *connection* $\nabla^\perp$ *in its normal bundle.* In a way, the simplest situation in which h and $\nabla^\perp$ do both show up, concerns surfaces M^2 in $\mathbb{E}^4$. In *Bernard Rouxel*'s research essay [**58**], (wherein remarkable connections are shown between recent and classic beings and notions belonging to the geometrical world of submanifolds), *among the first geometric studies of submanifolds with co-dimension* > 1 *and related to their normal spaces*, are mentioned *K. Kommerell*'s 1897 thesis at Tubingen, on the *characteristic conic in the normal plane of a surface* M^2 *in* $\mathbb{E}^4$, and, *generalisations* thereof for submanifolds M^n in $\mathbb{E}^{n+m}$ by *D. Perepelkine*. The characteristic conic F in the normal plane $T_p^\perp M^2$ at a point p of a surface M^2 in $\mathbb{E}^4$ is defined as *the locus in* $T_p^\perp M^2$ *of the intersections of* $T_p^\perp M^2$ *with its neighbouring normal planes*, and, the *center* of F is the endpoint of the *mean curvature vector* or the *Bompiani vector* $\vec{H}$ of M^2 in $\mathbb{E}^4$ at p, –thus, *minimal surfaces* M^2 in $\mathbb{E}^4$ are characterised by their characteristic conics being centered on these surfaces– . Kommerell also found that the *surfaces R of Riemann* are characterised by their *characteristic conics being circles centered on these surfaces* and discussed their link with holomorphic functions. And, finally, the better known *curvature ellipses G of surfaces M^2 in $\mathbb{E}^4$* are the *polar reciprocals* of Kommerell's characteristic conics *F with respect to the unit circles in the normal planes which are centered on these surfaces.*

From *Osserman*'s 1990 AMS Monthly article '*Curvature in the Eighties*' comes the following: "*The notion of curvature is one of the central concepts of differential geometry*; one could argue that it is *the* central one, *distinguishing the geometrical core of the subject from those aspects that are analytic, algebraic, or topological.* In the words of *Marcel Berger*, curvature is ≪*the Number 1 Riemmanian invariant and the most natural.* Gauss and then Riemann saw it instantly≫". A corresponding passage from Riemann's 'Habilitation-Schrift' concerning the *sectional, Riemannian curvature*, goes like this: "Um dem Krümmungsmass einer n fach ausgedehnte Mannigfaltigkeit in einem gegebenen Punkte und einer gegebenen durch ihn gelegten Flächenrichtung *eine greifbare Bedeutung* zu geben... ". And, *it would be well that the intrinsic as well as the extrinsic curvatures of submanifolds M^n in* $\mathbb{E}^{n+m}$, (and, also in other ambient spaces for that matter), *would be described as "haptic" as possible*; (actually, in several cases, this means nothing more than just to give the geometrical definitions of these curvatures). For instance, *for surfaces M^2 in $\mathbb{E}^3$*, besides treating *the Gauss curvature as the Jacobian of the Gauss map*, one might expect that the textbooks on elementary differential geometry would also explicitly treat *Sophie Germain*'s definition of her *mean curvature* H [**33**] and *Felice Casorati*'s definition of his *Casorati curvature* C [**7**]. And, at least *for submanifolds M^n in $\mathbb{E}^{n+m}$*, one might expect that the textbooks on submanifold theory would, as well as possible, treat the geometrical meaning of the extrinsic and intrinsic curvatures, say, directly or indirectly, eventually making use of suitable sections and/or projections of these submanifolds; (like, for instance, *a component* $R^{\perp}(\vec{t}_1, \vec{t}_2; \vec{n}_1, \vec{n}_2)$ *of the normal curvature tensor* $R^{\perp}$ of some M^n in $\mathbb{E}^{n+m}$ –$n \geq 2, m \geq 2$– , at some point p, being determined in terms of the area of the curvature ellipse at p of the surface which is the projection of the 2D normal section of M^n at p by the $(2+m)$D subspace $\mathbb{E}^{2+m} = (\vec{t}_1 \wedge \vec{t}_2) \oplus T_p^{\perp} M^n$ in $\mathbb{E}^{n+m}$ onto the 4D subspace $\mathbb{E}^4 = \vec{t}_1 \wedge \vec{t}_2 \wedge \vec{n}_1 \wedge \vec{n}_2$). By the way, for nD submanifolds, their *Casorati curvature* $C = ||h||^2/n$ likely is their *most valuable scalar valued extrinsic curvature*, certainly when intuitively, "visually" thinking about the meaning of being curved [**38, 72**]. Finally, "*Geometry being one topic, whether it is local or global*" (Chern), concerning *global results involving various curvatures of submanifolds, intrinsic ones and extrinsic ones*, one can find many beautiful examples in the already above indicated literature, to which, among the more recent such results, it seems well to add, in particular, *F. Dillen and W. Kühnel*'s results on *the total curvature of complete submanifolds of Euclidean spaces* [**25**] and *the solution of the Willmore conjecture* by *F.C. Marques and A. Neves* [**50**].

In the times before the (global and even local) isometric embedding theorems for Riemannian manifolds into Euclidean spaces had been proven, in several studies of Riemannian manifolds (M^n, g), it happened that some *intrinsic problems were worked out extrinsically*, by considering these manifolds, in one or other way, as submanifolds M^n in some Euclidean spaces $\mathbb{E}^{n+m}$, anyway. As such, in his treatment of the equivalence problem that he solved there independently from Christoffel and Lipschitz, in his 1871 master thesis at Kazan, *on 3D Riemannian manifolds*

(M^3, g), *F.M. Souvorov* obtained that *the critical values of the sectional curvatures* $K(p, \pi)$, at any point of M^3 and for any 2D plane π in T_pM^3, *occur in three mutually orthogonal planes*, (what, in the present special case of dimension 3, in fact amounts to finding, with the lines of intersection of these planes, *the intrinsic principal directions* of (M^3, g), i.e., *the three mutually orthogonal critical directions of its –later to come– Ricci tensor*), by considering these Riemannian manifolds (M^3, g) as hypersurfaces M^3 in $\mathbb{E}^4$, ($\sim$ " sacrificing the generality to the advantage of a more easy intuition"). And, end 1916, *Tullio Levi-Civita* [**49**] came to *his notion of parallel transport on Riemannian manifolds* (M^n, g) in a so nicely geometrically argumented way, by freeing the "formal analogon" of parallel transport in Euclidean spaces, as expressed on Riemannian manifolds when they are supposed to be submanifolds M^n in some $\mathbb{E}^{n+m}$, from all ingredients that refer to the ambient space, till retaining an expression only depending on the metrical fundamental form g; and, after having drawn some immediate geometrical consequences from this expression and after having made the link with Ricci's absolute differential calculus, he then gave *his "greifbare" interpretation of the Riemannian sectional curvature in terms of his parallelogramoids.* Around the same time, *J.A. Schouten* [**60**], in a purely intrinsic way, did arrive to *the equivalent notion of the Riemannian connection* and to *the interpretation of the Riemannian curvature as holonomy of tangent directions.* For the prehistory of the Levi-Civita parallel transport and for *Koszul*'s consideration of covariant differentiations as linear differential operators on tangent vector fields, (and the since then classical development of Riemannian geometry in *Koszul*'s ∇ *calculus*, which, as far as I know, was first done in *Katsumi Nomizu*'s doctoral thesis with Chern at Chicago, in the early 1950s), and *its impact on differential geometry in general, on other fields of mathematics and in theoretical physics*, see, i.p., [**27, 51**]. Of course, *the Riemannian metric g and the corresponding Riemannian connection ∇ on a proper Riemannian manifold (M^n, g)*, when –as the Theorem of Nash allows to do in many ways– considering this Riemannian manifold as a submanifold M^n of a Euclidean space $\mathbb{E}^{n+m}$, are nothing else than *the metric and the connection induced on M^n starting from the Euclidean metric and from the directional derivatives on the ambient space* $\mathbb{E}^{n+m}$.

10. The present contemplation of submanifolds will be ended in this last section, by briefly giving *some illustrations of just a few results from the geometry of submanifolds.* These results are mainly taken from the research work of *Professor Bang-Yen Chen*, who, for about half a century now, very rightfully is being considered as *a prime personification of submanifold theory.*

In the late 1970s, *B.-Y. Chen and T. Nagano* created their so-called (M_+, M_-) *theory*, which, by submanifold theory, provided a completely new approach to the study of the compact Cartan symmetric spaces; for surveys of this theory, see [**11, 15**]. Next is mentioned one application of this theory, namely, the determination of the 2-*rank of compact Lie groups.* The 2-rank $r_2(G)$ of a compact Lie group is *the maximal possible rank of the* 2-*subgroups of* G, and *A. Borel and J.-P. Serre* (in 1953) and *D.Quillen* (in 1971) had been able to determine this rank for some particular groups G. In their (M_+, M_-) theory, Chen and Nagano, for any connected Riemannian manifold M, introduced the 2-*number* $\#_2M$ as *the supremum of the cardinal numbers $\#A_2$ of all its antipodal subsets A_2*, (A_2 are the subsets of M for which all pairs of points are antipodal in M), which is a *Riemannian invariant.* And, for *compact symmetric spaces* M, they gave a method to determine

$\#_2 M$ by using pairs of totally geodesic (M_+, M_-) submanifolds of M. Thus, based on *Their Theorem which, for compact Lie groups G, links their geometric invariant* $\#_2 G$ *to Borel-Serre's algebraic invariant* $r_2(G)$: $r_2(G) = \log_2(\#_2 G)$, they showed *how to find the 2-rank of any compact Lie group.*

In his 1968 *Kansas Lecture Notes*, *Chern* asked to look for *intrinsic geometric conditions on Riemannian manifolds* (M^n, g) *that would prevent their minimal immersability in Euclidean spaces* $\mathbb{E}^{n+m}$, the only known such condition at that time being *a non-negative definite Ricci tensor* (and, besides, there is the topological condition of being *compact*). From the *Preface* of *Chen*'s book **[8]**, in this respect, comes the following :"In order to overcome such difficulties, *one needs to introduce new types of Riemannian invariants, different in nature from ≪classical≫ invariants.* Moreover, one also needs to establish general optimal relationships between the main extrinsic invariants with the new intrinsic invariants of the submanifolds. These were exactly the author's original motivations in early 1990s to introduce his so-called δ-curvature invariants on Riemannian manifolds. The δ-curvatures are very different in nature from the ≪classical≫ scalar and Ricci curvatures; simply due to the fact that both *scalar and Ricci curvatures are ≪total sums≫ of sectional curvatures on a Riemannian manifold*; in contrast, the author's δ-invariants are obtained from the scalar curvature by *throwing away a certain amount of sectional curvatures*". In particular, for a given Riemannian manifold (M^n, g) *these relationships do give certain controls on the possible shapes that can be assumed in some ambient Euclidean spaces* when according to the Theorem of Nash (M^n, g) would be isometrically embedded in there (globally or locally); in a way, thus are diminished, amongst others, *Yau*'s complaints vis-à-vis Nash's Theorem as formulated in his contribution to the 1991 book on '*Mathematics today and tomorrow*'. As first special case of the δ-curvatures, the *Chen curvature* $\delta(2)$ of a Riemannian manifold (M^n, g) is defined by $\delta(2) = \tau - K_{min}$, –at any point of M^n– , whereby, –at that point– , τ is the *scalar curvature* and K_{min} is the *infimum of the sectional curvatures* of (M^n, g). For $\delta(2)$, the above mentioned optimal relationship with the *scalar mean curvature* H, –at that point– , on (M^n, g), as a submanifold in some Euclidean space $\mathbb{E}^{n+m}$, is given by $\delta(2) \leq (n-2)n^2H^2/2(n-1)$, *and "=" holds if and only if the second fundamental form has a precisely known expression.* It is obvious how this *Theorem of Chen* gives an answer to Chern's above question. *For a given Riemannian manifold* (M^n, g), *among all its possible "isometric realisations" as submanifolds in Euclidean spaces* $\mathbb{E}^{n+m}$, *the ones for which, at every point, the equality "=" holds in the general soft inequality* $\leq$ *of the above Theorem of Chen, are those which give to* (M^n, g) *precisely that shape in its ambient space, for which the corresponding submanifold tension* H^2 *is as small as possible*; and, these are called the $\delta(2)$ *Chen ideal submanifolds.* Among the geometric properties of such Chen ideal submanifolds, one has the following ones: intrinsically, *their Weyl conformal curvature tensor is pseudo symmetric*; extrinsically, they are Chen submanifolds in the sense that *their allied mean curvature vector field vanishes identically*, or, equivalently, *their mean curvature vector field determines a first principal normal vector field*; and, relating their extrinsic and intrinsic qualities: *their extrinsic principal tangent directions coincide with their principal Ricci directions*, **[21, 22]**. And, with respect to his contribution to the book '*Chern –A Great Geometer of the Twentieth Century*', in which *M. Berger* wrote that: "*... knowing* K *is equivalent to knowing* R*...* (K standing for the *sectional curvatures* and R

for the *Riemannian curvature tensor* of Riemannian spaces, this is a *Theorem of E. Cartan*; L.V.). But the relations between K and R are subtle and still not completely understood (e.g. *what the critical planes of* K *are, and how they are distributed*)", the above $\delta(2)$-*Theorem of Chen* concretely deals with this problem for K_{min}, while other results of this kind, moreover deal with the 3*D tangent subspaces with minimal scalar curvature*, etc., besides that also consideration has been given to the situation of maximal curvatures.

From the beginning of *Chapter 6* of *Chen*'s book [**9**] is taken the following: "The main objects of studies in *algebraic geometry* are *algebraic varieties.* Because an algebraic variety is defined by using algebraic equations, one can define the *degree* of an algebraic variety by its algebraic structure. The concept of degree plays a fundamental role in algebraic geometry. On the other hand, according to Nash's embedding theorem, every Riemannian manifold can be realized as a Riemannian submanifold of some Euclidean space with sufficiently high dimension. However, *one lacks the notion of degree for Euclidean submanifolds.* Inspired by the above simple observations, the author introduced in the late 1970s the notions of ≪*order*≫ and ≪*type*≫ for submanifolds of Euclidean spaces. By applying the notion of order, the author was able to establish some *sharp estimates of the total mean curvature for closed Euclidean submanifolds in terms of their orders.* Moreover, by utilising the notions of type, the notions of finite type submanifolds and of finite type maps were naturally introduced. The family of submanifolds of finite type ... contains many important families of submanifolds; including all minimal submanifolds of Euclidean spaces, all minimal submanifolds of hyperspheres, all parallel submanifolds as well as all equivariantly immersed compact homogeneous submanifolds. Just like minimal submanifolds, submanifolds of finite type can be described by a *spectral variational principle*, namely as *critical points of directional deformations*'. Via his theory of finite type submanifolds, an important application that Bang-Yen Chen made of his δ-invariants is the derivation of *new intrinsic results*, for instance, *estimates on the first non-zero eigenvalue of their Laplacian for compact homogeneous Riemannian manifolds* (M^n, g) *with irreducible isotropy action*; see also [**8**]. And, these kind of results, as far as I know, are *first "materialisations"* in the sense of the following quote from *M. Gromov*'s contribution to the 1985 Astérisque book '*Elie Cartan et les mathématiques d'aujourd'hui*': ∼"*This question* (*the isometric immersion problem for Riemannian manifolds* (M^n, g); L.V.) *was raised by Schläfli in* 1873 *and one probably believed at that time that the existence of an isometric immersion of* (M^n, g) *in some Euclidean space might be helpful in the study of the intrinsic geometry of* (M^n, g). *... this belief has not been materialized yet ...*".

Since the beginning 1970s, Chen authored and co-authored many articles on *totally umbilical-* and on *quasi-umbilical submanifolds*; cfr. *Chapters* 5 *and* 6 in his 1973 *'Geometry of Submanifolds'* and *Chapter* 7 in his 1981 *'Geometry of Submanifolds and Its Applications'* [**12, 13**]. Inspired, on the one hand, by *Katsumi Nomizu*'s 1968 *Tôhoku* article on *Cartan- and Szabó symmetric spaces* (locally- and semi symmetric spaces), and, on the other hand, by the 1979 *Soochow* article of *Bang-Yen Chen* on *totally umbilical submanifolds of Cartan symmetric spaces* and by *Zbigniew Olszak*'s independent study of this same topic, in 1983 *Adamów and Deszcz* [**1**] proved that *extrinsic spheres in Szabó symmetric spaces are pseudo symmetric*, and, it further turned out that *extrinsic spheres in pseudo symmetric* *spaces also are pseudo symmetric spaces.* These last results became the real starting point of systematic studies of the latter kind of symmetric spaces, mainly in the school of Wroclaw, and, most in particular by *Ryszard Deszcz*, whence such spaces also became called *Deszcz symmetric spaces* [**77**]. Hereafter, we consider Riemannian manifolds (M^n, g) of dimensions $n \geq 3$. The most possible symmetric such spaces are the *real space forms*, or the *CC spaces* $M^n(c)$, the *spaces with constant Riemannian sectional curvatures* $K = c$, ($c = 0$ or $c > 0$ or $c < 0$): these are the spaces (M^n, g) that are *homogeneous and isotropic*, (at every point they are geometrically the same in all directions and their geometric properties are the same at all points). By the *lemma of Schur, the real space forms are the spaces* (M^n, g) *for which* the sectional curvatures $K(p, \pi)$ at every point p of M^n are independent of the 2-planes π in T_pM, or, still, the spaces for which *the sectional curvature function* K *at every point is isotropic.* The *locally Euclidean* or *locally flat spaces* $M^n(0)$ –in *H. Weyl*'s definition of *what is symmetry*– are characterised by the property that, *at all their points, all tangent directions remain invariant under their parallel transport fully around all infinitesimal co-ordinate parallelograms cornered at those points*; (at a point p and for such parallelograms tangent to a plane π at p, according to *Schouten*, it is precisely $K(p, \pi)$ that measures eventual changes of directions under such transports: "*K constitutes the holonomy of tangent directions*"). When applying *projective transformations* to locally flat spaces $M^n(0)$, CC spaces $M^n(c)$ of any possible c can be obtained, while *the class of the real space forms is closed under projective transformations.* Concerning the symmetry of the sectional curvatures $K(p, \bar{\pi})$, for tangent 2-planes $\bar{\pi}$ at p, under the same transformation of parallel transport as before, one has the following: "*L constitutes the holonomy of sectional curvatures K*", whereby L denotes the *Haesen or double sectional curvature* of (M^n, g); the double sectional curvatures $L(p, \pi, \bar{\pi})$ measure the eventual changes of the value of $K(p, \bar{\pi})$ when moving $\bar{\pi}$ parallely around the above parallelograms [**39**]). In contrast to K, there is no Lemma of Schur for L; so, for the *Deszcz symmetric spaces*, that is for *the spaces* (M^n, g) *for which the double sectional curvature function* L *at every point is isotropic*, $L(p, \pi, \bar{\pi})$ being independent of the planes π and $\bar{\pi}$ in general still changes from point to point, (and if not one speaks of *Deszcz symmetric spaces of constant type*). For a survey on Deszcz symmetry, see [**37**]. *The Szabó symmetric spaces are characterised as the spaces* (M^n, g) *with vanishing Haesen curvature*, ($L = 0$). When applying *projective transformations* to Szabó symmetric spaces, Deszcz symmetric spaces with non-zero curvature functions L can be obtained, while *the class of the Deszcz symmetric spaces is closed under projective transformations.* In the context of some former comments on human vision and some above quotes of Helmholtz and Gauss concerning their studies on Riemannian geometry and inner geometry of surfaces, respectively, *"for beings like us", the locally conformally Euclidean or locally conformally flat spaces certainly do form one of the most important special classes*, say $\mathcal{C}$, *of Riemannian manifolds.* As is well known, *within* $\mathcal{C}$, *the spaces which are intrinsically isotropic*, (that is, the manifolds (M^n, g) for which, *at every point, all directions have the same Ricci curvature*, or, still, *the Einstein spaces*), *are nothing but the real space forms. Hence, in case one would want to look for an extension of the class of the CC spaces by weakening their isotropy property in a reasonable way, –that is effectively, but not drastically– , it seems most natural to consider* the locally conformally Euclidean spaces which are Deszcz symmetric*: indeed, a Riemannian manifold* (M^n, g) *with vanishing conformal Weyl tensor is* Deszcz symmetric if and only if its Ricci tensor has at most two distinct eigenvalues *(or, so to say, if* (M^n, g) *is* "half and half" *Einstein*), [**23**]. Thus, when having as purpose to extend the class of the real space forms by weakening their isotropy property, the prime sort of Riemannian manifolds (M^n, g) to consider, very well may be the *locally conformally Euclidean quasi Einstein spaces*, (these latter space's *Ricci tensor* having *two distinct eigenvalues*, one, with multiplicity 1, and the other, with multiplicity $n-1$). From the *intrinsic point of view*, these Riemannian manifolds are the *QCC spaces*, *the spaces of "quasi constant sectional curvatures"* –as they later were called– , while, from the *extrinsic point of view*, these spaces are *proper quasi umbilical hypersurfaces* M^n *in CC spaces* $M^n(c)$, and, actually, –apart from the above considerations on Deszcz symmetry, via which *Mira Petrović-Torgašev and I* came to these spaces in the first place– , this was essentially obtained at MSU in 1972 by *Chen-Houh-Yano* and *Chen-Yano*, respectively; see [**12**, **40**]. In particular, for such quasi umbilical hypersurfaces M^n in Euclidean spaces $\mathbb{E}^{n+1}$, with *principal curvature* λ *of multiplicity* $n-1$ and with *principal curvature* μ *of multiplicity* 1, *the isotropic Haesen curvature* can actually be seen as $L = \lambda\mu$. To close this paragraph, next still do follow two remarks in this respect. Firstly, the above may be considered as some *alternative, metrical geometrical approach* to the following citation of *Thurston* [**65**]: " *What is geometry? Up till now, we have discussed three kinds of three-dimensional geometry: hyperbolic, Euclidean and spherical.* They have in common the property of being as uniform as possible: their isometries can move any point to any other point (*homogeneity*), and can take any orthonormal frame in the tangent space at a point to any other orthonormal frame at that point (*isotropy*). *There are more possibilities if we remove the isotropy condition, allowing the space to have a grain, so to speak, so that certain directions are geometrically distinguished from others*". Finally, when normally adapting what came before to the pseudo Riemannian setting, one may be happy hereby, in particular, to meet the *Lorentzian Friedmann-Lemaître-Robertson-Walker warped product space-times* in the above context; see also [**10**].

The last illustration of some of Professor Chen's work in submanifold theory that will be given here, concerns "*the differential geometry of the position vector field*", on which field of research [**14**] is a recent survey article. In a way, herewith we go *back to the technical manner* in which *Euler* and later in particular *Gauss, described surfaces* M^2 *in* $\mathbb{E}^3$. Whereas Chen made the following study for submanifolds M^n in $\mathbb{E}^{n+m}$ of all dimensions n and with all codimensions m, in 2001, and did it also for the indefinite situation, in 2003, hereafter however, *we will restrict to just discuss the case of surfaces* M^2 *in* $\mathbb{E}^3$. Then, as *immediate application in biology*, we will try to show that *many of the surface shapes that we see in nature, are the result of one (1) reasonable process of "growth out of a pole" of a diversity of really living beings*, (the "living" being stressed here a bit to make a distinction with some geometrical things that a.o. PapaCartan also looked at as real beings). And that this collection of often so beautiful shapes which emerge from a purely geometrical basic study of surfaces M^2 in $\mathbb{E}^3$, –as has been tried above to focus our attention on maybe a bit more than usual– , certainly played some rôle in *the development of geometry as the science of human vision*, seemed a good reason to end here, where, in some way, it all began. The study of Chen involved, concerns his "*submanifolds of constant ratio*". Let X be the position vector field of a surface M^2 in $\mathbb{E}^3$. Then X has *a natural and canonical, orthogonal decomposition into a tangential component* X^T *and a normal component* $X^\perp$ *with respect to the surface*; (when "standing" on the surface at some point, this could be thought of as a *HV decomposition* of X at that point). And Chen said a surface M^2 in $\mathbb{E}^3$ to be *of constant ratio, when the ratio of the lengths* $\|X^T\|$ *and* $\|X^\perp\|$ *is constant*, (which, of course, a.o., is *equivalent* with the angles between the position vector and the tangent space and between the position vector and the normal space at any point of the surface being also constant). *Another characterisation* that he proved *for these surfaces is that* $\|\operatorname{grad}\rho\|$ *is constant*, whereby ρ is the "*polar radial function*", or, still, $\rho = \|X\|$; and he obtained a *complete, explicit classification* of all constant ratio surfaces M^2 in $\mathbb{E}^3$. In the unabridged editions of *D'Arcy Thompson*'s book '*On Growth and Form*', one can find the author's *basic principle of growth in a plane out of a pole*, or, out of any given point, O, which geometrically would lead *to curves* M^1 *with a position vector field* X *in a Euclidean plane with origin* O. It comes from the following passage of the section '*The equiangular spiral in its dynamical aspect*': " In mechanical structures, curvature is essentially a mechanical phenomenon. It is found in flexible structures as a result of bending, or it may be introduced into construction for the purpose of resisting such a bending-moment. *But neither shell, nor tooth, nor claw are flexible structures*; *they have not been bent into their peculiar curvature, they have grown into it.* We may for a moment, however, *regard the equiangular or logarithmic spiral of our shell from the dynamic point of view*, by *looking at growth itself as the force concerned.* In the growing structure, let *growth at a point* P be *resolved into a force* F *acting along the line joining* P *to a pole* O, *and a force* T *in a direction perpendicular to* OP; and *let the magnitude* (...) *of these rates of growth remain constant.* It follows that *the resultant of the forces* F *and* T (which is in the growing direction, that is, in the direction at P of the tangent to the curve $M^1 = (P)$ formed by this growing process; L.V.) *makes a constant angle with the radius vector.* But *a constant angle between tangent and radius vector* is a fundamental property of *the "equiangular" spiral*: the very property with which *Descartes* started his investigation (in 1638; L.V.), and that which gives its alternative name to the curve. (...). In the growth of a shell, we can conceive no simpler law than this, that it shall *widen and lengthen in the same proportions*: and *this simplest of laws is what Nature tends to follow.* The shell, like the creature within it, *grows in size but does not change its shape*; and *the existence of this constant relativity in growth, or constant similarity of form, is of the essence* ...". But, of course, a shell or *whatever that grows out of some pole in nature*, never grows into some curve in a plane, but *grows into some surface in space.* Let us consider *such a surface* M^2 *with position vector field* X *in a Euclidean space* $\mathbb{E}^3$, *with origin* O *at the pole of growth.* Then, our purpose being to apply D'Arcy Thompson's above principle for growth into planar curves, to some appropriate plane related to growth into surfaces M^2 in $\mathbb{E}^3$, we proceed as follows. *When* at every "point" X of M^2 in $\mathbb{E}^3$, *the polar distance* ρ *remains constant*, i.e. when $\|\operatorname{grad}\rho\| = 0$, *then* M^2 *is (part of) a sphere.* And, *whenever* $\|\operatorname{grad}\rho\| \neq 0$, *there is at* X *a well determined tangent direction in which the growth is extremal, in which* ρ *changes most*, namely, *the direction of* $\operatorname{grad}\rho$. Then applying the above principle of planar growth to take place in this well determined plane $X \wedge \operatorname{grad}\rho$, one readily sees the *equivalence with* M^2 *in* $\mathbb{E}^3$ *being a surface of constant ratio.* One may be referred to [**36**] for more information in this respect: for seeing that, when in particular looking at shells, *all the pretty different shapes of shells*, –Nautilus ones, the discoid and the turbinate ones– , *come out of this same growing process*; for seeing how often in nature occurring "irregularities" *of such Euclidean shapes, may well be described via Gielis transformations*; and, that *this all works well for* nD *growth out of a pole in* $(n+m)D$ *spaces.* And in view of the above, in [**36**] was formulated the following "*D'Arcy Thompson and Bang-Yen Chen's basic principle for natural growth out of pole: during the growth the instantaneous maximal change of the polar distance is constant*".

"*Go for the simple things!*" is what the masters always advice their pupils to do, which, for studies as for life in general, is very easy to say, but is most difficult to do. And this is so excellent! Only really talented people who had the great fortune to have been taught by really talented and devoted teachers, by long and serious thinking on their own about the essentials of the very basic fields of the sciences and/or the arts, have no guarantee, but have a chance of maybe, sometimes, eventually to see something of value in their field of work that was not seen before. And *Professor Chen* is one of the very rare, true geometers who has been having this experience many times indeed. But, as indicated by the successive citations of *Huntley* and *Bronowski* in Section 2, also people like most of us, by longly and seriously studying the works of the real masters of our fields, *we too can experience these masters' joy of creation and insight.* And, further, according to the quote of *Gabos* at the end of that section, *all of us can well experience the great joy of explaining what we know and understand to others.*

There is the geometry of submanifolds of Euclidean- and of pseudo Euclidean spaces, with in their intrinsic parts the geometry of Riemannian spaces and of pseudo Riemannian spaces and of metrically degenerate spaces, and there is also the geometry of the submanifolds of such spaces –including, for instance, the geometry of the submanifolds of hyperspheres in Euclidean spaces–, and there is the geometry of submanifolds of Kaehler spaces, like the complex projective spaces of constant holomorphic curvature, and there are also naturally herewith related geometries, of which those of line congruences and of other systems of lines in Euclidean- and in pseudo Euclidean spaces are first instances, and, moreover, there are many other fields of importance in geometry as a whole, of course, –other fields that originated in mathematical studies making generalisations and extensions in some of the already mentioned fields, or, other fields that originated in mathematical studies of phenomena or problems occurring in the natural sciences or in technology–. Reading and reflecting about geometry, in any of the former cases, and, reading and reflecting about algebra and analysis, certainly not in a lesser extent, it may be of interest to have well in mind the contents of the following part of *von Neumann*'s '*The Mathematician*': "The most vital characteristic fact about mathematics is, in my opinion, its quite peculiar relationship to the natural sciences, or, more generally, to any science which interprets experience on a higher than purely descriptive level. Most people, mathematicians and others, will agree that mathematics is not an empirical science, or at least that it is practiced in a manner which differs in several decisive respects from the techniques of the empirical sciences. And, yet, its development is very closely linked with the natural sciences. Some of the best inspirations of modern mathematics (I believe, the best ones) clearly originated in the natural sciences. (...). There is a peculiar duplicity in the nature of mathematics. One has to realize this duplicity, to accept it, and to assimilate it into one's thinking of the subject. This double face is the face of mathematics, and I do not believe that any simplified, unitarian view of the thing is possible without sacrificing the essence. (...). One expects a mathematical theorem or a mathematical theory not only to describe and to classify in a simple and elegant way numerous and a priori disparate special cases. One also expects ≪elegance≫ in its ≪architectural≫, structural makeup. Ease in stating the problem, great difficulty in getting hold of it and in all attempts of approaching it, then again some very surprising twist by which the approach, or some part of the approach, becomes easy, etc. (...). I think that it is a relatively good approximation to the truth –which is much too complicated to allow anything but approximations– that mathematical ideas originate in empirics, although the genealogy is sometimes long and obscure. But, once they are so conceived, the subject begins to live a peculiar life of its own and is better compared to a creative one, governed by almost entirely aesthetic motivations, than to anything else and in particular, to an empirical science. There is, however, a further point which, I believe, needs stressing. As a mathematical discipline travels far from this empirical source, or still more, if it is a second and third generation only indirectly inspired by ideas coming from ≪reality≫, it is beset with very grave dangers. It becomes more and more purely aestheticizing, more and more purely ≪l'art pour l'art≫. This needs not be bad, if the field is surrounded by correlated subjects, which still have closer empirical connections, or if the discipline is under influence of men with an exceptionally well-developed taste. But there is a grave danger that the subject will develop along the lines of least resistance, that the stream, so far from its source, will separate into a multitude of insignificant branches, and that the discipline becomes a disorganized mass of details and complexities. In other words, at a great distance from its empirical source, or after much ≪abstract≫ inbreeding, a mathematical subject is in danger of degeneration".

In any case, proceeding in a positive mood, in the words of *Polya*: "*Mathematics, taught and learned appropriately, improves the mind and implants good habits of thought*", and on the cover of his 1977 MAA-New Mathematical Library book "*Mathematical Methods in Science*", is printed *Galileo*'s well known quote from '*Saggiatore*': "*The great book of nature lies ever open before our eyes and the true philosophy is written in it* (...). *But we cannot read it unless we have first learned the language and the characters in which it is written.* (...). *It is written in the language of mathematics and the characters are triangles, circles and other geometric figures* (...)". And from *Galileo*'s '*Discoveries and Opinions*' comes the following: "*I believe that vision, the sense eminent above all others in the proportion of the finite to the infinite, the temporal to the instantaneous, the illuminated to the obscure –that vision, I say, is related to light itself*", while, as formulated by *Teilhard de Chardin*:"*The history of the living world can be summarised as the elaboration of ever more perfect eyes within a cosmos in which there is always more to be seen*".

Next come the closing sentences of *Poincaré*'s '*La valeur de la Science*': "*Ce n'est que par la Science et par l'Art que valent les civilisations.* On s'est étonné de cette formule: la Science pour la Science, et pourtant cela vaut bien la vie pour la vie, si la vie n'est que misère; et même le bonheur pour le bonheur, si l'on ne croit pas que tous les plaisirs sont de même qualité, si l'on ne veut pas admettre que le but de la civilisation soit de fournir de l'alcool aux gens qui aiment à boire. Tout action doit avoir un but. Nous devons souffrir, nous devons travailler, nous devons payer notre place au spectacle, mais c'est pour voir; ou tout au moins pour que d'autres voient un jour. Tout ce qui n'est pas pensée est le pur néant; puisque nous ne pouvons penser que la pensée et que tous les mots dont nous disposons pour parler des choses ne peuvent exprimer que des pensées; dire qu'il y a autre chose que la pensée, c'est donc une affirmation qui ne peut avoir du sens. Et cependant –étrange contradiction pour ceux qui croient au temps– l'histoire géologique nous montre que la vie n'est qu'un court épisode entre deux éternités de mort, et que, dans cet épisode même la pensée consciente n'a duré et ne durera qu'un moment. *La pensée n'est qu'un éclair au milieu d'une longue nuit. Mais c'est cet éclair qui est tout*".

"*Mathematics*" in its meaning of "*knowledge and understanding*" is at the basis of *wisdom.* To gain this knowledge and understanding is not at all an easy matter, rather, is much more like an *art. In Dutch*, this view led to the introduction of the term '*wiskunde*', instead of the "*Mathematik, les mathématiques, matematica, ...*" *in other languages.* The arts and the sciences are amongst the main characteristics of human civilisation. The real scientists and the real artists share an immense desire for understanding and beauty. The studies aiming for an ever better understanding, which invariably result in seeing ever more beauty, are amongst the most valuable human activities. The percentage of members of a human society who really experience great joy when observing the creations of real scientists and real artists is a scale to measure the degree of civilisation of this society. As last "finally" of this article, here is a quote in this respect from the first chapter '*Einblick*' of his book [**31**], in which Fischer announces the conclusion of '*Die andere Bildung*': "Wie kommt es, das trotz aller Bemühungen hierzulande (∗; L.V.) von einem ≪public understanding of science≫ keine Rede sein kann? In einem Ausblick wird am Ende aller Erklärungen und Darstellungen dargelegt, dass sich diese missliche Lage erst ändert, wen wir ≪Wissenschaft als Kunst denken≫, wie es Goethe einmal in seiner 'Farbenlehre' formuliert hat. *Wissenschaft wird erst verstanden*, so vermute und behaupte ich, *wenn sie wie ein Kunstwerk gestaltet wird*, *das sich eine bestimmte wahrnehmbare Form bekommen soll*, *die ein offenes Geheimnis tragen und zur Schau stellen kan.* Genau dan werden naturwissenschaftliche Bildung und der verantwortungsvolle Umgang mit dem grösten Kapital der Menschheit möglich sein"; (the above ∗ after Fischer's "*hierzulande*" maybe could be replaced by "*and almost everywhere*"?). And, "à bout de souffle", two more citations, respectively from *René Thom* and from *Blaise Pascal*, will definitely put an end to the present contemplation of submanifold theory: "*Comprendre signifie donc avant tout géométriser*" and "*Ce qui passe la géométrie nous surpasse*".

Acknowledgments

The author is very obliged to all members of our Section of Geometry who work on Submanifold Theory, for the time that they often make to have serious and less serious discussions on geometry, and also for their patience in listening to some of the author's monologues about matters that are really too far from the source.

CiT-PiT, 15-06-2019.

References

[1] A. Adamów and R. Deszcz, *On totally umbilical submanifolds of some class Riemannian manifolds*, Demonstratio Math. **16** (1983), no. 1, 39–59. MR723962

[2] D. Bao, S.-S. Chern, and Z. Shen, *An introduction to Riemann-Finsler geometry*, Graduate Texts in Mathematics, vol. 200, Springer-Verlag, New York, 2000. MR1747675

[3] M. Berger, *A panoramic view of Riemannian geometry*, Springer-Verlag, Berlin, 2003. MR2002701

[4] *Big History*, Dorling Kindersley LTD, London, 2016.

[5] J. Bronowski, *The Ascent of Man*, BBC Books, Reading, 2011.

[6] E. Cartan, *Géométrie euclidienne et géométrie riemannienne*, Scientia 25 (49):393, 1931.

[7] F. Casorati, *Mesure de la courbure des surfaces suivant l'idée commune* (French), Acta Math. **14** (1890), no. 1, 95–110, DOI 10.1007/BF02413317. Ses rapports avec lesmesures de courbure gaussienne et moyenne. MR1554792

[8] B.-Y. Chen, *Pseudo-Riemannian geometry, δ-invariants and applications*, World Scientific Publishing Co. Pte. Ltd., Hackensack, NJ, 2011. With a foreword by Leopold Verstraelen. MR2799371

[9] B.-Y. Chen, *Total mean curvature and submanifolds of finite type*, 2nd ed., Series in Pure Mathematics, vol. 27, World Scientific Publishing Co. Pte. Ltd., Hackensack, NJ, 2015. With a foreword by Leopold Verstraelen. MR3362186

[10] B.-Y. Chen, *Differential geometry of warped product manifolds and submanifolds*, World Scientific Publishing Co. Pte. Ltd., Hackensack, NJ, 2017. With a foreword by Leopold Verstraelen. MR3699316

[11] B.-Y. Chen, *A new approach to compact symmetric spaces and applications*, Katholieke Universiteit Leuven, Louvain, 1987. A report on joint work with Professor T. Nagano. MR952556

[12] B.-y. Chen, *Geometry of submanifolds*, Marcel Dekker, Inc., New York, 1973. Pure and Applied Mathematics, No. 22. MR0353212

[13] B.-y. Chen, *Geometry of submanifolds and its applications*, Science University of Tokyo, Tokyo, 1981. MR627323

[14] B.-Y. Chen, *Topics in differential geometry associated with position vector fields on Euclidean submanifolds*, Arab J. Math. Sci. **23** (2017), no. 1, 1–17, DOI 10.1016/j.ajmsc.2016.08.001. MR3589496

[15] B.-Y. Chen, *Two-numbers and their applications—a survey*, Bull. Belg. Math. Soc. Simon Stevin **25** (2018), no. 4, 565–596. MR3896273

[16] S. S. Chern, *Differential geometry: its past and its future*, Actes du Congrès International des Mathématiciens (Nice, 1970), Gauthier-Villars, Paris, 1971, pp. 41–53. MR0428217

[17] J. L. Coolidge, *A history of geometrical methods*, Dover Publications, Inc., New York, 1963. MR0160143

[18] A. Cuvellier, *Esthétique*, in *Grand Memento Encyclopédique Larousse. Tome 1*, Larousse, Paris, 1937.

[19] F.D., *Nous avons tous la bosse des math*, Dossier Mathématiques, Science et Avenir, 840, 2017, 36.

[20] *De Clootcrans Press,* Traiectum, The Netherlands, KoenderinkJan@gmail.com.

[21] S. Decu, A. Pantić, M. Petrović-Torgašev, and L. Verstraelen, *Ricci and Casorati principal directions of $\delta(2)$ Chen ideal submanifolds*, Kragujevac J. Math. **37** (2013), no. 1, 25–31. MR3073695

[22] S. Decu, B. Jahanara, M. Petrović-Torgašev, and L. Verstraelen, *On the Chen character of $\delta(2)$-ideal submanifolds*, Kragujevac J. Math. **32** (2009), 37–46. MR2584238

[23] J. Deprez, R. Deszcz, and L. Verstraelen, *Examples of pseudo-symmetric conformally flat warped products*, Chinese J. Math. **17** (1989), no. 1, 51–65. MR1007875
[24] F. Dillen, *Submanifolds in differential geometry*, Bull. Soc. Math. Belg. Sér. B **42** (1990), no. 2, 177–190. MR1061230
[25] F. Dillen and W. Kühnel, *Total curvature of complete submanifolds of Euclidean space*, Tohoku Math. J. (2) **57** (2005), no. 2, 171–200. MR2137465
[26] P. Dombrowski, *150 years after Gauss' "Disquisitiones generales circa superficies curvas"*, Astérisque, vol. 62, Société Mathématique de France, Paris, 1979. With the original text of Gauss. MR535996
[27] P. Dombrowski, *Differentialgeometrie* (German), Ein Jahrhundert Mathematik 1890–1990, Dokumente Gesch. Math., vol. 6, Friedr. Vieweg, Braunschweig, 1990, pp. 323–360. MR1085968
[28] P. Dombrowski, *Wege in euklidischen Ebenen/ Kinematik der Speziellen Relativiätstheorie (Eine Auswahl geometrischer Themen mit Beiträgen zu deren Ideen-Geschichte. Heinz Hopf, Willi Rinow, Erhard Schmidt)*, Springer, Berlin, 1999.
[29] R. Dontot, *Géométrie*, in *Grand Memento Encyclopédique Larousse. Tome 2*, Larousse, Paris, 1937.
[30] *Encyclopedia of Mathematical Sciences. Geometry I-VI,* (ed. R.V.Gambrelidze), Springer, Berlin, 1991-2001.
[31] E. P. Fischer, *Die andere Bildung*, Ullstein Verlag, Ulm, 2003.
[32] W. R. Fuchs, *Modern Mathematics,* Weidenfeld & Nicolson, London, 1967.
[33] S. Germain, *Mémoire sur la coubure des surfaces* (French), J. Reine Angew. Math. **7** (1831), 1–29, DOI 10.1515/crll.1831.7.1. MR1577823
[34] L. Godeaux, *Esquisses d'une histoire des Sciences Mathématiques en Belgique*, Collection Nationale, Bruxelles, 1943.
[35] T. Gowers, *Mathematics*, Very Short Introductions, vol. 66, Oxford University Press, Oxford, 2002. A very short introduction. MR2147526
[36] S. Haesen, A. I. Nistor, and L. Verstraelen, *On growth and form and geometry. I*, Kragujevac J. Math. **36** (2012), no. 1, 5–25. MR2990632
[37] S. Haesen and L. Verstraelen, *Natural intrinsic geometrical symmetries*, SIGMA Symmetry Integrability Geom. Methods Appl. **5** (2009), Paper 086, 15, DOI 10.3842/SIGMA.2009.086. MR2559675
[38] S. Haesen, D. Kowalczyk, and L. Verstraelen, *On the extrinsic principal directions of Riemannian submanifolds*, Note Mat. **29** (2009), no. 2, 41–53. MR2789830
[39] S. Haesen and L. Verstraelen, *Properties of a scalar curvature invariant depending on two planes*, Manuscripta Math. **122** (2007), no. 1, 59–72, DOI 10.1007/s00229-006-0056-0. MR2287700
[40] *Handbook of Differential Geometry I–II. (Eds. F.J.E. Dillen e.a.)*, Elsevier North-Holland, Amsterdam, 2000-2005.
[41] W. Gellert, H. Küstner, M. Hellwich, and H. Kästner (eds.), *Kleine Enzyklopädie: Mathematik* (German), VEB Bibliographisches Institut, Leipzig, 1977. Mit einer Einleitung von Hans Reichardt; Zehnte Auflage. MR0441622
[42] F. Klein, *Elementary Mathematics from an Advanced Standpoint. Vol. I. Arithmetic, Algebra, Analysis*, Dover Publications, New York, N. Y., 1945. MR0015349
[43] G. Kinckhuysen, *De Gront der Meetkonst– Algebra ofte Stelkonst*, Haarlem, 1660-1661.
[44] S. Kobayashi, K. Nomizu, *Foundations of Differential Geometry*, vol. I, vol. II, Wiley, New York, 1963 and 1969.
[45] J. Koenderink, *Color for the sciences,* The MIT Press, Cambridge, 2010.
[46] V. Kommerell, *Analytische Geometrie der Ebene und des Raumes, in 'Vorlesungen weber die Geschichte der Mathematik'* (ed. M. Cantor), Teubner, Berlin, 1908.
[47] W. Kühnel, *Differentialgeometrie*: *Kurven—Flächen—Mannigfaltigkeiten. [Curves—surfaces—manifolds]* (German), 4th ed., Vieweg Studium: Aufbaukurs Mathematik. [Vieweg Studies: Mathematics Course], Vieweg, Wiesbaden, 2008. MR2527182
[48] D. Laugwitz, *Bernhard Riemann 1826–1866*: *Wendepunkte in der Auffassung der Mathematik. [Turning points in the conception of mathematics]* (German, with German summary), Vita Mathematica, vol. 10, Birkhäuser Verlag, Basel, 1996. MR1487377

[49] T. Levi-Civita, *Notione de parallelismo in una varietà qualunque e conseguente spezificazione geometrica della curvatura riemanniana,* Rend. Circolo Mathematico di Palermo, 42 (1917), 173–205.

[50] F. C. Marques and A. Neves, *Min-max theory and the Willmore conjecture*, Ann. of Math. (2) **179** (2014), no. 2, 683–782, DOI 10.4007/annals.2014.179.2.6. MR3152944

[51] K. H. Parshall, *Reviews: Mathematics of the 19th Century: Mathematical Logic, Algebra, Number Theory Probability Theory*, Amer. Math. Monthly **101** (1994), no. 4, 369–374, DOI 10.2307/2975639. MR1542526

[52] O. Neugebauer, *The exact sciences in antiquity*, Dover, New York, 2016.

[53] B. Ons and P. Verstraelen, *A geometrical description of visual sensation II*, Kragujevac J. Math. **34** (2010), 5–24. MR2897891

[54] R. Osserman, *Curvature in the eighties*, Amer. Math. Monthly **97** (1990), no. 8, 731–756, DOI 10.2307/2324577. MR1072814

[55] D. Palmer, *A History of Earth in 100 Groundbraking Discoveries*, Firefly Books Ltd., Richmond Hill, 2011.

[56] K. Reich, *Die Entwicklung des Tensorkalküls* (German, with German summary), Science Networks. Historical Studies, vol. 11, Birkhäuser Verlag, Basel, 1994. Vom absoluten Differentialkalkül zur Relativitätstheorie. [From the absolute differential calculus to relativity theory]. MR1353442

[57] K. Reich, *Die Geschichte der Differentialgeometrie von Gauss bis Riemann (1828–1868)* (German), Arch. History Exact Sci. **11** (1973/74), 273–382, DOI 10.1007/BF00357392. MR0497714

[58] B. Rouxel, *Sur quelques propriétés anallagmatiques de l'espace euclidean* $\mathbb{E}^4$, *Mémoire couronné, Acad. Royale belge, 1982.*

[59] J. A. Schouten, D. J. Struik, *Einführung in die neuren Methoden der Differentialgeometrie,* vol. II., Noordhoff, Groningen, 1938.

[60] J.A. Schouten, *Die direkte Analysis zur neuren Relativitätstheorie*, Verhandelingen der Koninklijke Akademie van Wetenschappen te Amsterdam, Bd. 12, no. 6, (1918), (95 s.).

[61] I. Stewart, *Symmetry: a very short introduction*, Very Short Introductions, vol. 353, Oxford University Press, Oxford, 2013. MR3086949

[62] I. Stewart, *Taming the Infinite– The story of mathematics from the first numbers to chaos theory,* Quercus, London, 2008.

[63] D. J. Struik, *Outline of a history of differential geometry I and II.(International review devoted to the History of Science and Civilization) 19 and 20, 1933, 92-120 and 161-191; Lectures on Classical Differential Geometry*, Dover, New York, 1961.

[64] T. Gowers, J. Barrow-Green, and I. Leader (eds.), *The Princeton companion to mathematics*, Princeton University Press, Princeton, NJ, 2008. MR2467561

[65] W. Thurston, *The Geometry and Topology of* 3*-manifolds,* Lecture Notes, Princeton University Press, 1979.

[66] K. Trenčevski, *Geometrical interpretation of the principal directions and principal curvatures of submanifolds*, Differ. Geom. Dyn. Syst. **2** (2000), no. 1, 50–58. MR1794323

[67] K. Trenčevski, *On the osculating spaces of submanifolds in Euclidean spaces*, Kragujevac J. Math. **36** (2012), no. 1, 45–49. MR2990635

[68] K. Ueno, *An introduction to algebraic geometry*, Translations of Mathematical Monographs, vol. 166, American Mathematical Society, Providence, RI, 1997. Translated from the 1995 Japanese original by Katsumi Nomizu. MR1454617

[69] B. L. van der Waerden, *Ontwakende Wetenschap. Egyptische, Babylonische en Griekse Wiskunde* (Dutch), P. Noordhoff, Groningen, 1950. MR0038286

[70] B. L. van der Waerden, *Geometry and algebra in ancient civilizations*, Springer-Verlag, Berlin, 1983. MR706258

[71] O. Veblen, J.H.C. Whitehead, *The Foundation of Differential Geometry,* Cambridge University Press, Cambridge, 1932.

[72] L. Verstraelen, *Geometry of submanifolds I. The first Casorati curvature indicatrices*, Kragujevac J. Math. **37** (2013), no. 1, 5–23. MR3073694

[73] L. Verstraelen, *A concise mini history of geometry*, Kragujevac J. Math. **38** (2014), no. 1, 5–21, DOI 10.5937/KgJMath1401005V. MR3229448

[74] L. Verstraelen, *Psychology and geometry I. On the geometry of the human kind*, Filomat **29** (2015), no. 3, 545–552, DOI 10.2298/FIL1503545V. MR3359447

[75] L. Verstraelen, *Psychology and geometry II. On the theater in which we have the impression to be on stage*, Riemannian Geometry and Applications—Proceedings RIGA 2014, Editura Univ. Bucur., Bucharest, 2014, pp. 213–229. MR3329948

[76] L. Verstraelen, *Natural extrinsic geometrical symmetries—an introduction*, Recent advances in the geometry of submanifolds—dedicated to the memory of Franki Dillen (1963–2013), Contemp. Math., vol. 674, Amer. Math. Soc., Providence, RI, 2016, pp. 5–16. MR3565240

[77] L. Verstraelen, *Comments on pseudo-symmetry in the sense of Ryszard Deszcz*, Geometry and topology of submanifolds, VI (Leuven, 1993/Brussels, 1993), World Sci. Publ., River Edge, NJ, 1994, pp. 199–209. MR1315102

[78] P. Vincensini, *La géométrie différentielle au XIXème siècle. Avec quelques réflexions générales sur les mathématiques* (French), Scientia (Milano) **107** (1972), 617–696. With an English translation. MR0504956

[79] H. Weyl, *Relativity as a stimulus in mathematical research,* Proc. American Philosophical Society, 93(7), 1949, 535–541.

KU Leuven, Department of Mathematics, Celestijnenlaan 200B - Box 2400, 3001 Leuven, Belgium

Email address: `leopold.verstraelen@kuleuven.be`

Contemporary Mathematics
Volume **756**, 2020
https://doi.org/10.1090/conm/756/15196

Spectral gaps on complete Riemannian manifolds

Nelia Charalambous, Helton Leal, and Zhiqin Lu

This paper is dedicated to Professor B.Y. Chen on his 75th Birthday

Abstract. In this short note, we survey some basic results related to the New Weyl criterion for the essential spectrum. We then use the language of Gromov-Hausdorff convergence to prove a spectral gap theorem.

1. Introduction

Let X be a complete noncompact Riemannian manifold of dimension n and denote by Δ the Laplacian acting on smooth functions with compact support $\mathcal{C}_0^\infty(X)$. It is well known that the self-adjoint extension of Δ on $L^2(X)$ exists, and is a unique nonpositive definite and densely defined linear operator.

The spectrum of $-\Delta$, denoted by $\sigma(-\Delta)$, consists of all $\lambda \in \mathbb{C}$ for which $\Delta + \lambda I$ fails to be invertible. The essential spectrum of $-\Delta$, $\sigma_{\rm ess}(-\Delta)$, consists of the cluster points in the spectrum and of isolated eigenvalues of infinite multiplicity. The pure point spectrum is defined by

$$\sigma_{\rm pp}(-\Delta) = \sigma(-\Delta)\backslash\sigma_{\rm ess}(-\Delta).$$

The spectral structure of a noncompact complete manifold is in general more complex than in the compact case. For a compact Riemannian manifold, by the Hodge Theorem, all spectral points of the Laplacian belong to the pure point spectrum. However, interestingly enough, while for most compact manifolds it is impossible to accurately compute the pure point spectrum, for a complete noncompact Riemannian manifold it is possible to locate the essential spectrum of the Laplacian in a large class of manifolds.

In this note, we will first survey some major results about the essential spectrum of a complete Riemannian manifold. Then we will use the language of Gromov-Haudorff to discuss a spectra gap phenomenon similar to a recent result of Schoen-Tran [**25**].

2. The New Weyl criterion

Donnelly pioneered the study of the essential spectrum using the Weyl criterion. In 1981, he proved the following result [**11**]

Key words and phrases. Essential spectrum, spectral gap, Laplacian, Gromov-Hausdorff convergence.

The third author is partially supported by the NSF grant DMS-19-08513.

The second author is partially supported by the CNPq-Brazil Doctorate Abroad scholarship.

THEOREM 2.1. *Let $\sigma > 0$ be a positive number. If there exists an infinite dimensional subspace G in the domain of Δ such that*

$$\|\Delta u + \lambda u\|_{L^2} \leq \sigma \|u\|_{L^2} \tag{1}$$

for all $u \in G$, then

$$\sigma_{\text{ess}}(-\Delta) \cap (\lambda - \sigma, \lambda + \sigma) \neq \emptyset.$$

The functions u are referred to as the *approximate eigenfunctions* corresponding to the eigenvalue λ. The above criterion is simple to apply and has directed the study of the essential spectrum of the Laplacian for the last three decades. A related result of the above is as follows: let u be a nonzero smooth function with compact support. If (1) is satisfied, then

$$\sigma(-\Delta) \cap (\lambda - \sigma, \lambda + \sigma) \neq \emptyset.$$

Using this criterion, it is not hard to prove that $\mathbb{R}^n$ has essential spectrum $[0, \infty)$. Moreover, with additional assumptions on the curvature and geometry of the manifold we can locate the essential spectrum (see for example [**8**, **12**, **14**, **15**, **18**, **29**]) by comparing the manifold to the n-dimensional Euclidean space.

The main difficulty in applying Donnelly's Weyl criterion stems from the fact that it requires canonical smooth functions on a manifold. There are many canonical smooth functions on a manifold, including the heat kernel and the Green's function. However, on a general manifold we do not have an explicit expression for these functions. It is possible to give upper and / or lower bounds for these functions, but those require another canonical function, namely the distance function on the manifold. Due to the presence of cut-loci, the distance function is in general not smooth. It is however Lipschitz and locally L^1 (cf. Cheeger [**6**, Chap 4], also see [**21**, **28**]). Thus in order to use the Weyl criterion, we must be in a setting where the distance function is smooth, or the manifold has a pole.

In [**4**] the following example is given which illustrates the non-regular nature of the distance function. Take $M = S^1 \times (-\infty, \infty)$, letting (θ, x) be the coordinates. Then the radial r function which gives the distance of (θ, x) to the point $(0, 0)$ is

$$r(\theta, x) = \sqrt{x^2 + (\min(\theta, 2\pi - \theta))^2}.$$

A straightforward computation gives

$$\Delta r = -\frac{2\pi}{\sqrt{x^2 + \pi^2}} \delta_{\{\theta = \pi\}} + \text{ a smooth function,}$$

where $\delta_{\{\theta=\pi\}}$ is the Delta function along the submanifold $\{\theta = \pi\}$. Therefore Δr is not locally L^2.

As was observed in [**28**] and [**4**], although in most problems the ideal space to work with is the L^2 function space, in comparison to L^q spaces, this is not the case when considering the spectrum of the Laplacian. On a Riemannian manifold, most of the approximate eigenfunctions we can write out explicitly must be related to the distance function. As the above example illustrates however, the Laplacian of the distance function is locally bounded in L^1, but not in L^2, thus making it easier to compute the L^1 spectrum of the Laplacian instead of the L^2 spectrum.

The failure of the L^2 integrability of the Laplacian of the distance function was one of the main difficulties in applying the classical criterion above. In fact, it was not possible to prove that the L^2 essential spectrum of the Laplacian on a manifold with nonnegative Ricci curvature is $[0, \infty)$ by directly computing the L^2 spectrum via the classical Weyl criterion. Around 1999, two papers addressed this difficulty. Donnelly [**12**] proved that the essential spectrum of the Laplacian is $[0,\infty)$ over manifolds with nonnegative Ricci curvature and maximal volume growth. Under his assumptions, the cone of the manifold at infinity is locally Euclidean and the result follows from spectrum continuity. On the other hand, J-P. Wang [**28**], by employing the seminal theorem of K. T. Sturm [**27**], removed the maximal volume growth condition. Wang's result confirmed the conjecture that the spectrum of manifolds with nonnegative Ricci curvature is $[0,\infty)$. In [**21**], Lu-Zhou gave a technical generalization of Wang's result which includes the case of manifolds of finite volume. We further elaborate on Sturm's result and its consequences in Section 3.

What was lacking from these results, was the direct relationship between the L^2 spectrum and L^p spaces. In [**4**], we found such a link.

THEOREM 2.2. *Let X be a Riemannian manifold and let Δ be the Laplacian. Assume that for $\lambda \in \mathbb{R}^+$, there exists a nonzero function u in the domain of Δ such that*

$$\|u\|_{L^\infty} \cdot \|\Delta u + \lambda u\|_{L^1} \leq \delta \|u\|_{L^2}^2 \tag{2}$$

for some positive number $\delta > 0$. Then

$$\sigma(-\Delta) \cap (\lambda - \varepsilon, \lambda + \varepsilon) \neq \emptyset,$$

where

$$\varepsilon = \min(1, (\lambda + 2)\delta^{1/3}).$$

Moreover,

$$\sigma_{\text{ess}}(-\Delta) \cap (\lambda - \varepsilon, \lambda + \varepsilon) \neq \emptyset,$$

if for any compact subset K of X, there exists a nonzero function u in the domain of Δ satisfying (2) whose support is outside K.

Using this theorem, we are able to compute the spectrum of a Ricci nonnegative manifold directly, using functions constructed from the distance function. The above theorem also allowed us to find the most general conditions possible so that the spectrum of the Laplacian on functions is maximal, in other words it is $[0,\infty)$.

THEOREM 2.3. *Let X be a complete noncompact Riemannian manifold. Take a fixed point x_o, and let $r(x) = d(x, x_o)$ be the radial distance to x_o. Assume that the radial Ricci curvature away from x_o is asymptotically nonnegative, in other words, there exists a continuous positive function $\delta(r)$ on $\mathbb{R}^+$ such that*

(i). $\lim_{r\to\infty} \delta(r) = 0$ *and*

(ii). $\mathrm{Ric}_X(\partial r, \partial r) \geq -(n-1)\delta(r)$ *away from the cut-locus of x_o.*

If the volume of the manifold is finite we additionally assume that its volume does not decay exponentially at x_o. Then the L^2 spectrum of the Laplacian is $[0,\infty)$.

3. On a theorem of Sturm

In 1993, Sturm [**27**] proved an interesting result, which relates the L^2 spectrum to the L^p spectrum of the Laplacian.

The L^p spectrum, denoted by $\sigma_p(-\Delta)$ for $p \geq 1$, is defined as the set of complex numbers λ such that the operator $\Delta + \lambda I$ fails to be invertible on the space $L^p(X)$. Note that unlike the case of L^2 spectrum, the L^p spectrum may contain complex numbers. For example, over real hyperbolic space, for $p \neq 2$ the L^p spectrum is a parabolic region in complex plane depending on p and the order of the form [**9**].

We say that a noncompact Riemannian manifold has uniformly subexponential volume growth, if for any $\varepsilon > 0$, there is a constant $C = C(\varepsilon)$, depending only on ε such that

$$\mathrm{vol}(B_x(R)) \leq C(\varepsilon)\,\mathrm{vol}(B_x(1))e^{\varepsilon R}.$$

for any $x \in X$, and $R \geq 1$, where $B_x(R)$ denotes the geodesic ball of radius R centered at x.

It should be noted that a manifold with finite volume does *not* necessarily have uniformly subexponential volume growth. For example, let X be a noncompact hyperbolic space with finite volume. Assume that X has a cusp at infinity. Then for $x_n \to \infty$, $\mathrm{vol}(B_{x_n}(1))$ has exponential decay with respect to the distance to a fixed point x_o. On the other hand, if X has uniformly subexponential volume growth, then for any $\varepsilon > 0$, there exists $C(\varepsilon) > 0$ such that

$$\mathrm{vol}(B_{x_n}(R)) \leq C(\varepsilon)\ \mathrm{vol}(B_{x_n}(1))e^{\varepsilon R}$$

for any $R > 1$. If we take $R = d(x_n, x_o) + 2$, then we have

$$\mathrm{vol}(B_{x_o}(1)) \leq \mathrm{vol}(B_{x_n}(R)) \leq C(\varepsilon)\ \mathrm{vol}(B_{x_n}(1))e^{\varepsilon d(x_n, x_o)}$$

which is a contradiction.

THEOREM 3.1 (Sturm). *Assume that X is a complete Riemannian manifold of uniformly subexponential volume growth. Assume furthermore that the Ricci curvature of X has a lower bound. Then for any $p \geq 1$, $\sigma_p(-\Delta) = \sigma_2(-\Delta)$, that is, all the L^p spectra coincide.*

Using the above result, J-P. Wang [**28**] proved that

THEOREM 3.2. *Let X be a complete Riemannian manifold and denote by $r(x)$ the radial distance to a fixed point $x_o \in X$. Suppose that*

$$\mathrm{Ric}_X(x) \geq -\frac{\delta}{r(x)^2},$$

where $\delta = \delta(n) > 0$ is a constant that depends only on the dimension n. Then, for all $p \geq 1$, $\sigma_p(-\Delta) = [0, \infty)$.

In [**21**], the following generalization of Wang's result has been proved, also by applying Sturm's Theorem

THEOREM 3.3. *Let X be a complete Riemannian manifold and again denote by $r(x)$ the radial distance to a fixed point $x_o \in X$. Let $\varepsilon(r)$ be a positive function such that $\varepsilon(r) \to 0$ as $r \to \infty$. Assume that*

$$\mathrm{Ric}_X(x) \geq -\varepsilon(r(x)),$$

Then for all $p \geq 1$, $\sigma_p(-\Delta) = [0, \infty)$.

One can compare the consequences of Sturm's Theorem about the spectrum to those of Theorem 2.2. When we assume that the volume has uniformly subexponential growth and the Ricci curvature is bounded below, then we can get information about all of the L^p spectra. On the other hand, by applying Theorem 2.2, we can obtain stronger information for just the L^2 spectrum without any assumptions on the smooothness of the distance function.

4. Spectral continuity and a gap theorem

In the previous sections we have studied results that allow us to find large sets of noncompact manifolds whose essential spectrum is a connected subset of the real line. There are however many known cases where the essential spectrum has an arbitrary number of gaps [**1, 19, 20, 22, 25**]. In the last part of the paper we are interested in further exploring this set of manifolds. We will first turn our attention to spectral continuity and then study the evolution of the spectrum of a manifold under Gromov-Hausdorff convergence. We will then use these ideas to prove the existence of gaps in the essential spectrum of a periodic manifold, which is close in spirit to a recent result by Schoen and Tran [**25**].

The first natural case to consider is the evolution of eigenvalues under the continuous deformation of a manifold or its Riemannian metric. Dodziuk proved the following result in [**10**].

THEOREM 4.1 (Dodziuk). *Let X be a compact manifold and let g_t be a family of Riemannian metrics on X. Assume that*

$$g_t \to g$$

in the C^0 topology. Then the spectrum (eigenvalues) of g_t converges to the spectrum of g.

In the recent paper [**5**] (also see [**23**] for related results), we generalized spectral continuity to the case when the quadratic forms of two self-adjoint operators are ε-close.

Let $\mathcal{H}$ be a Hilbert space with two inner products $(\cdot,\cdot)_0$ and $(\cdot,\cdot)_1$. We consider two densely defined nonnegative operators H_0 and H_1 on $\mathcal{H}$ that are self-adjoint with respect to the inner products $(\cdot,\cdot)_0$ and $(\cdot,\cdot)_1$ respectively. Let Q_0, Q_1 be their respective quadratic forms and denote the two norms on $\mathcal{H}$ by $\|\cdot\|_0$ and $\|\cdot\|_1$. Note that both Q_0 and Q_1 are nonnegative.

We denote the domain of the Friedrichs extension of H_0 and H_1 by $\mathfrak{Dom}(H_0)$ and $\mathfrak{Dom}(H_1)$ respectively. We assume that there exists a dense subspace $\mathcal{C} \subset \mathcal{H}$ such that $\mathcal{C}$ is contained in $\mathfrak{Dom}(H_0) \cap \mathfrak{Dom}(H_1)$ (in the case of the Laplacian, $\mathcal{C}$ will be the space of smooth functions/forms with compact support).

DEFINITION 1. We say that the operators H_0, H_1 are ε-close, if there exists a positive constant $0 < \varepsilon < 1$ such that for all $u \in \mathcal{C}$ the following two inequalities hold

$$(1-\varepsilon)\,\|u\|_0^2 \le \|u\|_1^2 \le (1+\varepsilon)\,\|u\|_0^2; \tag{3}$$

$$(1-\varepsilon)\,Q_0(u,u) \le Q_1(u,u) \le (1+\varepsilon)\,Q_0(u,u). \tag{4}$$

We note that if H_0, H_1 are ε-close, then for any $u, v \in \mathcal{C}$

$$|(u,v)_1 - (u,v)_0| \le \varepsilon(\|u\|_0\,\|v\|_0); \tag{5}$$

$$|Q_1(u,v) - Q_0(u,v)| \le \varepsilon\,[Q_0(u,u)\,Q_0(v,v)]^{1/2}. \tag{6}$$

Moreover, it can be shown that the resolvents of the two operators are also ε close.

In [**5**] we showed that two ε-close operators have nearby spectra. This result has an important application in the context of the Laplacian over a Riemannian manifold with two ε-close metrics over it. In particular, it allows us to prove the following theorem which holds even in the noncompact case, thus generalizing the result of Dodziuk.

THEOREM 4.2. *Let X^n be an orientable manifold, and let g_0, g_1 be two smooth complete Riemannian metrics on X that are ε-close for some $0 < \varepsilon < 1/2$.*

Fix $A > 0$. Then for any $\lambda \in \sigma(k, \Delta_1) \cap [0, A]$

$$\operatorname{dist}(\lambda, \sigma(k, \Delta_0)) < c(A, n)\, \varepsilon^{\frac{1}{3}}$$

for some constant $c(A)$ depending only on A. A similar result holds for the essential spectra of the operators. In particular,

$$d_{\mathfrak{h}}(\sigma(k, \Delta_1), \sigma(k, \Delta_0)) = o(1),$$

where $o(1) \to 0$, as $\varepsilon \to 0$.

In the above theorem $d_{\mathfrak{h}}$ denotes the pointed Gromov-Hausdorff distance between the spectra as subsets of the real line, and with a common fixed point $\{-1\}$.

In the setting of a family of compact Riemannian manifolds which is convergent in the Gromov-Hausdorff sense, we have the following important results due to Fukaya [**16**] and Cheeger-Colding [**7**].

THEOREM 4.3 (Fukaya). *Let X_t be a family of compact Riemannian manifolds which is Gromov-Hausdorff convergent to a compact metric space X. We assume that X is not a point. Assume that the curvatures of the manifolds X_t are uniformly bounded. Then the eigenvalues of X_t converge to those of X.*

Cheeger and Colding generalized the above theorem and proved

THEOREM 4.4 (Cheeger-Colding). *Let X_t be a family of compact Riemannian manifolds which is Gromov-Hausdorff convergent to a compact metric space X. We assume that X is not a point. Assume that the Ricci curvatures of the manifolds X_t are uniformly bounded below. Then the eigenvalues of X_t converge to those of X.*

There is no known common generalization of the results of Dodziuk and Fukaya-Cheeger-Colding. In this paper we will study a special case, which will allow us to find manifolds with gaps in the L^2 essential spectrum.

Let $(X_1, g_1), (X_2, g_2)$ be two complete Riemannian manifolds. Let $x_1 \in X_1$ and $x_2 \in X_2$ be two fixed points on the manifolds respectively.

Let

$$N = S^{n-1} \times (-2, 2)$$

be the product manifold equipped with the metric $g_N = \epsilon^2 g_0$, where g_0 is the standard product metric.

For any $\varepsilon > 0$, we construct the manifold X_ε by glueing the three manifolds X_1, X_2, N in the following way.

Let $f_1 : S^{n-1} \times (-2, -1) \to X_1$ be the function

$$f_1(\theta, t) = \exp_{x_1}(-t\varepsilon\theta),$$

where $\exp_{x_1}$ is the exponential map from $T_{x_1}X_1 \to X_1$. In particular $\exp_{x_1}(0) = x_1$. Similarly, let $f_2 : S^{n-1} \times (1, 2) \to X_2$ be the function

$$f_2(\theta, t) = \exp_{x_2}(t\varepsilon\theta),$$

where $\exp_{x_2}$ is the exponential map from $T_{x_2}X_2 \to X_2$. In particular $\exp_{x_2}(0) = x_2$.

It is clear that f_i $(i = 1,2)$ are diffeomorphisms between their domains and ranges. Let X_ε denote the composite manifold defined by (X_1, X_2, N, f_1, f_2), such that

$$X_\varepsilon = (X_1 \backslash B_{x_1}(\varepsilon)) \cup (X_2 \backslash B_{x_2}(\varepsilon)) \cup N/\sim, \tag{7}$$

where we identify f_i with their images respectively for $i = 1, 2$. Roughly speaking, X_ε is constructed from X_1, X_2 by removing two balls of radius ε and adding a neck connecting them.

Abusing notation, we will identify g_i with $f_i^*(g_i)$ for $i = 1, 2$ on the sets where they are defined.

We construct the metric g_ε on X_ε as follows. Outside the neck region N, we do not change the metric, i.e., $g_\varepsilon = g_i$ on $X_i \setminus N$. On N, let ρ_0, ρ_1, ρ_2 be a partition of unity for $\mathbb{R}$ in the following sense. ρ_i are nonnegative smooth functions with compact support. ρ_0 is identically 1 on $[-1, 1]$ and $\mathrm{supp}(\rho_0) \subset [-2, 2]$; ρ_1 is identically 1 on $(-\infty, -2]$ and ρ_2 is identically 1 on $[2, \infty)$, and

$$\rho_0 + \rho_1 + \rho_2 = 1.$$

We define

$$g_\varepsilon = \rho_0 g_N + \rho_1 g_1 + \rho_2 g_2.$$

We use M to denote the manifold obtained by taking $X_1 = X_2 = \mathbb{R}^n$, $x_1 = x_2 = 0 \in \mathbb{R}^n$ and $\varepsilon = 1$. Let $x_0 = (1, 0) \in N$ be a fixed middle point of M. In order words, M consists of two copies of $\mathbb{R}^n$ joint by a tube of radius 1 and length 2.

We prove that

PROPOSITION 4.1. *Let X_1, X_2 be two compact Riemannian manifolds. Using the above notations, we have*

(1) *$(X_\varepsilon, g_\varepsilon)$ is Gromov-Hausdorff convergent to the metric space X_0, which is the union $X_1 \cup X_2$ with x_1 identified with x_2 ;*
(2) *Let x_0 be a reference point in the middle of the neck N. The pointed Riemannian manifolds $(X_\varepsilon, \varepsilon^{-2} g_\varepsilon, x_0)$ are Gromov-Hausdorff convergent to (M, x_0).*

PROOF. For (1), consider the Gromov-Hausdorff approximations $\varphi : X_\varepsilon \to X_0$ and $\psi : X_0 \to X_\epsilon$ defined by

$$\varphi(x) = \begin{cases} x & \text{if } x \in X_1 \cup X_2 \\ x_1 = x_2 & \text{if } x \in N/(X_1 \cup X_2) \end{cases}$$

and

$$\psi(x) = \begin{cases} x & \text{if } x \neq x_1 \text{ and } x \neq x_2 \\ x_0 & \text{if } x = x_1 = x_2 \end{cases}$$

It's easy to see that

$$|d_{X_0}(\varphi(x), \varphi(y)) - d_{X_\varepsilon}(x, y)| \leq 4\varepsilon, |d_{X_\varepsilon}(\psi(a), \psi(b)) - d_{X_0}(a, b)| \leq 4\varepsilon$$

$\forall x, y \in X_\varepsilon, a, b, \in X_0$, where L_ε is the length of the neck N. As $\lim\limits_{x \to 0} L_\varepsilon = 0$, X_ε is Gromov-Hausdorff convergent to X_0.

Note that in this setting the collar region (where the cylinder is glued to the manifold) shrinks to a point.

For (2), it is easy to see convergence on the neck $N\setminus(X_1\cup X_2)$, so let's consider the congergence $(X_i, \varepsilon^{-2}g_i) \to \mathbb{R}^n$. Let $\delta > 0$ be less than the injectivity radius of (X_i, g_i) at x_i. We know that

$$d(\exp_{x_i})_x = I + o(\delta^2),$$

where $\exp_{x_i}$ is the exponential map with respect to the metric g_i, $d(x, x_i) < \delta$.

Given $R > 0$, choose $\varepsilon < \frac{\delta}{R}$ so that the ball $B_{x_i}(R) \subset (X_i, \varepsilon^{-2}g_i)$ is a subset of $B_{x_i}(\delta) \subset (X_i, g_i)$, and therefore

$$|d(exp_{x_i})^* g - I| = o(\delta^2)$$

in $B_{x_i}(R) \subset (X_i, \varepsilon^{-2}g_i)$, which proves the convergence.

□

LEMMA 1. *The Sobolev constants for both $(X_\varepsilon, g_\varepsilon)$ and $(X_\varepsilon, \varepsilon^{-2}g_\varepsilon)$ are uniformly bounded.*

PROOF. The limit of $(X_\varepsilon, \varepsilon^{-2}g_\varepsilon, x_\varepsilon)$ is the space M, which is obtained by connecting two copies of $\mathbb{R}^n$ by a neck with fixed size. By continuity, in order to prove the uniform bound for the Sobolev space, it suffices to prove the Sobolev inequality on the limiting space.

Let f be a smooth function of compact support on the limiting space M. As before, we can write

$$f = f_1 + f_2 + f_3$$

where f_1, f_2 has their support within one copy of $\mathbb{R}^n$ and f_3 has its support within a fixed geodesic ball. Since on Euclidean space we have uniform Sobolev constants, we have

$$\left(\int |f_i|^{\frac{2n}{n-2}}\right)^{\frac{n-2}{n}} \leq C \int |\nabla f_i|^2$$

for $i = 1, 2$. On the other hand, we have the usual Sobolev inequality on a compact manifold, thus we have

$$\left(\int |f_3|^{\frac{2n}{n-2}}\right)^{\frac{n-2}{n}} \leq C' \int |\nabla f_3|^2$$

with a possibly different Sobolev constant C'. Combining the above we get

$$\left(\int |f_i|^{\frac{2n}{n-2}}\right)^{\frac{n-2}{n}} \leq max(C, C') \sum_{i=1}^{3} \int |\nabla f_i|^2 \leq C\left(\int |\nabla f|^2 + |f|^2\right).$$

Since the Sobolev constant is independent of scaling, we have also proved the existence of a uniform Sobolev constant for $(X_\varepsilon, g_\varepsilon)$. □

In what follows we prove the main technical theorem of this paper.

THEOREM 4.5. *Let X_1, X_2 be two compact Riemannian manifolds and take $\lambda \notin \mathrm{Spec}(X_1) \cup \mathrm{Spec}(X_2)$. Consider the manifold $(X_\varepsilon, g_\varepsilon)$ defined above. Set $2\delta = dist(\lambda, \mathrm{Spec}(X_1) \cup \mathrm{Spec}(X_2))$ and take $\lambda' \in (\lambda - \delta, \lambda + \delta)$. Then, for any $\varepsilon > 0$ small enough, $\lambda' \notin \mathrm{Spec}(X_\varepsilon)$.*

PROOF. We will prove the theorem by contradiction. Assume that for any $\varepsilon > 0$, there is a $\lambda_\varepsilon \in (\lambda - \delta, \lambda + \delta)$ such that λ_ε is an eigenvalue of X_ε. Let f_ε be the corresponding eigenfunction. Then

$$\Delta_\varepsilon f_\varepsilon + \lambda_\varepsilon f_\varepsilon = 0.$$

By a standard Moser iteration argument using the uniform Sobolev inequality, we have

$$\|f_\varepsilon\|_{L^\infty}^2 \leq C \int_{X_\varepsilon} \|f_\varepsilon\|^2.$$

As a result, if we normalize the L^2 norm of f_ε to be 1 and given that $\mathrm{vol}(X_\varepsilon)$ is uniformly bounded, we get that the f_ε are uniformly bounded in L^∞. At the same time must have a sequence ε_i such that $\lambda_{\varepsilon_i} \to \lambda_0 \in [\lambda - \delta, \lambda + \delta]$. By passing to a subsequence if necessary, we must have $f_{\varepsilon_i} \to \xi$, converging to an L^2 function ξ on X_0. Let $\xi_i = \xi|_{X_i \setminus \{x_i\}}$ for $i = 1, 2$. By the above argument, since ξ is bounded, it follows that each ξ_i extends to a smooth function on X_i. Moreover, we have

$$\Delta \xi_i + \lambda_0 \xi_i = 0$$

for $i = 1, 2$ with $\lambda_0 \in [\lambda - \delta, \lambda + \delta]$. Since each of the f_ε is bounded, with L^2 norm equal to 1 for all ε, at least one of the ξ_i is not zero. This contradicts the assumption on λ_ε. □

REMARK 1. The above theorem can be interpreted as a spectral continuity result: let $\lambda_k(X_\varepsilon)$ be the k-th eigenvalue of X_ε. Then a subsequence $\lambda_k(X_{\varepsilon_i})$ is convergent to the corresponding eigenvalue of the limit space.

REMARK 2. Since the Ricci curvature of X_ε has no lower bound, Theorem 4.5 is not a special case of the theorem of Cheeger-Colding. It is neither a special case of the theorem of Dodziuk because the limit space is singular.

Let X be a fixed compact manifold. Let $x_1, x_2 \in X$ two distinct points. Construct a metric space by first making $\mathbb{Z}$ copies of X, and labelling them X_j. Then glue the point x_2 of X_j onto the point x_1 of X_{j+1} for each $j \in \mathbb{Z}$. Denote by M the metric space obtained through this gluing process.

DEFINITION 2. A smoothing X_eps of M is a smooth manifold constructed as in (7) at each x_1, x_2. Oviously, under the Gromov-Hausdorff convergence, we have

$$\lim_{\varepsilon \to 0} X_\varepsilon = M.$$

Similar to to Lemma 1, we have

LEMMA 2. *The Sobolev constant for X_ε is independent to ε.*

Using Theorem 4.5 we can prove the following.

THEOREM 4.6. *Let X_ε be a smoothing of M constructed in a similar process as in the beginning of the section. That is, X_ε is smooth and the Gromov-Hausdorff limit of X_ε is M. Then, for ε small enough, the essential spectrum of X_ε has gaps.*

PROOF. Note that since X_ε is a periodic manifold, its spectrum must coincide with its essential spectrum.

Our proof is a generalization of the method used in the proof of Theorem 4.5. Let $\lambda_\varepsilon \in \mathrm{Spec}(X_\varepsilon, g_\varepsilon)$. We shall prove that a subsequence λ_{ε_i} should be convergent to $\lambda_0 \in \mathrm{Spec}(X)$. Fix $\delta > 0$. Let f_ε be the approximating eigenfunction by the Weyl criterion such that

$$\|\Delta f_\varepsilon + \lambda_\varepsilon f_\varepsilon\|_{L^2} \leq \delta \|f_\varepsilon\|_{L^2}.$$

It is not difficult to see that there exist λ'_ε such that $\lambda_\varepsilon - \lambda'_\varepsilon = o(1)$ and for which we can find a function f'_ε that is the Dirichlet eigenfunction corresponding to λ'_ε on the support of f_ε. That is,

$$\Delta f'_\varepsilon + \lambda'_\varepsilon f'_\varepsilon = 0.$$

We normalize f'_ε so that the L^2 norm of f'_ε on K is 1. By using the uniform Sobolev inequality, we can prove that the f'_ε are uniformly bounded.

Let y_ε be the maximum point of f_ε. By the periodic property of X_ε, by translating if necessary, we may assume that y_ε is within a fixed copy of X. Normalizing f'_ε so that the maximum of it is 1. Then a subsequence of f'_ε will be convergent to a *non-zero* function f_0 on the copy X which, by elliptic estimates, must be smooth on the smooth part of M. Since f is also bounded, f must be an eigenfunction of X. Therefore $\lambda_0 \in \operatorname{Spec}(X)$.

□

REMARK 3. It should be noted that the convergence rate depends on λ. For larger λ the rate of convergence may in fact be slower. It would therefore be interesting to know whether there is an infinite number of gaps in the spectrum of X_ε for ε small enough.

References

[1] C. Anné, G. Carron, and O. Post, *Gaps in the differential forms spectrum on cyclic coverings*, Math. Z. **262** (2009), no. 1, 57–90, DOI 10.1007/s00209-008-0363-0. MR2491601

[2] G. A. Baker and J. Dodziuk, *Stability of spectra of Hodge-de Rham Laplacians*, Math. Z. **224** (1997), no. 3, 327–345, DOI 10.1007/PL00004293. MR1439194

[3] N. Charalambous, *On the equivalence of heat kernel estimates and logarithmic Sobolev inequalities for the Hodge Laplacian*, J. Differential Equations **233** (2007), no. 1, 291–312, DOI 10.1016/j.jde.2006.10.007. MR2290281

[4] N. Charalambous and Z. Lu, *On the spectrum of the Laplacian*, Math. Ann. **359** (2014), no. 1-2, 211–238, DOI 10.1007/s00208-013-1000-8. MR3201899

[5] N. Charalambous and Z. Lu, *The spectrum of continuously perturbed operators and the Laplacian on forms*, Differential Geom. Appl. **65** (2019), 227–240, DOI 10.1016/j.difgeo.2019.05.002. MR3948872

[6] J. Cheeger, *Degeneration of Riemannian metrics under Ricci curvature bounds*, Lezioni Fermiane. [Fermi Lectures], Scuola Normale Superiore, Pisa, 2001. MR2006642

[7] J. Cheeger and T. H. Colding, *On the structure of spaces with Ricci curvature bounded below. III*, J. Differential Geom. **54** (2000), no. 1, 37–74. MR1815411

[8] Z. H. Chen and Z. Q. Lu, *Essential spectrum of complete Riemannian manifolds*, Sci. China Ser. A **35** (1992), no. 3, 276–282. MR1183713

[9] E. B. Davies, B. Simon, and M. Taylor, *L^p spectral theory of Kleinian groups*, J. Funct. Anal. **78** (1988), no. 1, 116–136, DOI 10.1016/0022-1236(88)90135-8. MR937635

[10] J. Dodziuk, *Eigenvalues of the Laplacian on forms*, Proc. Amer. Math. Soc. **85** (1982), no. 3, 437–443, DOI 10.2307/2043863. MR656119

[11] H. Donnelly, *On the essential spectrum of a complete Riemannian manifold*, Topology **20** (1981), no. 1, 1–14, DOI 10.1016/0040-9383(81)90012-4. MR592568

[12] H. Donnelly, *Exhaustion functions and the spectrum of Riemannian manifolds*, Indiana Univ. Math. J. **46** (1997), no. 2, 505–527, DOI 10.1512/iumj.1997.46.1338. MR1481601

[13] H. Donnelly, *Spectrum of the Laplacian on asymptotically Euclidean spaces*, Michigan Math. J. **46** (1999), no. 1, 101–111, DOI 10.1307/mmj/1030132362. MR1682891

[14] J. F. Escobar, *On the spectrum of the Laplacian on complete Riemannian manifolds*, Comm. Partial Differential Equations **11** (1986), no. 1, 63–85, DOI 10.1080/03605308608820418. MR814547

[15] J. F. Escobar and A. Freire, *The spectrum of the Laplacian of manifolds of positive curvature*, Duke Math. J. **65** (1992), no. 1, 1–21, DOI 10.1215/S0012-7094-92-06501-X. MR1148983

[16] K. Fukaya, *Collapsing of Riemannian manifolds and eigenvalues of Laplace operator*, Invent. Math. **87** (1987), no. 3, 517–547, DOI 10.1007/BF01389241. MR874035

[17] D. Gilbarg and N. S. Trudinger, *Elliptic partial differential equations of second order*, Classics in Mathematics, Springer-Verlag, Berlin, 2001. Reprint of the 1998 edition. MR1814364

[18] J. Y. Li, *Spectrum of the Laplacian on a complete Riemannian manifold with nonnegative Ricci curvature which possess a pole*, J. Math. Soc. Japan **46** (1994), no. 2, 213–216, DOI 10.2969/jmsj/04620213. MR1264938

[19] F. Lledó and O. Post, *Existence of spectral gaps, covering manifolds and residually finite groups*, Rev. Math. Phys. **20** (2008), no. 2, 199–231, DOI 10.1142/S0129055X08003286. MR2400010

[20] J. Lott, *On the spectrum of a finite-volume negatively-curved manifold*, Amer. J. Math. **123** (2001), no. 2, 185–205. MR1828220

[21] Z. Lu and D. Zhou, *On the essential spectrum of complete non-compact manifolds*, J. Funct. Anal. **260** (2011), no. 11, 3283–3298, DOI 10.1016/j.jfa.2010.10.010. MR2776570

[22] O. Post, *Periodic manifolds with spectral gaps*, J. Differential Equations **187** (2003), no. 1, 23–45, DOI 10.1016/S0022-0396(02)00006-2. MR1946544

[23] O. Post, *Spectral convergence of quasi-one-dimensional spaces*, Ann. Henri Poincaré **7** (2006), no. 5, 933–973, DOI 10.1007/s00023-006-0272-x. MR2254756

[24] M. Reed and B. Simon, *Methods of modern mathematical physics. IV. Analysis of operators*, Academic Press [Harcourt Brace Jovanovich, Publishers], New York-London, 1978. MR0493421

[25] R. Schoen and H. Tran, *Complete manifolds with bounded curvature and spectral gaps*, J. Differential Equations **261** (2016), no. 4, 2584–2606, DOI 10.1016/j.jde.2016.05.002. MR3505201

[26] R. Schoen and S.-T. Yau, *Lectures on differential geometry*, Conference Proceedings and Lecture Notes in Geometry and Topology, I, International Press, Cambridge, MA, 1994. Lecture notes prepared by Wei Yue Ding, Kung Ching Chang [Gong Qing Zhang], Jia Qing Zhong and Yi Chao Xu; Translated from the Chinese by Ding and S. Y. Cheng; With a preface translated from the Chinese by Kaising Tso. MR1333601

[27] K.-T. Sturm, *On the L^p-spectrum of uniformly elliptic operators on Riemannian manifolds*, J. Funct. Anal. **118** (1993), no. 2, 442–453, DOI 10.1006/jfan.1993.1150. MR1250269

[28] J. Wang, *The spectrum of the Laplacian on a manifold of nonnegative Ricci curvature*, Math. Res. Lett. **4** (1997), no. 4, 473–479, DOI 10.4310/MRL.1997.v4.n4.a4. MR1470419

[29] D. T. Zhou, *Essential spectrum of the Laplacian on manifolds of nonnegative curvature*, Internat. Math. Res. Notices **5** (1994), 209 ff., approx. 6 pp., DOI 10.1155/S1073792894000231. MR1270134

Department of Mathematics and Statistics, University of Cyprus, Nicosia, 1678, Cyprus

Email address, Nelia Charalambous: `nelia@ucy.ac.cy`

Department of Mathematics, University of California, Irvine, Irvine, CA 92697, USA

Email address, Helton: `hleal@uci.edu`

Department of Mathematics, University of California, Irvine, Irvine, CA 92697, USA

Email address, Zhiqin Lu: `zlu@uci.edu`

Contemporary Mathematics
Volume **756**, 2020
https://doi.org/10.1090/conm/756/15197

On stability and index of minimal submanifolds

Hang Chen

ABSTRACT. Minimal submanifolds are critical points of the volume functional. It plays an important role in the geometry of submanifolds and has been widely and deeply studied. In this paper, we survey the stability and the index of minimal submanifolds in several special ambient spaces. In the first part, we mainly summarize known results on the existence and the classification of closed stable submanifolds of arbitrary codimension in compact rank one symmetric spaces and product manifolds. In the second part, we exhibit results on minimal hypersurfaces with lower index in a sphere, a projective space and the product manifold of two spheres, including a new result of the author.

1. Introduction

Let Σ^n be an n-dimensional immersed submanifold in an $(n+p)$-dimensional Riemannian manifold M^{n+p}. For simplicity, in this paper we always assume that Σ is closed except special declaration. It is well known that Σ is minimal if and only if it is a critical point of the volume functional. Precisely, if we consider a variation Σ_t $(-\epsilon < t < \epsilon)$ of Σ such that $\Sigma_0 = \Sigma$, and let $V(t)$ denote the volume of Σ_t, then

$$V'(0) = -\int_\Sigma \langle n\mathbf{H}, \eta\rangle, \tag{1.1}$$

here $\mathbf{H}$ is the mean curvature vector of Σ in M, and η is the normal variational vector field on Σ.

The study of minimal submanifolds is one of the most important branches in the geometry of submanifolds. The topics on minimal submanifolds include the existence and nonexistence of the immersions, pinching theorems, the classification, the stability and Morse index and so on, and a lot of results have been obtained on these topics. There are several nice books for references, see [**22, 26, 40, 78**].

In this paper, we mainly review the stability and Morse index, which arise from the second variation of the volume functional $V(t)$, for several special ambient spaces.

In Section 2, we consider the case of arbitrary codimension. We briefly recall the second variation formula, the definitions of stability and Morse index of minimal submanifolds, and then we survey some known results on the existence and

2010 *Mathematics Subject Classification.* Primary 53C20, 53C42.
Key words and phrases. Minimal submanifolds, Stability, Index.
The author was partially supported by NSFC Grant No. 11601426.

classification of closed stable submanifolds in the rank one symmetric spaces and certain product manifolds.

In Section 3, we consider the hypersurface case (i.e. codimension one). At first, we review the index jump phenomenon of closed minimal hypersurfaces in a sphere. Next, we review classification results on minimal hypersurfaces with index one in a projective space. At last, we review the index of minimal hypersurfaces in product manifolds. In particular, we show a new result on minimal hypersurfaces with index one in $\mathbb{S}^n(1) \times \mathbb{S}^1(r)$, which extends part of Urbano's result [**74**] from 2-dimension to the higher dimension.

In Section 4, we prove our new theorem in Section 3, which is inspired by Urbano's proof in 2-dimension case. The proof depends on a decomposition of a Killing vector field on the ambient space. In the end, we propose some questions which can be considered in the future.

2. Minimal submanifolds of arbitrary codimension

Taking the second order derivative of $V(t)$ respect to t, at the critical points, we have the second variation formula (cf. [**22, 68**])

$$V''(0) = -\int_\Sigma \langle \mathfrak{J}\eta, \eta \rangle, \tag{2.1}$$

where $\mathfrak{J}$ is the Jacobi operator given by

$$\mathfrak{J} = \Delta^\perp + \mathfrak{B} + \mathfrak{R}, \tag{2.2}$$

$$\Delta^\perp \eta = \sum_{i=1}^n (\nabla^\perp_{e_i}\nabla^\perp_{e_i} - \nabla^\perp_{\nabla_{e_i}e_i})\eta, \quad \mathfrak{B}(\eta) = \sum_{i=1}^n h(e_i, A_\eta e_i), \quad \mathfrak{R}(\eta) = \sum_{i=1}^n (\bar{R}(\eta, e_i)e_i)^\perp,$$

where $\{e_i\}_{1\le i\le n}$ is a tangent orthonormal frame on Σ, ∇ is the Levi-Civita connection on Σ, $\nabla^\perp$ is the normal connection, h is the second fundamental form of Σ, A_η is the shape operator associated to η, $\bar{R}$ is the curvature tensor of M and $\perp$ denotes normal component.

It is well known that (cf. [**68**]) $\mathfrak{J}$ is a strongly elliptic operator acting on the sections of the normal bundle $\Gamma(T^\perp\Sigma)$, and it has distinct, real eigenvalues $\{\lambda_i\}$ such that

$$\lambda_1 < \lambda_2 < \cdots < \lambda_i < \cdots \longrightarrow +\infty,$$

and the dimension of each eigenspace E_{λ_i} is finite.

The (Morse) index and the nullity of Σ are defined respectively by

$$\mathrm{Ind}(\Sigma) = \sum_{\lambda<0} \dim E_\lambda, \quad \mathrm{Null}(\Sigma) = \dim E_0.$$

A natural and important question is whether the minimal submanifold yields a local minimum of the volume functional or not, namely, the minimal submanifold is stable or not. Recall that Σ is called *stable* if and only if

$$Q(\eta) := -\int_\Sigma \langle \mathfrak{J}\eta, \eta \rangle \ge 0, \ \forall \eta \in \Gamma(T^\perp\Sigma).$$

We can easily see that Σ is stable if and only if $\mathrm{Ind}(\Sigma) = 0$.

2.1. Minimal submanifolds in a sphere. The simplest case is that the ambient space is a standard sphere. Since J. Simons' pioneering work [**68**], there have been a lot of results on this case, such as pinching theorems ([**11**, **21**, **30**, **34**, **39**, **42**, **64**, **68**, **81**], etc.), characterization and classification theorems ([**6**–**9**, **79**], etc.). On the index and nullity, Simons [**68**] proved that

THEOREM 2.1 (Theorem 5.1.1 in [**68**]). *For a closed minimal submanifold Σ^n immersed in a standard sphere $\mathbb{S}^{n+p}$, we have*

$$\mathrm{Ind}(\Sigma) \geq p; \quad \mathrm{Null}(\Sigma) \geq (n+1)p. \tag{2.3}$$

Moreover, $\mathrm{Ind}(\Sigma) = p \Longleftrightarrow \mathrm{Null}(\Sigma) = (n+1)p \Longleftrightarrow \Sigma = \mathbb{S}^n$*(totally geodesic).*

This theorem means that $(0,p)$ is a first forbidden zone for $\mathrm{Ind}(\Sigma)$. Later, Ejiri [**30**] proved that, if Σ^2 is a closed minimal surface of genus 0, fully immersed in $\mathbb{S}^{2m}(1)(m \geq 2)$, then $\mathrm{Ind}(\Sigma) \geq 2(m(m+2)-3)$. This partially improves Simons' result. In general case, El Soufi (Theorem 2.1 in [**31**]) proved that, under the same assumptions of Theorem 2.1, $\mathrm{Ind}(\Sigma) \geq n+p+1$ for any non-totally geodesic Σ, which means that $(p, n+p+1)$ is a second forbidden zone for $\mathrm{Ind}(\Sigma)$.

2.2. Existence and nonexistence of stable submanifolds. As a direct corollary of Theorem 2.1, there is no stable closed minimal submanifold in $\mathbb{S}^{n+p}$. The existence and classification of closed stable minimal submanifolds in compact rank one symmetric spaces have been solved completely:

THEOREM 2.2. *Let Σ^m be an m-dimensional closed minimal submanifold immersed in M. Then Σ is stable if and only if Σ is one of the cases listed in the following table.*

	Lawson and Simons [**41**]	Ohnita [**54**]		
M	$P^n(\mathbb{C})$	$P^n(\mathbb{R})$	$P^n(\mathbb{H})$	$P^2(\mathbf{Cay})$
Σ	*complex submanifold*	$P^l(\mathbb{R})$	$P^l(\mathbb{H})$	$P^1(\mathbf{Cay})$

Here $1 \leq l \leq n-1$, $m = l$ for $\Sigma^m = P^l(\mathbb{R})$ and $m = 4l$ for $\Sigma^m = P^l(\mathbb{H})$. Noting that $P^n(\mathbb{C})$ is 1/4-pinched (i.e. the sectional curvature is in $[1/4, 1]$), based on Theorem 2.1 and Theorem 2.2, Lawson and Simons proposed the following conjecture:

CONJECTURE 2.3 ([**41**]). *There are no closed stable minimal submanifolds in any compact, simply connected, strictly 1/4-pinched (i.e. the sectional curvature is in $(1/4, 1]$) Riemannian manifold.*

This conjecture is open, but known results support this conjecture. Shen and Xu [**66**] proved that there are no closed stable minimal submanifolds in a closed simply connected 0.77-pinched Riemannian manifold. Howard and Wei [**37**] verified the above conjecture for several classes of positively curved manifolds. On the other hand, when M is a submanifold of the Euclidean space $\mathbb{R}^N$, some nonexistence results of stable minimal submanifolds of M were obtained by different people under different assumptions (e.g. δ-pinched hypersurface in $\mathbb{R}^N$, algebraic conditions on the second fundamental form of M in $\mathbb{R}^N$), and in some results the pinching constant δ can be less than 1/4 (cf. [**37**, **38**, **43**, **54**, **65**, **68**]).

2.3. Minimal manifolds in a product manifold. It is much harder when considering minimal submanifolds of codimension > 1 in ambient spaces different from ones mentioned above. Only a few particular cases have been studied, see the following papers and the references therein: [**5, 12, 13**] for the existence and nonexistence, [**4, 14–18, 71, 75**] for the classifications, [**3, 10, 51–53, 73**] for the second variation, the eigenvalue estimates and the stabilities.

Recently, Torralbo and Urbano [**70**] proved a classification theorem for stable compact minimal submanifolds of a Riemannian product of a sphere $\mathbb{S}^m(r)$ and any Riemannian manifold M.

THEOREM 2.4 (Theorem 1 in [**70**]). *Let M be any Riemannian manifold and $\Phi = (\phi, \psi) : \Sigma^n \to \mathbb{S}^m(r) \times M$ be a minimal immersion of a compact n-manifold Σ, $n \geq 2$, satisfying either $m \geq 3$ or $m = 2$ and Φ is a hypersurface. Then, Φ is stable if and only if*

(1) *$\Sigma = \mathbb{S}^m(r)$ and $\Phi(\Sigma)$ is a slice $\mathbb{S}^m(r) \times \{q\}$ with q a point of M.*
(2) *Σ is a covering of M and $\Phi(\Sigma)$ is a slice $\{p\} \times M$ with p a point of $\mathbb{S}^m(r)$.*
(3) *$\psi : \Sigma \to M$ is a stable minimal submanifold and $\Phi(\Sigma)$ is $\{p\} \times \psi(\Sigma)$ with p a point of $\mathbb{S}^m(r)$.*
(4) *$\Sigma = \mathbb{S}^m(r) \times \hat{\Sigma}$, $\Phi = Id \times \psi$, and $\psi : \hat{\Sigma} \to M$ is a stable minimal submanifold.*

Afterward, Xianfeng Wang and the author [**20**] generalized Torralbo and Urbano's result to a Riemannian product of a δ-pinched hypersurface M_1 in the Euclidean space and an arbitrary Riemannian manifold M_2.

THEOREM 2.5 ([**20**]). *Let $\Phi = (\phi, \psi) : \Sigma^n \to M = M_1 \times M_2$ be a minimal immersion of an n-dimensional ($n \geq 2$) compact manifold Σ into M, where M_1 is a compact connected hypersurface in $\mathbb{R}^{m_1+1}$ ($m_1 \geq 3$) and M_2 is any Riemannian manifold. Assume that the sectional curvature K_{M_1} of M_1 satisfies*

$$\frac{1}{\sqrt{m_1 - 1}} \leq K_{M_1} \leq 1.$$

Then, Φ is stable if and only if

(1) *$\Sigma = M_1$ and $\Phi(\Sigma)$ is a slice $M_1 \times \{p_2\}$ with p_2 a point of M_2.*
(2) *Σ is a covering of M_2 and $\Phi(\Sigma)$ is a slice $\{p_1\} \times M_2$ with p_1 a point of M_1.*
(3) *$\psi : \Sigma \to M_2$ is a stable minimal submanifold and $\Phi(\Sigma)$ is $\{p_1\} \times \psi(\Sigma)$ with p_1 a point of M_1.*
(4) *$\Sigma = M_1 \times \hat{\Sigma}$, $\Phi = Id \times \psi$, and $\psi : \hat{\Sigma} \to M_2$ is a stable minimal submanifold.*

In particular, when the ambient space is an m-dimensional ($m \geq 3$) compact hypersurface M in $\mathbb{R}^{m+1}$, if the sectional curvature K_M of M satisfies

$$\frac{1}{\sqrt{m + 1}} \leq K_M \leq 1,$$

then there exist no stable compact minimal submanifolds in M.

A surprising fact is that the pinching constant δ is a strictly monotonically decreasing function of m_1 ($= \dim M_1$) converging to 0 as $m_1 \to \infty$.

One basic tool of studying submanifolds in $M_1 \times M_2$ is analyzing the *almost product structure*, which is a linear isometry $\bar{P}$ defined by (cf. Chapter 8 in [**80**])

$$\bar{P}(v_1, v_2) = (v_1, -v_2), \quad v_i \in TM_i, i = 1, 2.$$

Besides, the proof of Theorem 2.5 (and Theorem 2.4) depends on the following criteria.

LEMMA 2.6 ([**20**]). *Let Σ be a closed minimal submanifold in $M = M_1 \times M_2$, where M_1 is an m_1-dimensional compact submanifold in a real space form $\mathbb{R}^{N_1}(c)$ $(c \geq 0)$, and M_2 is any Riemannian manifold. Let B_1 denote the second fundamental form of the immersion $M_1 \to \mathbb{R}^{N_1}(c)$, and define*

$$F = \sum_{i,\beta} \Big(\langle B_1(e^1_\beta, e^1_\beta), B_1(e^1_i, e^1_i) \rangle - 2\|B_1(e^1_i, e^1_\beta)\|^2 + c|e^1_i|^2|e^1_\beta|^2 - 2c\langle e^1_i, e^1_\beta \rangle^2 \Big).$$

If $F > 0$ for any local orthonormal frames $\{e_i\}_{1 \leq i \leq n}$ tangent to Σ and $\{e_\alpha\}_{n+1 \leq \alpha \leq n+p}$ normal to Σ, then Σ is unstable. If Σ is stable, then $\int_\Sigma F \leq 0$.

If M_2 vanishes to a point in Lemma 2.6, then the ambient space M is an m-dimensional submanifold of a real space form $\mathbb{R}^N(c)(c \geq 0)$. Let B denote the second fundamental form of the immersion $M \to \mathbb{R}^N(c)$, then we have

$$F = \sum_{i,\beta} \{ \langle B(e_i, e_i), B(e_\beta, e_\beta) \rangle - 2\|B(e_i, e_\beta)\|^2 \} + n(m-n)c,$$

so Lemma 2.6 recovers the well-known results due to Lawson-Simons [**41**] $(c > 0)$ and Y. Xin [**77**] $(c = 0)$, which can be applied to deal with the cases mentioned in Section 2.1.

3. Minimal hypersurfaces with lower index

3.1. Two-sided and one-sided hypersurfaces. Now let's consider the hypersurface case. Let Σ^n be a minimal hypersurface immersed in M^{n+1}, and we denote the immersion by $\Phi : \Sigma \to M$. Then the immersion Φ (or the hypersurface Σ itself) is called *two-sided* (resp. *one-sided*) if the normal bundle is trivial (resp. nontrivial).

It is valuable to point out, the two-sidedness and one-sidedness are related to the orientability and non-orientability of the hypersurface Σ, but they are not intrinsic properties of Σ and depend on the ambient manifold M. When M is orientable, the two-sidedness is equivalent to the orientability of Σ. However, when M is non-orientable, there are many possibilities. For example, taking $M = P^2(\mathbb{R}) \times \mathbb{R}$, then $P^1(\mathbb{R}) \times \mathbb{R} \subset P^2(\mathbb{R}) \times \mathbb{R}$ is an orientable one-sided surface; $P^2(\mathbb{R}) \times \{0\} \subset P^2(\mathbb{R}) \times \mathbb{R}$ is a non-orientable two-sided surface; $\mathbb{S}^2$ is orientable and $\Phi(\mathbb{S}^2) = P^2(\mathbb{R}) \times \{0\}$ is two-sided.

The two-sidedness case is simpler. When Φ is two-sided, there is a global unit normal vector field ν on Σ, so each normal vector field $\eta \in \Gamma(T^\perp\Sigma)$ corresponds to a function $f \in C^\infty(\Sigma)$ by $\eta = f\nu$, and the Jacobi operator (2.2) becomes a Schrödinger operator acting on $C^\infty(\Sigma)$:

$$J = \Delta + |A|^2 + \overline{\mathrm{Ric}}^M(\nu, \nu), \tag{3.1}$$

where $|A|^2$ is the squared norm of the second fundamental form, and $\overline{\mathrm{Ric}}^M$ is the Ricci curvature of M . Now Φ (or Σ) is *stable* if and only

$$Q(f) := -\int_\Sigma fJf \geq 0, \forall f \in C^\infty(\Sigma). \tag{3.2}$$

In complete non-compact or one-sided case, it is more subtle even for stable surfaces in a 3-dimensional manifold, and we refer the readers to [**2, 28, 32, 33,**

45, 47, 49, 58–62, 67] and the references therein. For this reason, in the rest of this section, we mainly focus on closed two-sided minimal hypersurfaces with lower index in a sphere, a real projective space and the product manifold of two spheres. For general ambient spaces, the existence of (embedded) minimal hypersurfaces are usually obtained by min-max theory, and the reader can refer [**25, 44, 48, 69, 82**] and the references therein.

3.2. Minimal hypersurfaces in a sphere. When the ambient space M is the unit sphere $\mathbb{S}^{n+1}(1)$, $J = \Delta + |A|^2 + n$, which follows that $\lambda_1^J \leq -n$. Actually, we have

THEOREM 3.1 ([**56,68,76**]). *Let Σ^n be a closed orientable minimal hypersurface in $\mathbb{S}^{n+1}(1)$, then either*

(1) $\lambda_1^J = -n$ *and* Σ *is totally geodesic; or*
(2) $\lambda_1^J \leq -2n$, *and* $\lambda_1^J = -2n$ *if and only if* Σ *is a Clifford torus.*

Recently, Qing-Ming Cheng and Daguang Chen [**19**] obtained an optimal upper bound of λ_1^J only depending on the mean curvature H and the dimension n for CMC hypersurfaces in $\mathbb{S}^{n+1}(1)$, which recovers Theorem 3.1 when $H = 0$.

As mentioned after Theorem 2.1 in Section 2.1, if Σ is not totally geodesic, then $\mathrm{Ind}(\Sigma) \geq n+2$. In fact, there is a better estimate for hypersurfaces.

THEOREM 3.2 (cf. [**31,72**]). *Let Σ^n be a closed orientable minimal hypersurface in $\mathbb{S}^{n+1}(1)$ and not totally geodesic, then*

$$\mathrm{Ind}(\Sigma) \geq n+3. \tag{3.3}$$

Theorem 3.2 is proved by Urbano [**72**] for $n = 2$ and by El Soufi [**31**] for general n. The key step of the proof is that, the vector space $V = \{f_a := \langle \nu, a\rangle \mid a \in \mathbb{R}^{n+2}\}$ has dimension $n+2$, and each non-zero f_a is an eigenfunction of J associated to the eigenvalue $-n$. On the other hand, $-n$ is not the first eigenvalue of J (since λ_1^J is simple, or cf. Theorem 3.1). Hence $\mathrm{Ind}(\Sigma) \geq n+3$.

The lower bound in Theorem 3.2 is optimal, as each Clifford torus

$$C_{k,n-k} := \mathbb{S}^k(\sqrt{k/n}) \times \mathbb{S}^{n-k}(\sqrt{(n-k)/n}) \to \mathbb{S}^{n+1}(1), \quad 1 \leq k \leq n-1 \tag{3.4}$$

has the index $n+3$. For this reason, we have the following conjecture, which has been proposed for a long time.

CONJECTURE 3.3. *Let Σ^n be minimal in $\mathbb{S}^{n+1}$ and not totally geodesic. Then*

$$\mathrm{Ind}(\Sigma) = n+3. \tag{3.5}$$

if and only if Σ *is a minimal Clifford torus.*

One feasible way to prove this conjecture is to show $|A|^2 = n$, which implies the conclusion by a very well-known result due to Chern, do Carmo and Kobayashi [**23**]. Urbano [**72**] proved it for $n = 2$. The first step, which is easier, is to show $\lambda_2^J = -2$. The second step, which is more complicated, is to show each coordinate function is the eigenfunction of J associated to λ_2^J. In this step, we need a technique of constructing test functions orthogonal to the first eigenfunction by conformal transformations on a sphere (cf. [**46**]); another critical ingredient is Gauss-Bonnet theorem for $n = 2$ so that we can determine the genus of the surface.

For $n \geq 3$, Conjecture 3.3 remains open. So far, we only know that it is true for general n under some additional hypotheses. All these hypotheses are used to derive $|A|^2 = n$ in the processes of the proofs, for instance,

- $|A|^2$ is constant (Guadalupe, Brasil Jr. and Delgado [**36**]);
- The group $O_\Sigma(n+2) := \{\gamma \in O(n+2) \mid \gamma(\Sigma) = \Sigma\}$ fixes only the origin in $\mathbb{R}^{n+2}$ (in particular, Σ has antipodal symmetry) (Perdomo [**55**]);
- $|A|^2 \geq n$ (A. Barros and P. Sousa [**27**]).

Very recently, Perdomo [**57**] proved that, if Σ is a closed minimal rotational hypersurface in $\mathbb{S}^{n+1}$, neither totally geodesic nor a product of two spheres, then $\mathrm{Ind}(\Sigma) \geq 3n+5$.

At last, we mention that, the lower bound in Theorem 3.2 was improved by Savo [**63**] when the first Betti number $b_1(\Sigma)$ (for $n \geq 3$) or the genus $g(\Sigma)$ (for $n = 2$) of Σ is sufficiently large, due to his remarkable comparison theorem between the spectrum of the Jacobi operator and that of the Hodge-Laplacian on 1-forms in the same paper. Savo proved that (Corollary 1.2 and Theorem 1.3 in [**63**]), for a closed, orientable minimal hypersurface Σ in $\mathbb{S}^{n+1}$,

$$\mathrm{Ind}(\Sigma) \geq \begin{cases} \dfrac{b_1(\Sigma)}{\binom{n+2}{2}} + n + 2 & \text{for } n \geq 3 \text{ and } b_1(\Sigma) \geq 1; \\[2ex] \dfrac{g(\Sigma)}{2} + 4 & \text{for } n = 2 \text{ and } g(\Sigma) \geq 1. \end{cases}$$

Ambrozio, Carlotto and Sharp [**1**] extended and generalized Savo's method, and the improved technique can be used to prove an inequality of the form $\mathrm{Ind}(\Sigma) \geq Cb_1(\Sigma)$ for minimal Σ in some other ambient spaces (e.g. compact symmetric spaces), see [**1, 24, 35, 50**].

3.3. Minimal hypersurfaces in the projective spaces. Considering the real projective space $P^{n+1}(\mathbb{R})$ as the quotient $\mathbb{S}^{n+1}(1)/\{\pm\}$, the Jacobi operator is still $J = \Delta + |A|^2 + n$. However, the classification of minimal hypersurfaces with some fixed lower index in $P^{n+1}(\mathbb{R})$ is very different from the ones in $\mathbb{S}^{n+1}(1)$. Due to Ohnita [**54**] (cf. Theorem 2.2), a closed minimal hypersurface with index 0 in $P^{n+1}(\mathbb{R})$ must be $P^n(\mathbb{R})$. A classification of orientable minimal surfaces with index one in $P^3(\mathbb{R})$ can be obtained from the work of Ritoré and Ros [**58**]. Later, do Carmo, Ritoré and Ros [**29**] obtained the complete classification of closed two sided minimal hypersurfaces with index one in $P^{n+1}(\mathbb{R})$.

THEOREM 3.4 ([**29**]). *The only closed two-sided minimal hypersurfaces with index one in $P^{n+1}(\mathbb{R})$ are the totally geodesic spheres and the minimal Clifford hypersurfaces.*

Moreover, the only embedded ones are the Clifford hypersurfaces.

We remark that the Clifford hypersurface mentioned in the above theorem actually is $C_{k,n-k}/\{\pm\}$ minimally embedded $P^{n+1}(\mathbb{R}) = \mathbb{S}^{n+1}(1)/\{\pm\}$ induced by $C_{k,n-k}$ minimally embedded in $\mathbb{S}^{n+1}(1)$, which is also called a Clifford hypersurface (see [**29**]).

3.4. Minimal hypersurfaces in a product manifold. Theorem 2.4 and Theorem 2.5 give classifications of closed minimal hypersurfaces with index 0 in certain product manifolds. A next natural question is whether we can classify the minimal hypersurfaces with index one. For surfaces, Urbano [**74**] proved that

THEOREM 3.5 (Theorem 4.8 in [**74**]). *Let $\Phi : \Sigma \to \mathbb{S}^2(1) \times \mathbb{S}^1(r), r \geq 1$ be a minimal immersion of a closed surface Σ.*

(1) *If* Σ *is orientable of genus* g*, then* $\mathrm{Ind}(\Sigma) \geq \frac{2g-1}{5}$*. Moreover,* $\mathrm{Ind}(\Sigma) = 1$ *if and only if* $r = 1$ *and* Φ *is the totally geodesic embedding* $\mathbb{S}^1(1) \times \mathbb{S}^1(1) \subset \mathbb{S}^2(1) \times \mathbb{S}^1(1)$*.*

(2) *If* Σ *is nonorientable, then* $\mathrm{Ind}(\Sigma) \geq \frac{g-1}{5}$*, where* g *is the genus of the two-fold oriented covering of* Σ*.*

Urbano also classified the stable orientable complete minimal surfaces of $P^2(\mathbb{R}) \times \mathbb{R}$ and $P^2(\mathbb{R}) \times \mathbb{S}^1(r)$ (see Theorem 4.10 and Theorem 4.12 in [**74**]).

In the higher dimension case, Ambrozio, Carlotto and Sharp [**1**] proved that if Σ is a closed, orientable (embedded) minimal hypersurface of $\mathbb{S}^n(1) \times \mathbb{S}^1(1)$ $(n \geq 3)$, then $\mathrm{Ind}(\Sigma) \geq \frac{2}{(n+3)(n+2)} b_1(\Sigma)$. But when n is large enough and $b_1(\Sigma)$ is small, we cannot get more information except $\mathrm{Ind}(\Sigma) \geq 0$ or 1. If $\mathrm{Ind}(\Sigma) = 0$, then $\Sigma = \mathbb{S}^n(1) \times \{q\}$ by Theorem 2.4. If $\mathrm{Ind}(\Sigma) = 1$, comparing Theorem 3.5, we have the following theorem.

THEOREM 3.6. *For* $n \geq 3, r \geq \frac{1}{\sqrt{n-1}}$*, let* $\Phi : \Sigma \to \mathbb{S}^n(1) \times \mathbb{S}^1(r)$ *be a minimal immersion of an orientable closed hypersurface* Σ*. Then* $\mathrm{Ind}(\Sigma) = 1$ *if and only if* $r = \frac{1}{\sqrt{n-1}}$ *and* Φ *is the totally geodesic embedding* $\Sigma = \mathbb{S}^{n-1}(1) \times \mathbb{S}^1(\frac{1}{\sqrt{n-1}}) \subset \mathbb{S}^n(1) \times \mathbb{S}^1(\frac{1}{\sqrt{n-1}})$*.*

4. Proof of Theorem 3.6

In this section, we will prove Theorem 3.6. The proof depends on a decomposition of a Killing vector field on $\mathbb{S}^n(1) \times \mathbb{S}^1(r)$, and the main idea follows from Urbano's proof for the case $n = 2$ [**74**].

4.1. Structure of $\mathbb{S}^n(1) \times \mathbb{S}^1(r)$. Consider the immersion

$$\mathbb{S}^n(1) \times \mathbb{S}^1(r) \to \mathbb{S}^{n+2}(\sqrt{1+r^2}) \to \mathbb{R}^{n+1} \times \mathbb{R}^2.$$

For any vector $a \in \mathbb{R}^{n+1}, b \in \mathbb{R}^2$, we identify a and b with $(a, 0)$ and $(0, b) \in \mathbb{R}^{n+1} \times \mathbb{R}^2$ respectively. Let $\{\tilde{e}_1, \cdots, \tilde{e}_n\}$ be an orthonormal frame on $\mathbb{S}^n(1)$, and ξ be the unit vector field on $\mathbb{S}^1(r)$. Then $\{\tilde{e}_1, \cdots, \tilde{e}_n, \tilde{e}_{n+1} := \xi\}$ is an orthonormal frame on $\mathbb{S}^n(1) \times \mathbb{S}^1(r)$.

Let x and y denote the position vector of $\mathbb{S}^n(1)$ in $\mathbb{R}^{n+1}$ and $\mathbb{S}^1(r)$ in $\mathbb{R}^2$ respectively. Then $N_1 = \frac{1}{\sqrt{1+r^2}}(rx, -\frac{1}{r}y)$ is the unit normal vector of $\mathbb{S}^n(1) \times \mathbb{S}^1(r)$ in $\mathbb{S}^{n+2}(\sqrt{1+r^2})$, and $N_2 = \frac{1}{\sqrt{1+r^2}}(x, y)$ is the unit normal vector of $\mathbb{S}^{n+2}(\sqrt{1+r^2})$ in $\mathbb{R}^{n+3}$. Let $\tilde{h}$ be the second fundamental form of $\mathbb{S}^n(1) \times \mathbb{S}^1(r)$ in $\mathbb{S}^{n+2}(\sqrt{1+r^2})$, $\bar{h}$ be the second fundamental form of $\mathbb{S}^{n+2}(\sqrt{1+r^2})$ in $\mathbb{R}^{n+3}$. Denote

$$\tilde{h}_{AB} = \langle \tilde{h}(\tilde{e}_A, \tilde{e}_B), N_1 \rangle, \quad \bar{h}_{AB} = \langle \bar{h}(\tilde{e}_A, \tilde{e}_B), N_2 \rangle \quad (1 \leq A, B \leq n+1),$$

A direct computation gives

$$(\tilde{h}_{AB}) = \frac{1}{\sqrt{1+r^2}} \begin{pmatrix} -rI_n & 0 \\ 0 & \frac{1}{r} \end{pmatrix}, \quad (\bar{h}_{AB}) = -\frac{1}{\sqrt{1+r^2}} I_{n+1}. \tag{4.1}$$

4.2. Structure of hypersurfaces in $\mathbb{S}^n(1) \times \mathbb{S}^1(r)$. Let Σ be a hypersurface in $\mathbb{S}^n(1) \times \mathbb{S}^1(r)$. Assume $\{e_1, \cdots, e_n\}$ is an orthonormal frame on Σ and ν is the unit normal of Σ in $\mathbb{S}^n(1) \times \mathbb{S}^1(r)$. Denote by h the second fundamental form of Σ in $\mathbb{S}^n(1) \times \mathbb{S}^1(r)$. Set $h_{ij} = \langle h(e_i, e_j), \nu \rangle$, then $H = \frac{1}{n} \sum_{i=1}^n h_{ii}$ is the mean curvature of Σ in $\mathbb{S}^n(1) \times \mathbb{S}^1(r)$.

We decompose ξ into the tangent part and normal part to Σ as $\xi = T + C\nu$, where $T = \sum_{i=1}^n T_i e_i$. Then $\nu = \sum_{i=1}^n a_i \tilde{e}_i + C\xi$ for some a_i such that $\sum_{i=1}^n a_i^2 = 1 - C^2 = |T|^2$. From (4.1), we have

$$\sum_{i=1}^n \bar{h}_{ii} = \langle \bar{h}(e_i, e_i), N_2 \rangle = -\frac{n}{\sqrt{1+r^2}};$$

$$\begin{aligned}\sum_{i=1}^n \tilde{h}_{ii} &= \operatorname{tr}(\tilde{h}_{AB}) - \langle \tilde{h}(\nu,\nu), N_1 \rangle \\ &= \frac{1}{\sqrt{1+r^2}}\Big(-nr + \frac{1}{r}\Big) - \frac{1}{\sqrt{1+r^2}}\Big(-r\sum_{i=1}^n a_i^2 + \frac{C^2}{r}\Big) \\ &= \frac{1}{\sqrt{1+r^2}}\Big(\frac{1}{r} - (1+r^2)|T|^2\Big).\end{aligned}$$

Using the relation between the Laplacian of coordinate functions and the mean curvature vector for a submanifold in Euclidean space, we have

$$\begin{aligned}\Delta(x,y) &= nH\nu + \frac{1}{\sqrt{1+r^2}}\Big(\frac{1}{r} - (1+r^2)|T|^2\Big)N_1 - \frac{n}{\sqrt{1+r^2}}N_2 \\ &= nH\nu + \Big((-n+|T|^2)x, -\frac{|T|^2}{r^2}y\Big).\end{aligned} \tag{4.2}$$

Now we are ready to prove Theorem 3.6.

Proof of Theorem 3.6. Since Σ is minimal, i.e. $H = 0$, from (4.2) and (3.1), we have

$$\int_M \langle y,b\rangle J\langle y,b\rangle = \int_M \Big(\frac{|T|^2}{r^2} - (n-1)|T|^2 - |A|^2\Big)\langle y,b\rangle^2 \le 0 \tag{4.3}$$

for any fixed vector $b \in \mathbb{R}^2$. Here we used $\overline{\mathrm{Ric}}^M(\nu,\nu) = (n-1)|\sum_{i=1}^n a_i\tilde{e}_i|^2 = (n-1)|T|^2$.

Consider $V = \{\langle y,b\rangle | b \in \mathbb{R}^2\}$. If $\dim V < 1$, then there is a vector $b \neq 0$ such that $\langle y,b\rangle = 0$. So y is constant and then $\Phi(\Sigma) = \mathbb{S}^n \times \{q\}$ is a slice of $\mathbb{S}^n \times \mathbb{S}^1(r)$, which is stable, a contradiction. So $\dim V = 2$. By the assumption $\mathrm{Ind}(\Sigma) = 1$, we have $|A|^2 = 0$ and $\frac{1}{r^2} = n-1$. So Φ is a finite covering of the totally geodesic hypersurface $\mathbb{S}^{n-1}(1) \times \mathbb{S}^1(\frac{1}{\sqrt{n-1}})$.

On the other hand, a direct computation gives that, for an m-sheeted ($m \in \mathbb{Z}_+$) covering $\Sigma = \mathbb{S}^{n-1}(1) \times \mathbb{S}^1(mr) \to \mathbb{S}^{n-1}(1) \times \mathbb{S}^1(r)$, $J = \Delta + (n-1)$, and the first non-zero eigenvalue of Δ on Σ is

$$\lambda_1^\Delta = \min\Big\{n-1, \frac{1}{(mr)^2}\Big\}.$$

Since $\mathrm{Ind}(\Sigma) = 1$, we must have $\frac{1}{(mr)^2} \ge n-1$ so that the second eigenvalue λ_2^J of J is non-negative. Hence $m \le \frac{1}{\sqrt{n-1}r} = 1$ (since $\frac{1}{r^2} = n-1$) and then m must be 1. □

At last, we give some comments on Theorem 3.6.

(1) As mentioned in Remark 4.9 of [**74**], the method in the proof does not work for $r < 1/\sqrt{n-1}$, and any m-sheeted covering $\Sigma = \mathbb{S}^{n-1}(1) \times \mathbb{S}^1(mr) \to \mathbb{S}^{n-1}(1) \times \mathbb{S}^1(r)$ with $m \le \frac{1}{\sqrt{n-1}r}$ ($m \in \mathbb{Z}_+$) has index one.

(2) Theorem 3.6 implies that, when $r > 1/\sqrt{n-1}$, there are no closed orientable minimal hypersurfaces of index one in $\mathbb{S}^n \times \mathbb{S}^1(r)$. Hence, a natural question is that, can we find an optimal lower bound of the index of unstable closed orientable minimal Σ?

(3) It would be interesting to determine the minimal hypersurfaces with index one in general torus $\mathbb{S}^m(1) \times \mathbb{S}^l(r)(m, l \geq 2)$. Note that the decomposition of the Killing vector field ξ plays an important role in the above proof, but it doesn't work for this case.

Acknowledgments

The author was supported by NSFC Grant No. 11601426 and the Fundamental Research Funds for the Central Universities Grant No. 310201911cx013. The author would like to thank Prof. Guofang Wei, Prof. Haizhong Li and the anonymous referees for very helpful suggestions and comments.

References

[1] L. Ambrozio, A. Carlotto, and B. Sharp, *Comparing the Morse index and the first Betti number of minimal hypersurfaces*, J. Differential Geom. **108** (2018), no. 3, 379–410, DOI 10.4310/jdg/1519959621. MR3770846

[2] H.-D. Cao, Y. Shen, and S. Zhu, *The structure of stable minimal hypersurfaces in* $\mathbf{R}^{n+1}$, Math. Res. Lett. **4** (1997), no. 5, 637–644, DOI 10.4310/MRL.1997.v4.n5.a2. MR1484695

[3] I. Castro and F. Urbano, *Minimal Lagrangian surfaces in* $\mathbb{S}^2 \times \mathbb{S}^2$, Comm. Anal. Geom. **15** (2007), no. 2, 217–248. MR2344322

[4] B.-Y. Chen, F. Dillen, L. Verstraelen, and L. Vrancken, *An exotic totally real minimal immersion of* S^3 *in* $\mathbf{C}P^3$ *and its characterisation*, Proc. Roy. Soc. Edinburgh Sect. A **126** (1996), no. 1, 153–165, DOI 10.1017/S0308210500030651. MR1378838

[5] B.-Y. Chen, *The minimal imbedding of surfaces in Riemannian manifold* (English, with Chinese summary), Tamkang J. No. **7** (1968), 345–349. MR0246231

[6] B.-Y. Chen, *Minimal hypersurfaces in an m-sphere*, Proc. Amer. Math. Soc. **29** (1971), 375–380. MR0285999

[7] B.-Y. Chen, *Minimal submanifolds of a Euclidean hypersphere*, Yokohama Math. J. **19** (1971), 45–47. MR291960

[8] B.-Y. Chen, *Minimal surfaces in* S^m *with Gauss curvature* ≤ 0, Proc. Amer. Math. Soc. **31** (1972), 235–238. MR0286000

[9] B.-Y. Chen, *Minimal surfaces with constant Gauss curvature*, Proc. Amer. Math. Soc. **34** (1972), 504–508. MR0296828

[10] B.-Y. Chen, *On the first eigenvalue of Laplacian of compact minimal submanifolds of rank one symmetric spaces*, Chinese J. Math. **11** (1983), no. 4, 1–15. MR732862

[11] B.-Y. Chen, *Some pinching and classification theorems for minimal submanifolds*, Arch. Math. (Basel) **60** (1993), no. 6, 568–578. MR1216703

[12] B.-Y. Chen, *Some new obstructions to minimal and Lagrangian isometric immersions*, Japan. J. Math. (N.S.) **26** (2000), no. 1, 105–127, DOI 10.4099/math1924.26.105. MR1771434

[13] B.-Y. Chen, *On isometric minimal immersions from warped products into real space forms*, Proc. Edinb. Math. Soc. (2) **45** (2002), no. 3, 579–587, DOI 10.1017/S001309150100075X. MR1933740

[14] B.-Y. Chen, *Lagrangian minimal surfaces in Lorentzian complex plane*, Arch. Math. (Basel) **91** (2008), no. 4, 366–371, DOI 10.1007/s00013-008-2733-6. MR2447551

[15] B.-Y. Chen, *Minimal flat Lorentzian surfaces in Lorentzian complex space forms*, Publ. Math. Debrecen **73** (2008), no. 1-2, 233–248. MR2429037

[16] B.-Y. Chen, *A minimal immersion of the hyperbolic plane into the neutral pseudo-hyperbolic 4-space and its characterization*, Arch. Math. (Basel) **94** (2010), no. 3, 291–299, DOI 10.1007/s00013-009-0094-4. MR2602455

[17] B.-Y. Chen, *Classification of minimal Lorentz surfaces in indefinite space forms with arbitrary codimension and arbitrary index*, Publ. Math. Debrecen **78** (2011), no. 2, 485–503, DOI 10.5486/PMD.2011.4860. MR2796783

[18] B.-Y. Chen and L. Vrancken, *Lagrangian minimal isometric immersions of a Lorentzian real space form into a Lorentzian complex space form*, Tohoku Math. J. (2) **54** (2002), no. 1, 121–143. MR1878931

[19] D. Chen and Q.-M. Cheng, *Estimates for the first eigenvalue of Jacobi operator on hypersurfaces with constant mean curvature in spheres*, Calc. Var. Partial Differential Equations **56** (2017), no. 2, Art. 50, 12, DOI 10.1007/s00526-017-1132-x. MR3626321

[20] H. Chen and X. Wang, *On stable compact minimal submanifolds of Riemannian product manifolds*, J. Math. Anal. Appl. **402** (2013), no. 2, 693–701, DOI 10.1016/j.jmaa.2013.01.068. MR3029182

[21] H. Chen, and G. Wei, *Rigidity of minimal submanifolds in space forms*, J. Geom. Anal. (2020). https://doi.org/10.1007/s12220-020-00462-7

[22] S. S. Chern, *Minimal submanifolds in a Riemannian manifold*, University of Kansas, Department of Mathematics Technical Report 19 (New Series), Univ. of Kansas, Lawrence, Kan., 1968. MR0248648

[23] S. S. Chern, M. do Carmo, and S. Kobayashi, *Minimal submanifolds of a sphere with second fundamental form of constant length*, Functional Analysis and Related Fields (Proc. Conf. for M. Stone, Univ. Chicago, Chicago, Ill., 1968), Springer, New York, 1970, pp. 59–75. MR0273546

[24] O. Chodosh and C. Mantoulidis, *Minimal surfaces and the Allen-Cahn equation on 3-manifolds: index, multiplicity, and curvature estimates*, Ann. of Math. (2) **191** (2020), no. 1, 213–328, DOI 10.4007/annals.2020.191.1.4. MR4045964

[25] T. H. Colding and C. De Lellis, *The min-max construction of minimal surfaces*, Surveys in differential geometry, Vol. VIII (Boston, MA, 2002), Surv. Differ. Geom., vol. 8, Int. Press, Somerville, MA, 2003, pp. 75–107, DOI 10.4310/SDG.2003.v8.n1.a3. MR2039986

[26] T. H. Colding and W. P. Minicozzi II, *A course in minimal surfaces*, Graduate Studies in Mathematics, vol. 121, American Mathematical Society, Providence, RI, 2011. MR2780140

[27] A. A de Barros and P. A. A. Sousa, *Estimate for index of closed minimal hypersurfaces in spheres*, Kodai Math. J. **32** (2009), no. 3, 442–449. MR2582011

[28] M. do Carmo and C. K. Peng, *Stable complete minimal surfaces in* $\mathbf{R}^3$ *are planes*, Bull. Amer. Math. Soc. (N.S.) **1** (1979), no. 6, 903–906, DOI 10.1090/S0273-0979-1979-14689-5. MR546314

[29] M. do Carmo, M. Ritoré, and A. Ros, *Compact minimal hypersurfaces with index one in the real projective space*, Comment. Math. Helv. **75** (2000), no. 2, 247–254, DOI 10.1007/PL00000373. MR1774704

[30] N. Ejiri, *Compact minimal submanifolds of a sphere with positive Ricci curvature*, J. Math. Soc. Japan **31** (1979), no. 2, 251–256, DOI 10.2969/jmsj/03120251. MR527542

[31] A. El Soufi, *Applications harmoniques, immersions minimales et transformations conformes de la sphère* (French), Compositio Math. **85** (1993), no. 3, 281–298. MR1214448

[32] D. Fischer-Colbrie, *On complete minimal surfaces with finite Morse index in three-manifolds*, Invent. Math. **82** (1985), no. 1, 121–132, DOI 10.1007/BF01394782. MR808112

[33] D. Fischer-Colbrie and R. Schoen, *The structure of complete stable minimal surfaces in 3-manifolds of nonnegative scalar curvature*, Comm. Pure Appl. Math. **33** (1980), no. 2, 199–211, DOI 10.1002/cpa.3160330206. MR562550

[34] H. Gauchman, *Minimal submanifolds of a sphere with bounded second fundamental form*, Trans. Amer. Math. Soc. **298** (1986), no. 2, 779–791, DOI 10.2307/2000649. MR860393

[35] C. Gorodski, R. A. E. Mendes, and M. Radeschi, *Robust index bounds for minimal hypersurfaces of isoparametric submanifolds and symmetric spaces*, Calc. Var. Partial Differential Equations **58** (2019), no. 4, Art. 118, 25, DOI 10.1007/s00526-019-1552-x. MR3962913

[36] I. V. Guadalupe, A. Brasil Jr., and J. A. Delgado, *A characterization of the Clifford torus*, Rend. Circ. Mat. Palermo (2) **48** (1999), no. 3, 537–540. MR1731453

[37] R. Howard and S. W. Wei, *On the existence and nonexistence of stable submanifolds and currents in positively curved manifolds and the topology of submanifolds in Euclidean spaces*, Geometry and topology of submanifolds and currents, Contemp. Math., vol. 646, Amer. Math. Soc., Providence, RI, 2015, pp. 127–167, DOI 10.1090/conm/646/12978. MR3384978

[38] Z. Hu and G. Wei, *On the nonexistence of stable minimal submanifolds and the Lawson-Simons conjecture*, Colloq. Math. **96** (2003), no. 2, 213–223. MR2010356

[39] T. Itoh, *On Veronese manifolds*, J. Math. Soc. Japan **27** (1975), no. 3, 497–506, DOI 10.2969/jmsj/02730497. MR383309

[40] H. B. Lawson Jr., *Lectures on minimal submanifolds. Vol. I*, Monografías de Matemática [Mathematical Monographs], vol. 14, Instituto de Matemática Pura e Aplicada, Rio de Janeiro, 1977. MR527121

[41] H. B. Lawson Jr. and J. Simons, *On stable currents and their application to global problems in real and complex geometry*, Ann. of Math. (2) **98** (1973), 427–450, DOI 10.2307/1970913. MR324529

[42] H. Li, *Curvature pinching for odd-dimensional minimal submanifolds in a sphere*, Publ. Inst. Math. (Beograd) (N.S.) **53(67)** (1993), 122–132. MR1319765

[43] H. Li, *Some nonexistence theorems on stable minimal submanifolds*, Colloq. Math. **73** (1997), no. 1, 1–13, DOI 10.4064/cm-73-1-1-13. MR1436947

[44] H. Li and X. Zhou, *Existence of minimal surfaces of arbitrarily large Morse index*, Calc. Var. Partial Differential Equations **55** (2016), no. 3, Art. 64, 12, DOI 10.1007/s00526-016-1007-6. MR3509038

[45] P. Li and J. Wang, *Minimal hypersurfaces with finite index*, Math. Res. Lett. **9** (2002), no. 1, 95–103. MR1892316

[46] P. Li and S. T. Yau, *A new conformal invariant and its applications to the Willmore conjecture and the first eigenvalue of compact surfaces*, Invent. Math. **69** (1982), no. 2, 269–291, DOI 10.1007/BF01399507. MR674407

[47] F. J. López and A. Ros, *Complete minimal surfaces with index one and stable constant mean curvature surfaces*, Comment. Math. Helv. **64** (1989), no. 1, 34–43, DOI 10.1007/BF02564662. MR982560

[48] F. C. Marques and A. Neves, *Existence of infinitely many minimal hypersurfaces in positive Ricci curvature*, Invent. Math. **209** (2017), no. 2, 577–616. MR3674223

[49] W. H. Meeks III and H. Rosenberg, *Stable minimal surfaces in $M \times \mathbb{R}$*, J. Differential Geom. **68** (2004), no. 3, 515–534. MR2144539

[50] R. A. E. Mendes and M. Radeschi, *Virtual immersions and minimal hypersurfaces in compact symmetric spaces*, arXiv:1708.05881, 2017, to appear in Calc. Var. Partial Differential Equations.

[51] M. J. Micallef and J. G. Wolfson, *The second variation of area of minimal surfaces in four-manifolds*, Math. Ann. **295** (1993), no. 2, 245–267, DOI 10.1007/BF01444887. MR1202392

[52] S. Montiel and F. Urbano, *Second variation of superminimal surfaces into self-dual Einstein four-manifolds*, Trans. Amer. Math. Soc. **349** (1997), no. 6, 2253–2269, DOI 10.1090/S0002-9947-97-01933-8. MR1422905

[53] Y.-G. Oh, *Second variation and stabilities of minimal Lagrangian submanifolds in Kähler manifolds*, Invent. Math. **101** (1990), no. 2, 501–519, DOI 10.1007/BF01231513. MR1062973

[54] Y. Ohnita, *Stable minimal submanifolds in compact rank one symmetric spaces*, Tohoku Math. J. (2) **38** (1986), no. 2, 199–217, DOI 10.2748/tmj/1178228488. MR843807

[55] O. Perdomo, *Low index minimal hypersurfaces of spheres*, Asian J. Math. **5** (2001), no. 4, 741–749, DOI 10.4310/AJM.2001.v5.n4.a8. MR1913819

[56] O. Perdomo, *First stability eigenvalue characterization of Clifford hypersurfaces*, Proc. Amer. Math. Soc. **130** (2002), no. 11, 3379–3384, DOI 10.1090/S0002-9939-02-06451-1. MR1913017

[57] O. Perdomo, *Spectrum of the Laplacian and the Jacobi operator on rotational cmc hypersurfaces of spheres*, arXiv:1902.07348, 2019, to appear in Pacific J. Math.

[58] M. Ritoré and A. Ros, *Stable constant mean curvature tori and the isoperimetric problem in three space forms*, Comment. Math. Helv. **67** (1992), no. 2, 293–305, DOI 10.1007/BF02566501. MR1161286

[59] M. Ritoré and A. Ros, *The spaces of index one minimal surfaces and stable constant mean curvature surfaces embedded in flat three manifolds*, Trans. Amer. Math. Soc. **348** (1996), no. 1, 391–410, DOI 10.1090/S0002-9947-96-01496-1. MR1322955

[60] A. Ros, *One-sided complete stable minimal surfaces*, J. Differential Geom. **74** (2006), no. 1, 69–92. MR2260928

[61] M. Ross, *Complete nonorientable minimal surfaces in $\mathbf{R}^3$*, Comment. Math. Helv. **67** (1992), no. 1, 64–76, DOI 10.1007/BF02566489. MR1144614

[62] M. Ross, *Schwarz' P and D surfaces are stable*, Differential Geom. Appl. **2** (1992), no. 2, 179–195, DOI 10.1016/0926-2245(92)90032-I. MR1245555

[63] A. Savo, *Index bounds for minimal hypersurfaces of the sphere*, Indiana Univ. Math. J. **59** (2010), no. 3, 823–837, DOI 10.1512/iumj.2010.59.4013. MR2779062

[64] Y.-B. Shen, *Curvature pinching for three-dimensional minimal submanifolds in a sphere*, Proc. Amer. Math. Soc. **115** (1992), no. 3, 791–795, DOI 10.2307/2159229. MR1093604

[65] Y.-B. Shen and Q. He, *On stable currents and positively curved hypersurfaces*, Proc. Amer. Math. Soc. **129** (2001), no. 1, 237–246, DOI 10.1090/S0002-9939-00-05753-1. MR1784025

[66] Y.-B. Shen and H.-Q. Xu, *On the nonexistence of stable minimal submanifolds in positively pinched Riemannian manifolds*, Geometry and topology of submanifolds, X (Beijing/Berlin, 1999), World Sci. Publ., River Edge, NJ, 2000, pp. 274–283, DOI 10.1142/9789812792051_0023. MR1801920

[67] Y.-B. Shen and X.-H. Zhu, *On stable complete minimal hypersurfaces in* $\mathbf{R}^{n+1}$, Amer. J. Math. **120** (1998), no. 1, 103–116. MR1600268

[68] J. Simons, *Minimal varieties in riemannian manifolds*, Ann. of Math. (2) **88** (1968), 62–105, DOI 10.2307/1970556. MR233295

[69] A. Song, *Existence of infinitely many minimal hypersurfaces in closed manifolds*, arXiv:1806.08816 (2018).

[70] F. Torralbo and F. Urbano, *On stable compact minimal submanifolds*, Proc. Amer. Math. Soc. **142** (2014), no. 2, 651–658, DOI 10.1090/S0002-9939-2013-11810-1. MR3134005

[71] F. Torralbo and F. Urbano, *Minimal surfaces in* $\mathbb{S}^2 \times \mathbb{S}^2$, J. Geom. Anal. **25** (2015), no. 2, 1132–1156, DOI 10.1007/s12220-013-9460-3. MR3319965

[72] F. Urbano, *Minimal surfaces with low index in the three-dimensional sphere*, Proc. Amer. Math. Soc. **108** (1990), no. 4, 989–992. MR1007516

[73] F. Urbano, *Stability problems for minimal Lagrangian submanifolds*, Symposium on the Differential Geometry of Submanifolds, [s.n.], [s.l.], 2007, pp. 221–232. MR2509503

[74] F. Urbano, *Second variation of one-sided complete minimal surfaces*, Rev. Mat. Iberoam. **29** (2013), no. 2, 479–494, DOI 10.4171/RMI/727. MR3047425

[75] X. Wang, H. Li, and L. Vrancken, *Minimal Lagrangian isotropic immersions in indefinite complex space forms*, J. Geom. Phys. **62** (2012), no. 4, 707–723, DOI 10.1016/j.geomphys.2011.12.015. MR2888976

[76] C. Wu, *New characterizations of the Clifford tori and the Veronese surface*, Arch. Math. (Basel) **61** (1993), no. 3, 277–284. MR1231163

[77] Y. Xin, *An application of integral currents to the vanishing theorems*, Sci. Sinica Ser. A **27** (1984), no. 3, 233–241. MR763966

[78] Y. Xin, *Minimal submanifolds and related topics*, Nankai Tracts in Mathematics, vol. 16, World Scientific Publishing Co. Pte. Ltd., Hackensack, NJ, 2019. Second edition of [MR2035469]. MR3837570

[79] K. Yano and B.-Y. Chen, *Minimal submanifolds of a higher dimensional sphere*, Tensor (N.S.) **22** (1971), 369–373. MR0287495

[80] K. Yano and M. Kon, *Structures on manifolds*, Series in Pure Mathematics, vol. 3, World Scientific Publishing Co., Singapore, 1984. MR794310

[81] S. T. Yau, *Submanifolds with constant mean curvature. I, II*, Amer. J. Math. **96** (1974), 346–366; ibid. 97 (1975), 76–100, DOI 10.2307/2373638. MR370443

[82] X. Zhou, *Min-max hypersurface in manifold of positive Ricci curvature*, J. Differential Geom. **105** (2017), no. 2, 291–343. MR3606731

School of Mathematics and Statistics, Northwestern Polytechnical University, Xi' an 710129, People's Republic of China

Email address: chenhang86@nwpu.edu.cn

Contemporary Mathematics
Volume **756**, 2020
https://doi.org/10.1090/conm/756/15198

Maximally-warped metrics with harmonic curvature

Andrzej Derdzinski and Paolo Piccione

Abstract. We describe the local structure of Riemannian manifolds with harmonic curvature which admit a maximum number, in a well-defined sense, of local warped-product decompositions, and at the same time their Ricci tensor has, at some point, only simple eigenvalues. We also prove that, in every given dimension greater than two, the local-isometry types of such manifolds form a finite-dimensional moduli space, and a nonempty open subset of this moduli space is realized by locally irreducible complete metrics which are neither Ricci-parallel, nor – for dimensions greater than three – conformally flat.

Introduction

A Riemannian manifold is said to have *harmonic curvature* [**1**, Sect. 16.33] if

$$\operatorname{div} R = 0, \text{or, in local coordinates, } R_{ijl}{}^{k}{}_{,k} = 0, \tag{0.1}$$

R being the curvature tensor. We consider harmonic-curvature Riemannian manifolds (M, g) of dimensions $n \geq 3$ in which, with r denoting the Ricci tensor,

(0.2) r has n distinct eigenvalues at some point, and an open submanifold of (M, g) admits a nontrivial warped-product decomposition.

Such warped product decompositions have one-dimensional fibres (Corollary 1.3), and are in a one-to-one correspondence with certain one-dimensional Lie subalgebras of $\mathfrak{isom}(M', g')$, the Lie algebra of Killing fields on the Riemannian universal covering (M', g') of (M, g) (see Remark 3.2). The number γ of these subalgebras cannot exceed $n - 1$, cf. Corollary 4.2, and we refer to g as *maximally warped* if

$$\gamma = n - 1. \tag{0.3}$$

Our Theorem 5.6 describes the local structure of Riemannian manifolds (M, g) of dimensions $n \geq 3$, satisfying (0.1) – (0.3). Their local-isometry types turn out to form a $(2n - 3)$-dimensional moduli space (Remark 5.7), and we prove (in Theorem 7.3) that some nonempty open subset of the moduli space consists of local isometry types of such manifolds which in addition are

(0.4) complete, locally irreducible, and neither conformally flat (unless $n = 3$), nor Ricci-parallel.

2010 *Mathematics Subject Classification.* Primary 53C25; Secondary 53B20.

Both authors' research was supported in part by a FAPESP-OSU 2015 Regular Research Award (FAPESP grant: 2015/50265-6).

We do not know if any compact manifold can have the properties (0.1) – (0.3). However, we observe (see Theorem 1.4) that compact Riemannian manifolds with harmonic curvature that admit *global* nontrivial warped-product decompositions must have fibre dimensions greater than one and, consequently, cannot be *Ricci-generic* in the sense of satisfying the distinct-eigenvalues clause of (0.2).

Harmonicity of the curvature always follows if the metric is *Ricci-parallel*, or *locally reducible with harmonic-curvature factors*, or *conformally flat and of constant scalar curvature* while, in dimension three, harmonic curvature amounts to conformal flatness plus constancy of the scalar curvature [**1**, Sect. 16.35 and 16.4].

Compact Riemannian manifolds with (0.1) have been studied extensively. All their known examples (aside from the three classes italicized above), listed as 2,3,4 in [**1**, p. 432], admit – at least locally – nontrivial warped-product decompositions with a fibre of dimension greater than one. Consequently (see Corollary 1.3 below), they are not Ricci-generic. However, for examples 2 and 4 of [**1**, p. 432] the warped-product structure, rather than being an Ansatz, is a consequence of geometric conditions involving multiplicities of eigenvalues of the Ricci tensor [**2**] or self-dual Weyl tensor [**3**]. The following questions about compact Riemannian manifolds with harmonic curvature, lying outside of the three italicized classes, are thus open and natural: can they be Ricci-generic? must they, locally, have a warped-product decomposition and if not, how to describe those among them which have one?

Our Theorem 7.3 yields an affirmative answer to a weaker version of the first question in which completeness replaces compactness. On the other hand, the second question provides an obvious motivation for studying condition (0.3).

The authors wish to thank the referees for constructive comments, which have resulted in a substantial improvement of the presentation.

1. Preliminaries

Manifolds (always assumed connected), mappings and tensor fields are by definition C^∞-differentiable. By a *Codazzi tensor* [**1**, p. 435] on a Riemannian manifold one means a twice-covariant symmetric tensor field S with a totally symmetric covariant derivative ∇S. One then has two well-known facts [**1**, Sect. 16.4(ii)]:

$$\text{(1.1)}\qquad \begin{array}{rl} \text{i)} & \operatorname{div} R = 0 \text{ if and only if r is a Codazzi tensor,} \\ \text{ii)} & \text{the condition } \operatorname{div} R = 0 \text{ implies constancy of s,} \end{array}$$

s being the scalar curvature. As shown by DeTurck and Goldschmidt [**4**],

$$\text{(1.2)}\qquad \text{metrics with } \operatorname{div} R = 0 \text{ are real-analytic in suitable local coordinates.}$$

We call two (connected) real-analytic Riemannian manifolds *locally isometric* if they have open submanifolds that are both isometric to open submanifolds of a third such manifold. One easily sees that this is an equivalence relation. In view of the extension theorem for analytic isometries [**9**, Corollary 6.4 on p. 256], for two complete real-analytic Riemannian manifolds, being locally isometric to each other means the same as having isometric Riemannian universal coverings.

On a manifold with a torsion-free connection ∇, the Ricci tensor r satisfies the Bochner identity $\mathrm{r}(\cdot, v) + d[\operatorname{div} v] = \operatorname{div} \nabla v$, where v is any vector field. Its coordinate form $R_{jk}v^k = v^k{}_{,jk} - v^k{}_{,kj}$ arises via contraction from the Ricci identity $v^l{}_{,jk} - v^l{}_{,kj} = R_{jkq}{}^l v^q$. (We use the sign convention for R such that $R_{jk} = R_{jqk}{}^q$.)

Applied to the gradient v of a function ϕ on a Riemannian manifold, this yields

$$\mathrm{r}(\nabla\phi,\cdot) + d\Delta\phi = \mathrm{div}[\nabla d\phi]. \tag{1.3}$$

The *warped product* of Riemannian manifolds $(\overline{M},\overline{g})$ and (Σ,η) with the *warping function* $\phi : \overline{M} \to (0,\infty)$ is the Riemannian manifold

$$(M,g) = (\overline{M} \times \Sigma, \overline{g} + \phi^2\eta). \tag{1.4}$$

(The same symbols $\overline{g},\eta,\phi$ stand here also for the pullbacks of $\overline{g},\eta,\phi$ to the product $M = \overline{M}\times\Sigma$.) One calls $(\overline{M},\overline{g})$ and (Σ,η) the *base* and *fibre* of (1.4), and refers to (1.4) as *nontrivial* if ϕ is nonconstant. From now on we assume that $\dim\Sigma \geq 1$.

REMARK 1.1. As $\overline{g}+\phi^2\eta = \phi^2[\phi^{-2}\overline{g}+\eta]$, a warped product is nothing else than a Riemannian manifold conformal to a Riemannian product via multiplication by a positive function which is constant along one of the factor manifolds.

A proof of the following well-known lemma [**7**] is given in the Appendix.

LEMMA 1.2. *A warped product* (1.4) *with a nonconstant warping function* ϕ *has harmonic curvature if and only if the Levi-Civita connection* $\overline{\nabla}$ *of* $(\overline{M},\overline{g})$, *its Ricci tensor* $\overline{\mathrm{r}}$ *and the* $\overline{g}$*-gradient* $\overline{\nabla}\phi$ *of* ϕ *satisfy three conditions*:

(a) (Σ,η) *is an Einstein manifold, with some Einstein constant* κ.

(b) $\overline{\mathrm{r}} - p\phi^{-1}\overline{\nabla}d\phi$ *is a Codazzi tensor on* $(\overline{M},\overline{g})$, *where* $p = \dim\Sigma \geq 1$.

(c) $\phi^3\overline{\mathrm{div}}[\phi^{-1}\overline{\nabla}d\phi] = [(p-1)\Lambda - \kappa]d\phi + (1-p)\phi d\Lambda/2$, *with* $\Lambda = \overline{g}(\overline{\nabla}\phi,\overline{\nabla}\phi)$

and the $\overline{g}$*-divergence* $\overline{\mathrm{div}}$. *Then, at each point of* M, *for the Ricci tensor* r *of* g,

(d) *the space tangent to the fibre factor is contained in an eigenspace of* r.

One may rewrite (c) *as a requirement involving the* $\overline{g}$*-Laplacian* $\overline{\Delta}\phi$, *namely*,

(e) $\phi^2[\overline{\mathrm{r}}(\overline{\nabla}\phi,\cdot) + d\overline{\Delta}\phi] = [(p-1)\Lambda - \kappa]d\phi + (1-p/2)\phi d\Lambda$.

Finally, when $p = 1$, *and so* $\kappa = 0$, (c) *reads* $\overline{\mathrm{div}}[\phi^{-1}\overline{\nabla}d\phi] = 0$.

From (d) and, respectively, (c), we obtain two easy consequences:

COROLLARY 1.3. *In a warped-product Riemannian manifold with a nonconstant warping function, harmonic curvature, and a fibre of dimension greater than one, the Ricci tensor has, at every point, at least one multiple eigenvalue. The assumptions* (0.1) – (0.2) *thus imply one-dimensionality of the fibre for any warped-product decomposition in* (0.2).

THEOREM 1.4. *If a warped-product Riemannian manifold* (M,g) *has a compact base* $(\overline{M},\overline{g})$, *a nonconstant warping function, and harmonic curvature, then the Einstein constant* κ *of its fibre must be positive. Thus, the dimension of the fibre is greater than one and, in view of Corollary* 1.3, (M,g) *cannot satisfy the distinct-eigenvalues clause of* (0.2).

In fact, given a positive function ϕ on a Riemannian manifold $(\overline{M},\overline{g})$ and constants $\kappa, p\in\mathbb{R}$, let us set $v = \overline{\nabla}\phi$ and $w = \phi^2u - [(p-1)\Lambda-\kappa]v + (p-2)\phi\overline{\nabla}_v v$, where $u = \overline{\mathrm{div}}\overline{\nabla}v$ and $\Lambda = \overline{g}(v,v)$. Then the function $\phi^{p-4}\overline{g}(v,w)$ differs by a $\overline{g}$-divergence from $-[(p-1)\Lambda-\kappa]\phi^{p-4}\Lambda - \phi^{p-2}\overline{g}(\overline{\nabla}v,\overline{\nabla}v)$, while (c) reads $w = 0$, as $2\overline{\nabla}_v v = \overline{\nabla}\Lambda$. (An easy exercise.)

REMARK 1.5. The base and fibre factor distributions of any warped product are Ricci-orthogonal to each other. (See the equality $R_{ia} = 0$ in formula (A.3) of the Appendix.) Thus, if the base, or fibre, is one-dimensional, nonzero vectors tangent to it constitute eigenvectors of the Ricci tensor.

2. Vector fields

LEMMA 2.1. *Let a maximal integral curve* $(a_-, a_+) \ni t \mapsto x(t)$ *of a vector field* v *on a manifold* M, *with* $-\infty \leq a_- < a_+ \leq \infty$, *some* $t' \in (a_-, a_+)$, *and some compact set* $C \subseteq M$, *have the property that* $x(t) \in C$ *for all* $t \in [t', a_+)$ *or, respectively, for all* $t \in (a_-, t']$. *Then* $a_+ = \infty$ *or, respectively,* $a_- = -\infty$.

Consequently, a maximal integral curve of a vector field on a manifold, lying within a compact set, must be complete, that is, defined on $\mathbb{R}$.

PROOF. For a compactly supported function χ equal to 1 on an open set U containing C, the curve restricted to $[t', a_+)$, or to $(a_-, t']$, clearly remains half-maximal (not extendible beyond a_+, or a_-) when treated as an integral curve of χv. On the other hand, χv is complete due to compactness of its support. □

By a *section* of a locally trivial fibre bundle we mean, as usual, a submanifold Σ of the total space Q mapped diffeomorphically onto the base M by the bundle projection p. We also identify the section with the inverse $\psi : M \to \Sigma$ of the latter diffeomorphism, which makes it a mapping $\psi : M \to Q$ having $\mathrm{p} \circ \psi = \mathrm{Id}_M$. In the case of a vector bundle Q, a section ψ, and a zero $z \in M$ of ψ, the corresponding submanifold Σ of Q intersects the zero section M at z (that is, at the zero vector of the fibre Q_z), giving rise to the *differential* $\partial\psi_z$, defined to be the linear operator $T_zM \to Q_z$ obtained as the composite of the ordinary differential of $\psi : M \to Q$ at z (the inverse of $d\mathrm{p}_z : T_z\Sigma \to T_zM$), followed by the direct-sum projection $T_zQ = T_zM \oplus Q_z \to Q_z$. Relative to any local coordinates at z and a local trivialization of Q, the components of $A = \partial\psi_z$ form the matrix $[A_j^\lambda] = [\partial_j\psi^\lambda]$, with the partial derivatives of the components of ψ evaluated at z.

Two important examples are provided by zeros z of $\psi = v$, a vector field on M (with $Q = TM$) and of $\psi = df$, for a function $f : M \to \mathbb{R}$ (here $Q = T^*M$). In the former case, $A = \partial v_z$ (in coordinates: $A_j^k = \partial_j v^k$) is the infinitesimal generator of the one-parameter group of linear transformations of T_zM arising as the differentials, at the fixed point z, of the local diffeomorphisms forming the local flow of v. In the latter, $\partial df_z = \mathrm{Hess}_z f$, the Hessian of f at the critical point z.

Let v be a vector field on a manifold M, having a zero at $z \in M$, where one assumes M either to be an open submanifold of a vector space Y, or to have a submanifold N with $z \in N$ such that v is tangent to N at each point of N. In this way v, or the restriction of v to N, becomes a mapping $v : M \to Y$, or a vector field w on N. The equality $A_j^k = \partial_j v^k$ of the last paragraph, evaluated in coordinates for M which are linear functionals on Y or, respectively, in which N is defined by equating some coordinate functions to 0, clearly implies that

$$\text{(2.1)} \qquad \text{i) } \partial v_z = dv_z : Y \to Y, \text{ ii) } \partial w_z \text{ equals the restriction of } \partial v_z \text{ to } T_zN.$$

LEMMA 2.2. *Given a zero* $z \in M$ *of a vector field* v *on a manifold* M, *with the differential* $A = \partial v_z$, *let a function* $f : U \to \mathbb{R}$ *on a neighborhood* U *of* z *have* $df_z = 0$. *Then* $d\sigma_z = 0$ *and* $(u, u)_\sigma = 2(Au, u)_f$ *for the directional derivative* $\sigma = d_v f : U \to \mathbb{R}$, *all* $u \in T_zM$, *the Hessian* $(\,,\,)_f = \mathrm{Hess}_z f$, *and* $(\,,\,)_\sigma = \mathrm{Hess}_z\sigma$.

PROOF. With commas denoting, this time, partial derivatives relative to fixed local coordinates on a neighborhood of z, we have $\sigma = v^j f_{,j}$ as well as $\sigma_{,k} = v^j f_{,jk} + v^j{}_{,k} f_{,j}$ and $\sigma_{,kl} = v^j f_{,jkl} + v^j{}_{,l} f_{,jk} + v^j{}_{,k} f_{,jl} + v^j{}_{,kl} f_{,j}$. At z, both v^j and $f_{,j}$ vanish, while $v^j{}_{,k} = A_k^j$. This proves our claim. □

LEMMA 2.3. *Let $z \in N$ be a zero of a vector field w on a manifold N such that, for some $\varepsilon = \pm 1$, some Euclidean inner product $\langle,\rangle$ in T_zN, and $A = \varepsilon\partial w_z$, the bilinear form $\langle A\cdot,\cdot\rangle$ on T_zN is negative definite. In this case there exist arbitrarily small neighborhoods U of z with the following property: if a maximal integral curve $(a_-, a_+) \ni t \mapsto x(t)$ of w and $t' \in (a_-, a_+)$ satisfy the condition $x(t') \in U$, where $-\infty \le a_- < a_+ \le \infty$, then, denoting by $\pm$ the sign of ε, one has $a_\pm = \pm\infty$, and $x(t) \in U$ whenever $\varepsilon(t - t') \ge 0$.*

PROOF. We fix a Riemannian metric g on a neighborhood of z in N having $\langle,\rangle = g_z$. The required neighborhoods U of z are g-metric balls centered at z, small enough so as to have compact closures and be diffeomorphic images, under the g-exponential mapping at z, of the corresponding Euclidean balls around 0 in T_zN. This gives smoothness of the function $f : U \to \mathbb{R}$ such that $2f$ equals $\mathrm{dist}^2(z,\cdot)$, the squared g-distance from z, and using normal coordinates one obtains $\mathrm{Hess}_z f = \langle,\rangle$. If the g-metric ball U is sufficiently small, Lemma 2.2 for $v = \varepsilon w$ implies negativity of $\sigma = \varepsilon d_w f$ on $U \smallsetminus \{z\}$, as σ assumes at z the critical value 0 with a negative-definite Hessian. Our claim now easily follows from Lemma 2.1. □

REMARK 2.4. *The same* neighborhoods U of z will still satisfy the assertion of Lemma 2.3 if one replaces w by w/c for a constant $c > 0$ and $(a_-, a_+) \ni t \mapsto x(t)$ by $(ca_-, ca_+) \ni t \mapsto x(t/c)$.

REMARK 2.5. For any Killing field v on a Riemannian manifold (M, g), the pair $(v, \nabla v)$ constitutes a parallel section of the vector bundle $TM \oplus \mathfrak{so}(TM)$ endowed with a suitable linear connection [**5**, Remark 17.25 on p. 547]. Therefore,

(i) a Killing field on M is uniquely determined by its restriction to any nonempty open subset of M, while

(ii) assuming (M, g) to be simply connected and real-analytic, we conclude that any Killing field v on a nonempty connected open subset of M has a unique extension to a Killing field on M.

Given a nontrivial Killing vector field v on a Riemannian manifold and a function θ, the obvious equality $\pounds_{\theta v} g = \theta\pounds_v g + 2d\theta \odot g(v,\cdot)$ clearly implies that

(2.2) if θv is also a Killing field, θ must be constant, cf. Remark 2.5(i).

3. Integrable-complement Killing fields

This section presents a well-known correspondence – see, for instance, the Appendix in [**10**] – between warped product decompositions with a one-dimensional fibre and certain special Killing fields.

Let v a nontrivial Killing field on a Riemannian manifold (M, g) such that, on the dense (by Remark 2.5(i)) complement of its zero set, the distribution $v^\perp$ is integrable. In other words, locally, at points with $v \neq 0$, multiplying v by a suitable positive function one obtains a gradient vector field. Equivalently,

(3.1) the 1-form $g(v,\cdot)/g(v,v)$, defined wherever $v \neq 0$, is closed.

Namely, (3.1) is necessary: for $\xi = g(v,\cdot)$, due to skew-symmetry of $\nabla\xi$, the integrability condition $\xi \wedge d\xi = 0$ has the local-coordinate expression $\xi_{i,j}\xi_k + \xi_{j,k}\xi_i + \xi_{k,i}\xi_j = 0$, which transvected with v^k yields $v^k\xi_k\xi_{i,j} = v^k\xi_{k,j}\xi_i - v^k\xi_{k,i}\xi_j$, or

(3.2) $$2\beta\xi_{i,j} = \beta_{,j}\xi_i - \beta_{,i}\xi_j, \text{ where } \beta = v^k\xi_k = g(v,v).$$

Closedness of ξ/β amounts to symmetry of $\nabla(\xi/\beta)$, and so it now follows since (3.2) with $\xi_{i,j} = -\xi_{j,i}$ implies symmetry of $\beta^2(\xi_i/\beta)_{,j} = \beta\xi_{i,j} - \beta_{,j}\xi_i$ in i, j.

If v is a Killing field, $g(v, \dot{x})$ is constant along any geodesic $t \mapsto x = x(t)$, as $d[g(v, \dot{x})]/dt = g(\nabla_{\dot{x}} v, \dot{x}) = 0$. Then, with the orthogonal complement $v^\perp$ only defined away from the zero set of v, one easily sees that

(3.3) v is orthogonal to any geodesic passing through a zero of v, while whenever (3.1) holds,the distribution $v^\perp$ has totally geodesic leaves.

Remark 3.1. Local Killing fields v satisfying (3.1), outside of their zero sets, if treated as defined only up to multiplication by nonzero constants, stand in a natural one-to-one correspondence with local warped product decompositions of g that have a one-dimensional fibre. Here v is tangent to the fibre direction.

Namely, such a local decomposition is uniquely determined by the base and fibre factor distributions. Just one of them suffices, the other being its (necessarily integrable) orthogonal complement. That v locally spans the fibre factor distribution of a warped product follows from Remark 1.1 and the local version of de Rham's decomposition theorem: in view of (3.2), rewritten as $2\beta v^i{}_{,j} = v^i\beta_{,j} - \beta^{,i}\xi_j$, where $\beta = v^k\xi_k = g(v, v)$, and [**1**, Theorem 1.159], v is $\hat{g}$-parallel for the conformally related metric $\hat{g} = g/\beta$, with $d_v\beta = 2g(\nabla_v v, v) = 0$ due to skew-adjointness of ∇v. Conversely, for a warped product with a one-dimensional fibre, the required Killing field v comes from a local flow of local isometries of the fibre (cf. formula (A.2) in the Appendix), (2.2) implying uniqueness of v up to a constant factor.

Remark 3.2. From Remarks 2.5 and 3.1 it follows that, in the case of a real-analytic Riemannian manifold (M, g), denoting by $\mathfrak{isom}(M', g')$ the Lie algebra of Killing fields on the Riemannian universal covering (M', g') of (M, g), one has a natural bijective correspondence between the one-dimensional Lie subalgebras of $\mathfrak{isom}(M', g')$ spanned by Killing fields v satisfying (3.1), and the local warped product decompositions, with one-dimensional fibres, of g restricted to the dense open set where $v \neq 0$. As before, v is tangent to the fibre direction.

Lemma 3.3. *Let an open ball B around* 0 *in a Euclidean n-space, $n \geq 2$, admit a connection ∇ such that all line segments through* 0 *in B are ∇-totally geodesic and tangent at all points $x \in B \smallsetminus \{0\}$ to some codimension-one foliation $\mathcal{F}$ on $B \smallsetminus \{0\}$ having ∇-totally geodesic leaves. Then $n = 2$.*

Proof. Fix a leaf L of $\mathcal{F}$ and $x \in L$ such that the ∇-exponential mapping $\exp_x$ sends a Euclidean open ball B' centered at 0 in T_xB, diffeomorphically, onto a neighborhood $\exp_x(B')$ of 0 in B. Thus, $\exp_x(B' \cap T_xL) \smallsetminus J \subseteq L$, for $J = B \cap \{qx : q \in (-\infty, 0]\}$, and so $J \subseteq L$. (The leaves of $\mathcal{F}$ are locally closed, being, locally, the level sets of a submersion.) Hence $L \cup \{0\}$ is a smooth ∇-totally geodesic submanifold of B, with some tangent space V at 0, meaning in turn that $L \cup \{0\} = B \cap V$. Consequently, $n = 2$, for otherwise any two such codimension-one subspaces V of our Euclidean n-space would have a nontrivial intersection. ☐

When $\mathcal{F}$ is real-analytic, we can also obtain the above assertion by applying, to a sphere Σ around 0 in B, Haefliger's theorem [**6**] which states that a transversally orientable real-analytic codimension-one foliation may exist on a compact manifold Σ only if the fundamental group of Σ has an element of infinite order.

REMARK 3.4. Kobayashi [8] showed that the zero set of any Killing vector field on a Riemannian manifold (M, g) is either empty, or its connected components are mutually isolated totally geodesic submanifolds of even codimensions.

For a nontrivial Killing field v with (3.1), *the above codimensions must all equal* 2. This is immediate if one fixes a zero z of v and applies Lemma 3.3 to a small ball B in the normal space at z of the connected component through z such that $\exp_z$ maps B diffeomorphically onto a submanifold N of M, with ∇ and $\mathcal{F}$ denoting the $\exp_z$-pullback of the Levi-Civita connection of the submanifold metric h on N and, respectively, of the foliation on $N \smallsetminus \{z\}$ the leaves of which are intersections of N and the leaves of $v^\perp$, the latter defined wherever $v \neq 0$. (The local flow of v preserves N and h, and so v is tangent to N.) The restriction of v to N now constitutes an h-Killing field w having just one zero, at z, and satisfying (3.1) (for h, w rather than g, v), so that (3.3) allows us to use Lemma 3.3.

4. Multiply-warped metrics with $\operatorname{div} R = 0$

LEMMA 4.1. *Suppose that the Ricci tensor of a real-analytic Riemannian n-manifold (M, g) has n distinct eigenvalues at some point and, with the notation of Remark* 3.2, $\mathfrak{a}_2, \ldots, \mathfrak{a}_m$ *are distinct one-dimensional Lie subalgebras of* $\mathfrak{isom}(M', g')$ *spanned by Killing fields* $v_2, \ldots, v_m$ *such that each* $v = v_j$ *satisfies* (3.1). *Then* $m \leq n$, *and* $g(v_j, v_k) = 0$ *as well as* $[v_j, v_k] = 0$ *if* $j \neq k$. *Finally,* $g(\nabla_u v_j, v_k) = 0$ *whenever* $j, k, l \in \{2, \ldots, m\}$ *and* $u = v_l$.

PROOF. Remarks 3.2 and 1.5 imply that all v_j, wherever nonzero, are mutually nonproportional eigenvectors of the Ricci tensor, which makes them pointwise orthogonal to one another, as well as invariant, up to constant factors – by (2.2) – under each other's local flows. Thus, $m \leq n+1$ and, as $[v, w] = \pounds_v w$ for $v = v_j$ and $u = v_k$, one gets $[v_j, v_k] = cv_k$ with some constant c depending on j and k. Switching j and k, we see that $c = 0$. Now let $u = v_l$, $v = v_j$ and $u = v_k$, where $j, k, l \in \{2, \ldots, m\}$. We have $g(\nabla_u v, w) = 0$ if $u = w$ (due to the Killing property of v) and, therefore, also when $v = w$ (since u, v commute). Also, $g(\nabla_u v, w) = 0$ in the remaining case, with u, v different from w (and hence orthogonal to w): as a consequence of (3.3), outside of the zero set of w the distribution $w^\perp$ has totally geodesic leaves. This proves the final claim of the lemma, implying in turn that, if one had $m = n+1$, all v_j would be parallel, leading to flatness of g, and contradicting the Ricci-eigenvalues assumption. □

Due to DeTurck and Goldschmidt's real-analyticity theorem (1.2), we may combine Lemma 4.1 with Remark 3.2 and Corollary 1.3, obtaining

COROLLARY 4.2. *Under the assumptions* (0.1) – (0.2), *the integer* γ *defined in the Introduction does not exceed* $n-1$.

5. The local structure

Given an open interval $I \subseteq \mathbb{R}$, we introduce a Riemannian metric g on the open set $I \times \mathbb{R}^{n-1} \subseteq \mathbb{R}^n$, $n \geq 2$, by declaring its component functions in the Cartesian coordinates $x^1, x^2, \ldots, x^n$ to be

$$g_{kl} = 0 \text{ if } k \neq l, \; g_{11} = 1, \; g_{jj} = g_{jj}(t) \text{ for } t = x^1 \text{ and } j \geq 2, \tag{5.1}$$

where $I \ni t \mapsto (g_{22}(t), \ldots, g_{nn}(t)) \in (0, \infty)^{n-1}$ is any prescribed smooth curve. We also define the functions $y_2, \ldots, y_n$ and $\mathbf{y} = \operatorname{diag}(y_2, \ldots, y_n)$ of the variable $t \in I$, valued in $\mathbb{R}$ and, respectively, in the real vector space $\mathbb{E} \cong \mathbb{R}^{n-1}$ of all diagonal $(n-1) \times (n-1)$ matrices, by

$$2y_j g_{jj} = -\dot{g}_{jj} \text{ (no summation), with } ()^{\cdot} = d/dt. \tag{5.2}$$

REMARK 5.1. If $I = \mathbb{R}$ while $\mp y_j(t) \geq \delta$ whenever $\pm t$ is sufficiently large and positive, for both signs $\pm$, some constant $\delta > 0$, and all $j \geq 2$, then the above metric g is complete. In fact, (5.2) gives $\log g_{jj}(t) \to \infty$ as $|t| \to \infty$, so that $g_{jj}(t) \geq a$ with some constant $a \in (0, 1]$ and all $t \in \mathbb{R}$, which in turn gives $g \geq ag'$ (positive semidefiniteness of $g - ag'$) for the standard Euclidean metric g'. Completeness of g' now implies that of g, as g-bounded sets have compact closures due to the resulting inequality $\operatorname{dist} \geq a \operatorname{dist}'$ between distance functions.

Let us consider the second-order autonomous ordinary differential equation

$$\ddot{\mathbf{y}} - (\operatorname{tr}\mathbf{y} + \mathbf{y})\dot{\mathbf{y}} = (\operatorname{tr}\mathbf{y}^2)\mathbf{y} - (\operatorname{tr}\mathbf{y})\mathbf{y}^2 \tag{5.3}$$

imposed on a C^2 curve $I \ni t \mapsto \mathbf{y} \in \mathbb{E}$, in which $\mathbf{y}\dot{\mathbf{y}}$ and $\mathbf{y}^2 = \mathbf{y}\mathbf{y}$ represent diagonal-matrix products, while $\operatorname{tr}\mathbf{y}$ also denotes $\operatorname{tr}\mathbf{y}$ times the identity.

LEMMA 5.2. *For a metric g on $I \times \mathbb{R}^{n-1}$ defined by* (5.1) *and the corresponding curve $I \ni t \mapsto \mathbf{y} = \operatorname{diag}(y_2, \ldots, y_n)$ with* (5.2)*, at every point $(t, \mathbf{x}) \in I \times \mathbb{R}^{n-1}$, each of the coordinate vectors ∂_k, $k = 1, \ldots, n$, is an eigenvector of the Ricci tensor of g with an eigenvalue μ_k depending on t.*

(a) *Specifically, $\mu_1 = \operatorname{tr}\dot{\mathbf{y}} - \operatorname{tr}\mathbf{y}^2$ and $\mu_j = \dot{y}_j - y_j \operatorname{tr}\mathbf{y}$ if $j \geq 2$.*
(b) *The scalar curvature s of g equals $2\operatorname{tr}\dot{\mathbf{y}} - \operatorname{tr}\mathbf{y}^2 - (\operatorname{tr}\mathbf{y})^2$.*
(c) *$\partial_2, \ldots, \partial_n$ are g-Killing fields with integrable orthogonal complements.*
(d) *Given any fixed $\mathbf{x} \in \mathbb{R}^{n-1}$, the curve $I \ni t \mapsto (t, \mathbf{x})$ is a g-geodesic.*
(e) *g has harmonic curvature if and only if* (5.3) *holds.*

PROOF. We assume j, k, l to range over $\{2, \ldots, n\}$ and be mutually distinct. Repeated indices are *not* summed over. First, (c) is obvious as $g_{11}, g_{1j}, g_{jj}, g_{jk}$ only depend on $t = x^1$. Also, $\Gamma_{11}^1 = \Gamma_{11}^j = 0$, proving (d), while $\Gamma_{1j}^1 = \Gamma_{1j}^k = \Gamma_{jj}^j = \Gamma_{jj}^k = \Gamma_{jk}^j = \Gamma_{jk}^l = 0$ and $g^{jj}\Gamma_{jj}^1 = -\Gamma_{1j}^j = y_j$. Hence $R_{11} = \mu_1$ and $g^{jj}R_{jj} = \mu_j$ for μ_1, μ_j as in (a). This yields (a), and hence (b). (Each ∂_k spans the fibre direction of a warped product decomposition, and we may use Remark 1.5.) Next, $R_{11,j} = R_{1j,1} = R_{1j,k} = R_{jk,1} = R_{jk,j} = R_{jj,k} = R_{jk,l} = 0$. Finally, $g^{jj}R_{j1,j} = y_j(\mu_j - \mu_1)$ and $g^{jj}R_{jj,1} = \dot{\mu}_j$, so that (1.1.i) implies (e), □

We refer to a solution $I \ni t \mapsto \mathbf{y} \in \mathbb{E}$ of (5.3) as *maximal* if it cannot be extended to a larger open interval, and call it *Ricci-generic* whenever the n values $\mu_k = \mu_k(t)$ of Lemma 5.2(a) are all distinct at some $t \in I$ (or, equivalently, no two among the functions $\mu_1, \ldots, \mu_n$ coincide everywhere in I).

EXAMPLE 5.3. Two non-Ricci-generic maximal solutions of (5.3) are defined by $\mathbf{y} = -2\tanh nt$ and $\mathbf{y} = 2\tan nt$ (times the identity $\mathbf{1}$), with $I = \mathbb{R}$ or $I = (-\pi/(2n), \pi/(2n))$. In fact, $2\dot{\mathbf{y}} = n(\mathbf{y}^2 \mp 4)$ and so $\ddot{\mathbf{y}} = n\mathbf{y}\dot{\mathbf{y}}$, while for multiples $\mathbf{y}$ of $\mathbf{1}$ the right-hand side of (5.3) vanishes and $\operatorname{tr}\mathbf{y} + \mathbf{y} = n\mathbf{y}$.

EXAMPLE 5.4. Any solution $\mathbf{y} = \operatorname{diag}(y_2, \ldots, y_n)$ of (5.3), where $n \geq 2$, can be *trivially extended* to the solution $\operatorname{diag}(y_2, \ldots, y_n, 0, \ldots, 0)$ with a number $m > 0$ of additional zero components. The new metric defined using (5.1) – (5.2) is isometric to the Riemannian product of the original g and a flat metric on $\mathbb{R}^m$.

The set of maximal solutions of (5.3) is obviously preserved by the group K acting on it via replacement of $\mathbf{y}$ with $t \mapsto \pm\mathbf{y}(b \pm t)$, where $b \in \mathbb{R}$ and $\pm$ is either sign, combined with permutations of the components $y_2, \ldots, y_n$. We will use the term *K-equivalence* when two maximal solutions lie in the same K-orbit.

REMARK 5.5. Nonzero real numbers a act on maximal solutions $t \mapsto \mathbf{y}(t)$ of (5.3) by sending them to $t \mapsto a\mathbf{y}(at)$. (The new metric arising via (5.1) – (5.2) is isometric to g/a^2.) The group K defined above, obviously isomorphic to the direct product of the isometry group of $\mathbb{R}$ and the symmetric group S_{n-1}, along with the multiplicative group $\mathbb{R} \smallsetminus \{0\}$ acting as described here, together generate an action of a semidirect product of K and $(0, \infty)$.

THEOREM 5.6. *For any $n \geq 3$, the construction summarized by* (5.1) – (5.2) *provides a bijective correspondence between two sets consisting, respectively, of*

(i) *all K-equivalence classes of maximal Ricci-generic solutions to* (5.3), *and*
(ii) *all local-isometry types of Riemannian n-manifolds with* (0.1) – (0.3).

For the meaning of local-isometry types, see (1.2) *and the paragraph following it.*

PROOF. We need to show that the mapping from (i) to (ii) is: (A) well-defined, (B) injective, and (C) surjective.

Part (A) easily follows from Lemma 5.2 combined with the comment on g/a^2 in Remark 5.5, the latter applied to $a = \pm 1$. To obtain (B), note that the local-isometry type of a metric g arising from (5.1) – (5.3) determines the K-equivalence class of the maximal Ricci-generic solution $t \mapsto \mathbf{y}$ of (5.3). Namely, the g-Killing fields $\partial_2, \ldots, \partial_n$, valued in eigenvectors of the Ricci tensor of g (see Lemma 5.2), are – due to the Ricci-generic condition and (2.2) – unique up to permutations and multiplication by nonzero constants, which makes $y_2, \ldots, y_n$, defined by (5.1) with $g_{jj} = g(\partial_j, \partial_j)$, also unique up to permutations. The variable t, being an arc-length parameter of g-geodesics orthogonal to $\partial_2, \ldots, \partial_n$, cf. Lemma 5.2(d) and (5.1), is in turn unique up to substitutions by $b \pm t$, for constants b, as required.

Finally, to prove (C), we fix (M, g) of dimension $n \geq 3$ satisfying (0.1) – (0.3). Corollary 1.3 and (1.2), along with Remarks 3.2 and 2.5(ii), allow us to choose $\mathfrak{a}_2, \ldots, \mathfrak{a}_n$ and $v_2, \ldots, v_n$ as in Lemma 4.1 for $m = n$, and a point $x \in M$ at which all v_j are nonzero. (From now on j ranges over $\{2, \ldots, n\}$.) By the Lie-bracket assertion of Lemma 4.1, the local flow of each v_j preserves all v_j and, consequently, also a unit vector field v_1 on a neighborhood of x, orthogonal to all v_j. Since v_1 and all v_j commute with one another, they constitute the coordinate vector fields of a local coordinate system $x^1 = t, x^2, \ldots, x^n$ on a neighborhood of x, in which the metric g has the form (5.1) as a consequence of the last two lines of Lemma 4.1, with $m = n$. (In particular, the assertion $g(\nabla_u v_j, v_k) = 0$, for $u = v_l$ and $j, k, l \in \{2, \ldots, n\}$, applied to $j = k$, shows that $g_{jj} = g(v_j, v_j)$ only depend on the variable $t = x^1$.) Now Lemma 5.2(e) yields (C). $\square$

REMARK 5.7. The component version of (5.3) states that $\ddot{y}_j - (\operatorname{tr}\mathbf{y} + y_j)\dot{y}_j$ equals $y_j[\operatorname{tr}\mathbf{y}^2 - (\operatorname{tr}\mathbf{y})y_j]$. A solution $t \mapsto \mathbf{y}$ of (5.3) for $n \geq 3$, with *any* prescribed value at $t = 0$, may be chosen so as to make the values $\mu_1(0), \ldots, \mu_n(0)$ mutually distinct. (By Lemma 5.2(a), this amounts to using $\dot{\mathbf{y}}(0)$ that realizes $(\mu_2(0), \ldots, \mu_n(0))$ lying outside a finite union of specific hyperplanes in $\mathbb{E}$.) Consequently, the local-isometry types in Theorem 5.6(ii) form a *moduli space* of dimension $2n - 3$.

6. The scalar-curvature integral

Not surprisingly, in the light of (1.1.ii) and parts (b), (e) of Lemma 5.2,

$$\mathrm{s} = 2\,\mathrm{tr}\,\dot{\mathbf{y}} - \mathrm{tr}\,\mathbf{y}^2 - (\mathrm{tr}\,\mathbf{y})^2 \text{ is constant whenever } t \mapsto \mathbf{y} \text{ satisfies (5.3).} \tag{6.1}$$

LEMMA 6.1. *For any solution $I \ni t \mapsto \mathbf{y} \in \mathbb{E}$ of* (5.3) *defined on $\mathbb{R}$, and not identically equal to zero, one must have* $\mathrm{s} < 0$ *in* (6.1).

PROOF. Under the assumption that $\mathrm{s} \geq 0$, (6.1) gives $2\,\mathrm{tr}\,\dot{\mathbf{y}} \geq \mathrm{tr}\,\mathbf{y}^2 + (\mathrm{tr}\,\mathbf{y})^2$ for our solution $\mathbb{R} \ni t \mapsto \mathbf{y} \in \mathbb{E}$, and so $\mathrm{tr}\,\mathbf{y}$ is nondecreasing and nonconstant. Fixing $t' \in \mathbb{R}$ such that $\mathrm{tr}\,\mathbf{y}(t') \neq 0$, we define a constant $c > 0$ by $(n-1)c^2 = [\mathrm{tr}\,\mathbf{y}(t')]^2$. Depending on whether $\mathrm{tr}\,\mathbf{y}(t')$ is positive or negative, monotonicity of $\mathrm{tr}\,\mathbf{y}$ gives $(\mathrm{tr}\,\mathbf{y})^2 \geq (n-1)c^2$ on $[t', \infty)$ or, respectively, on $(-\infty, t']$. The Schwarz inequality $(\mathrm{tr}\,\mathbf{x})^2 \leq (n-1)\,\mathrm{tr}\,\mathbf{x}^2$ now shows that $\mathrm{tr}\,\mathbf{y}^2 \geq c^2$ on $[t', \infty)$, or on $(-\infty, t']$. The relation $2\,\mathrm{tr}\,\dot{\mathbf{y}} \geq \mathrm{tr}\,\mathbf{y}^2 + (\mathrm{tr}\,\mathbf{y})^2$ (see above) thus yields $2\,\mathrm{tr}\,\dot{\mathbf{y}} \geq c^2 + (\mathrm{tr}\,\mathbf{y})^2$, that is, $\dot{\alpha} \geq c^2$ on $[t', \infty)$ or $(-\infty, t']$, where $\alpha = 2\tan^{-1}(\mathrm{tr}\,\mathbf{y}/c)$. Consequently, $\alpha \to \pm\infty$ as $t \to \pm\infty$ for some sign $\pm$, contrary to boundedness of α. □

REMARK 6.2. A Riemannian manifold $(I \times \mathbb{R}^{n-1}, g)$ arising from (5.1) – (5.3), which makes it real-analytic, may be locally isometric to a compact (and hence complete) real-analytic Riemannian manifold, in the sense of the paragraph following (1.2), even if the solution $I \ni t \mapsto \mathbf{y} \in \mathbb{E}$ of (5.3) has no extension to one defined on $\mathbb{R}$. This is illustrated by the trivial extension (Example 5.4), with $m > 0$ additional zeros, of the solution $y_2(t) = 2\tan 2t$ of Example 5.3, for $n = 2$, further modified using $a = 1/2$ in Remark 5.5, so as to become $t \mapsto (\tan t, 0, \ldots, 0)$. Since the latter realizes (5.2) with $g_{22} = \cos^2 t$, it represents, locally, a product of the standard sphere S^2 with a flat torus T^m.

7. Completeness

Let $n \geq 3$. In the usual fashion, (5.3) is equivalent to the first-order system

$$\dot{\mathbf{y}} = \mathbf{p},\ \dot{\mathbf{p}} = (\mathrm{tr}\,\mathbf{y} + \mathbf{y})\mathbf{p} + (\mathrm{tr}\,\mathbf{y}^2)\mathbf{y} - (\mathrm{tr}\,\mathbf{y})\mathbf{y}^2. \tag{7.1}$$

Solutions $t \mapsto \mathbf{y}$ of (5.3) thus correspond to integral curves $t \mapsto (\mathbf{y}, \mathbf{p})$ of the vector field v on $\mathbb{E} \times \mathbb{E}$ represented by (7.1), and expressed as

$$(\mathbf{y}, \mathbf{p}) \mapsto v_{(\mathbf{y},\mathbf{p})} = (\mathbf{p}, (\mathrm{tr}\,\mathbf{y} + \mathbf{y})\mathbf{p} + (\mathrm{tr}\,\mathbf{y}^2)\mathbf{y} - (\mathrm{tr}\,\mathbf{y})\mathbf{y}^2) \tag{7.2}$$

when identified with a mapping $\mathbb{E} \times \mathbb{E} \to \mathbb{E} \times \mathbb{E}$. This v has an obvious curve $\mathbb{R} \ni q \mapsto q(\mathbf{1}, \mathbf{0})$ of zeros, where $\mathbf{1} \in \mathbb{E}$ is the identity. Evaluating the differentials of $v : \mathbb{E}\times\mathbb{E} \to \mathbb{E}\times\mathbb{E}$ at $q(\mathbf{1}, \mathbf{0})$, and of the function $\mathbb{E}\times\mathbb{E} \ni (\mathbf{y}, \mathbf{p}) \mapsto \mathrm{s} = 2\,\mathrm{tr}\,\mathbf{p} - \mathrm{tr}\,\mathbf{y}^2 - (\mathrm{tr}\,\mathbf{y})^2 \in \mathbb{R}$, cf. (6.1), at any $(\mathbf{y}, \mathbf{p}) \in \mathbb{E} \times \mathbb{E}$, we obtain $dv_{q(\mathbf{1},\mathbf{0})}(\hat{\mathbf{y}}, \hat{\mathbf{p}}) = (\hat{\mathbf{p}}, nq\hat{\mathbf{p}} + q^2\,\mathrm{tr}\,\hat{\mathbf{y}} - (n-1)q^2\hat{\mathbf{y}})$ and $d\mathrm{s}_{(\mathbf{y},\mathbf{p})}(\hat{\mathbf{y}}, \hat{\mathbf{p}}) = 2[\mathrm{tr}\,\hat{\mathbf{p}} - \mathrm{tr}\,\mathbf{y}\hat{\mathbf{y}} - (\mathrm{tr}\,\mathbf{y})\,\mathrm{tr}\,\hat{\mathbf{y}}]$. When $q \neq 0$, the linear endomorphism $dv_{q(\mathbf{1},\mathbf{0})}$ of $\mathbb{E} \times \mathbb{E}$ is diagonalizable, with the eigenvalues $0, nq, (n-1)q, q$ of multiplicities $1, 1, n-2, n-2$, the eigenspace for each of the four eigenvalues λ consisting of all $(\hat{\mathbf{y}}, \hat{\mathbf{p}})$ such that $\hat{\mathbf{p}} = \lambda\hat{\mathbf{y}}$ and either $\hat{\mathbf{y}}$ equals a multiple of the identity (for $\lambda \in \{0, nq\}$), or $\mathrm{tr}\,\hat{\mathbf{y}} = 0$ (if $\lambda \in \{(n-1)q, q\}$).

On the other hand, s has no critical points in $\mathbb{E} \times \mathbb{E}$, and v is tangent to the level sets of s. The latter sets are codimension-one real-analytic submanifolds of $\mathbb{E} \times \mathbb{E}$, and those among them intersecting the curve $\mathbb{R} \ni q \mapsto q(\mathbf{1}, \mathbf{0})$ correspond, by (6.1), to $\mathrm{s} = -n(n-1)q^2$, that is, to all nonpositive values of s. If we fix $q \neq 0$, the tangent space at $z = q(\mathbf{1}, \mathbf{0})$ of the hypersurface N given by $\mathrm{s} = -n(n-1)q^2$, equal to the kernel of $ds_{q(\mathbf{1},\mathbf{0})}$, coincides, due to dimensional reasons, with the span of the eigenspaces of $dv_{q(\mathbf{1},\mathbf{0})}$ for the three nonzero eigenvalues $nq, (n-1)q, q$. (See the preceding paragraph and the above formula for $ds_{(\mathbf{y},\mathbf{p})}(\hat{\mathbf{y}},\hat{\mathbf{p}})$.) From (2.1) it now follows that ∂w_z, for the vector field w on N arising as the restriction of v, is diagonalizable, with positive (or, negative) eigenvalues. Thus, as $z = q(\mathbf{1},\mathbf{0})$,

(7.3) our z, N, w and $\varepsilon = -\operatorname{sgn} q$ satisfy the hypothesis of Lemma 2.3.

REMARK 7.1. Whenever $c \in \mathbb{R} \smallsetminus \{0\}$, the assignment $(\mathbf{y},\mathbf{p}) \mapsto (c\mathbf{y}, c^2\mathbf{p})$ is a diffeomorphism $F_c : \mathbb{E}\times\mathbb{E} \to \mathbb{E}\times\mathbb{E}$, sending our vector field v to v/c, and pulling the function s back to c^2s. Using our N given by $\mathrm{s} = -n(n-1)q^2$ we obtain a diffeomorphism $(0,\infty)\times N \ni (c,x) \mapsto F(c,x) = F_c(x)$ onto the open set in $\mathbb{E}\times\mathbb{E}$ on which $\mathrm{s} < 0$, as one sees defining its inverse by $F^{-1}(x') = (c,x)$, if $\mathrm{s}(x') < 0$, with c, x such that $n(n-1)(cq)^2 = -\mathrm{s}(x')$ and $x = F(1/c, x')$.

In the next theorem, we fix an integer $n \geq 3$, again denoting by $\mathbb{E}$ the space of all diagonal $(n-1)\times(n-1)$ matrices, and by $\mathbf{1}\in\mathbb{E}$ the identity.

THEOREM 7.2. *For any $(\xi,\zeta) \in \mathbb{R}\times(0,\infty)$, every maximal solution $t \mapsto \mathbf{y}$ of (5.3) with $(\mathbf{y}(0),\dot{\mathbf{y}}(0))$ sufficiently close to $(\xi\mathbf{1},-\zeta\mathbf{1})$ in $\mathbb{E}\times\mathbb{E}$ has the domain $\mathbb{R}$, and the metric g on $\mathbb{R}^n$ defined by (5.1) – (5.2) is complete.*

PROOF. The solution $\mathbb{R} \ni t \mapsto \mathbf{y}_{1,0}(t) = -2\tanh nt$ (times the identity $\mathbf{1}$) of Example 5.3 leads, via Remark 5.5, to further solutions $t \mapsto \mathbf{y}_{a,b}(t) = a\mathbf{y}_{1,0}(at+b)$, where $a, b \in \mathbb{R}$ and $a \neq 0$. Suitably chosen and fixed such a, b clearly realize, at $t = 0$, any initial data $(\xi\mathbf{1},-\zeta\mathbf{1}) = (\mathbf{y}_{a,b}(0),\dot{\mathbf{y}}_{a,b}(0))$ for $(\xi,\zeta)\in\mathbb{R}\times(0,\infty)$. Setting $x_{a,b}(t) = (\mathbf{y}_{a,b}(t),\dot{\mathbf{y}}_{a,b}(t))$ and $z_\pm = \mp 2|a|(\mathbf{1},\mathbf{0})$ we get $x_{a,b}(t) \to z_\pm$ as $t \to \pm\infty$. In the discussion preceding (7.3), applied to $q = \mp 2|a|$, both choices of the sign $\pm$ lead to the same N, given by $\mathrm{s} = -n(n-1)q^2$, and the same w, while $z_+, z_- \in N$ are two different zeros of w. Using (7.3) we now choose neighborhoods $U_\pm$ of $z_\pm$ in N satisfying the assertion of Lemma 2.3 for $x(t)$ equal to our $x_{a,b}(t)$, and $t'_\pm \in \mathbb{R}$ with $x_{a,b}(t'_\pm) \in U_\pm$. Since $z_\pm = \mp 2|a|(\mathbf{1},\mathbf{0})$, we may also require that

(7.4) $\pm y_j > |a|$ whenever $(y_2,\dots,y_n,p_2,\dots,p_n) \in U_\pm$ and $j \in \{2,\dots,n\}$.

By continuity, $x(t'_\pm) \in U_\pm$ for some neighborhood U_0 of $x_{a,b}(0)$ in N and all integral curves $t \mapsto x(t) \in N$ of w with $x(0) \in U_0$. The image of $(0,\infty)\times U_0$ under the diffeomorphism F of Remark 7.1 is now a neighborhood of $x_{a,b}(0) = (\xi\mathbf{1},-\zeta\mathbf{1})$ in $\mathbb{E}\times\mathbb{E}$, the existence of which constitutes our assertion: according to Remark 7.1, this F-image equals the union of $F_c(U_0)$ over $c > 0$, and each F_c maps N diffeomorphically onto the s-preimage of the value $-n(n-1)(cq)^2$, while the push-forward, under $F_c : N \to F_c(N)$, of w, obtained by restricting v to N, is the restriction of v/c to $F_c(N)$. However, the discussion preceding (7.3), and (7.3) itself, apply to every $q \neq 0$, and the use of v/c rather than v makes no difference (Remark 2.4). Now (7.4) combined with Remark 5.1 yields completeness of g. □

Our next result shows that the examples arising from Lemma 5.2(e) are not generally Ricci-parallel, or locally reducible, or (when $n \geq 4$) conformally flat.

THEOREM 7.3. *The local-isometry types of Riemannian n-manifolds satisfying (0.1) – (0.4) form a set with a nonempty interior in the $(2n-3)$-dimensional moduli space of Remark 5.7.*

PROOF. According to Theorem 5.6, the local-isometry types of all n-dimensional (M,g) with (0.1) – (0.3) arise from (5.1) when one chooses a maximal Ricci-generic solution $I \ni t \mapsto \mathbf{y} \in \mathbb{E}$ of (5.3), and then fixes a smooth curve $I \ni t \mapsto (g_{22}(t),\dots,g_{nn}(t)) \in (0,\infty)^{n-1}$ satisfying (5.2). Restricting our discussion to the case where $0 \in I$, and then parametrizing such solutions (allowed, this time, not to be Ricci-generic) by their initial data at $t=0$, we identify them with points of a specific Euclidean space, and completeness of g is guaranteed by Theorem 7.2 once one assumes (as we do from now on) that the initial data range over a certain nonempty open subset of the latter space. Now, as in Remark 5.7, if $n \geq 3$, we can make the Ricci eigenvalue functions $\mu_1(0),\dots,\mu_n(0)$ of Lemma 5.2 mutually distinct (which leads to Ricci-genericity) just by ensuring that $(\mu_2(0),\dots,\mu_n(0))$ does not lie within a specific finite union of hyperplanes in $\mathbb{E}$. However, rather than using any prescribed $\mathbf{y}(0)$, cf. Remark 5.7, let us require $y_1(0),\dots,y_n(0)$ to be all nonzero. This amounts to imposing on the solution $t \mapsto \mathbf{y}$ of (5.3) a *further open condition* implying (see the proof of Lemma 5.2) that $R_{j1,j}(0) \neq 0$, and so g is not Ricci-parallel. In the proof of Lemma 5.2 we also saw that $\Gamma_{jj}^1(0) \neq 0$ and, consequently, g cannot be locally reducible. (If it were, the Ricci eigenvector fields $\partial_1,\dots,\partial_n$ of Lemma 5.2, with distinct eigenvalue functions $\mu_1,\dots,\mu_n$, would each be tangent to one or the other parallel factor distribution, giving $\Gamma_{jj}^1 = 0$ with some $j = 2,\dots,n$.) For $k \neq j$, one easily verifies that $g^{jj}g^{kk}R_{jkjk} = -y_jy_k$. Therefore, if W denotes the Weyl tensor, $(n-1)(n-2)g^{jj}g^{kk}W_{jkjk} = 2\operatorname{tr}\dot{\mathbf{y}} - \operatorname{tr}\mathbf{y}^2 - (\operatorname{tr}\mathbf{y})^2 + (n-1)[(y_j+y_k)\operatorname{tr}\mathbf{y} - (n-2)y_jy_k - \dot{y}_j - \dot{y}_k]$, where $\dot{y}_j$ appears with the coefficient $3-n$. An enhanced version of the last open condition thus precludes conformal flatness of our examples when $n \geq 4$. $\square$

Appendix: Warped products with harmonic curvature

For the reader's convenience, we gather here some facts that are well known [**7**] and easily verified. The repeated indices are always summed over. In (1.4) we set $m = \dim \overline{M}$ and $p = \dim \Sigma$, assuming that $mp \geq 1$ and $\phi : \overline{M} \to (0,\infty)$ is nonconstant. Thus, $\dim M = n$ with $n = m + p \geq 2$. We use product coordinates x^λ in M, consisting of local coordinates x^i for $\overline{M}$ and x^a for Σ, declaring

$$\text{(A.1)} \qquad \lambda,\mu,\nu \in \{1,\dots,n\},\ i,j,k \in \{1,\dots,m\},\ a,b,c \in \{m+1,\dots,n\}$$

to be our index ranges. Therefore, $\overline{g}_{ij}$ as well as $\theta = \log\phi$ depend only on the variables x^k, and η_{ab} only on x^c, that is, $\partial_a\overline{g}_{ij} = \partial_a\theta = \partial_i\eta_{ab} = 0$. Furthermore,

$$\text{(A.2)} \qquad g_{ij} = \overline{g}_{ij},\ g_{ia} = g_{ai} = 0,\ g_{ab} = e^{2\theta}\eta_{ab}.$$

For the Christoffel symbols $\Gamma_{\lambda\mu}^\nu, \overline{\Gamma}_{ij}^k, H_{ab}^c$ of $g,\overline{g},\eta$, their Ricci-tensor components $R_{\lambda\mu}, \overline{R}_{ij}, P_{ab}$, and the components $\overline{\nabla}_i\overline{\nabla}_j\theta$ of the $\overline{g}$-Hessian of θ, one has

$$g^{ij} = \overline{g}^{ij}, g^{ia} = g^{ai} = 0, g^{ab} = e^{-2\theta}\eta^{ab}, \Gamma_{ij}^k = \overline{\Gamma}_{ij}^k,\ \Gamma_{ia}^k = \Gamma_{ij}^a = 0, \Gamma_{ia}^b = \delta_a^b\theta_{,i},$$
$$\Gamma_{ab}^i = -e^{2\theta}\eta_{ab}\theta^{,i},\ \Gamma_{ab}^c = H_{ab}^c, R_{ij} = \overline{R}_{ij} - p[\overline{\nabla}_i\overline{\nabla}_j\theta + \theta_{,i}\theta_{,j}],$$

while, in terms of the $\overline{g}$-Laplacian $\overline{\Delta}$,

$$\text{(A.3)} \qquad R_{ia} = 0,\ R_{ab} = P_{ab} - p^{-1}e^{(2-p)\theta}[\overline{\Delta}e^{p\theta}]\eta_{ab}.$$

The components $R_{\lambda\mu,\nu}, \overline{\nabla}_i\overline{R}_{jk}, D_cP_{ab}$ of the covariant derivatives of the Ricci tensors of $g,\overline{g},\eta$ satisfy, with the usual conventions $\theta_{,i} = \partial_i\theta$ and $\theta^{,i} = \overline{g}^{ij}\partial_j\theta$, the relations

$$\text{(A.4)}\quad \begin{aligned} R_{jk,i} &= \overline{\nabla}_i\overline{R}_{jk} - p[\overline{\nabla}_i\overline{\nabla}_j\overline{\nabla}_k\theta + \overline{\nabla}_i(\theta_{,j}\theta_{,k})],\ R_{ij,a} = R_{aj,i} = 0,\\ R_{ib,a} &= e^{2\theta}(p^{-1}e^{-p\theta}[\overline{\Delta}e^{p\theta}]\theta_{,i} + [\overline{R}_{ij} - p\overline{\nabla}_i\overline{\nabla}_j\theta - p\theta_{,i}\theta_{,j}]\theta^{,j})\eta_{ab} - \theta_{,i}P_{ab},\\ R_{ab,i} &= -p^{-1}e^{2\theta}(e^{-p\theta}\overline{\Delta}e^{p\theta})_{,i}\eta_{ab} - 2\theta_{,i}P_{ab},\ R_{ab,c} = D_cP_{ab}. \end{aligned}$$

Let (a) – (e) refer to parts of Lemma 1.2, which we now proceed to prove. First,

(f) $R_{ab,i} = R_{ib,a}$ for all i, a, b as in (A.1) if and only if one has (a) and (e).

In fact, it suffices to verify (f) on the dense set $(U \cup U') \times \Sigma \subseteq M$, for the interior U of the zero set of $d\theta$ in $\overline{M}$ and the subset U' on which $d\theta \neq 0$. On U, according to (A.4), $R_{ab,i} = 0 = R_{ib,a}$ since $\overline{\Delta}e^{p\theta} = 0$. Similarly, on U', the equality $R_{ab,i} = R_{ib,a}$ amounts, by (A.4), to the condition $P_{ab} = \kappa\eta_{ab}$, for a function κ on Σ which must be constant, as it depends only on the variables x^j that are local coordinates in $\overline{M}$. Formulae (A.4) also show that κ is characterized by the relation $-\kappa e^{-2\theta}d\theta = p^{-1}(d[e^{-p\theta}\overline{\Delta}e^{p\theta}] + e^{-p\theta}[\overline{\Delta}e^{p\theta}]d\theta) + \overline{\mathrm{r}}(\overline{\nabla}\theta, \cdot) - p\overline{g}(\overline{\nabla}\theta, \overline{\nabla}\theta)d\theta - pd[\overline{g}(\overline{\nabla}\theta, \overline{\nabla}\theta)]/2$ which, rewritten in terms of $\phi = e^\theta$, becomes (e).

The equivalence of (e) and (c) is in turn obvious from (1.3). Next, by (A.4),

(g) $R_{jk,i} = R_{ik,j}$ for all i, j, k with (A.1) if and only if (b) holds,

since $\phi^{-1}\overline{\nabla}d\phi = \overline{\nabla}d\theta + d\theta \otimes d\theta$. The main claim of Lemma 1.2 is thus immediate: harmonicity of the curvature amounts to the Codazzi equation for the Ricci tensor, cf. (1.1).i), while (A.4 clearly reduces the latter to the cases (f) – (g).

Finally, (d) follows from (a) and (A.3).

References

[1] A. L. Besse, *Einstein manifolds*, Ergebnisse der Mathematik und ihrer Grenzgebiete (3) [Results in Mathematics and Related Areas (3)], vol. 10, Springer-Verlag, Berlin, 1987. MR867684

[2] A. Derdziński, *On compact Riemannian manifolds with harmonic curvature*, Math. Ann. **259** (1982), no. 2, 145–152, DOI 10.1007/BF01457307. MR656660

[3] A. Derdziński, *Riemannian metrics with harmonic curvature on 2-sphere bundles over compact surfaces* (English, with French summary), Bull. Soc. Math. France **116** (1988), no. 2, 133–156. MR971557

[4] D. DeTurck and H. Goldschmidt, *Regularity theorems in Riemannian geometry. II. Harmonic curvature and the Weyl tensor*, Forum Math. **1** (1989), no. 4, 377–394, DOI 10.1515/form.1989.1.377. MR1016679

[5] F. J. E. Dillen and L. C. A. Verstraelen (eds.), *Handbook of differential geometry. Vol. I*, North-Holland, Amsterdam, 2000. MR1736851

[6] A. Haefliger, *Structures feuilletées et cohomologie à valeur dans un faisceau de groupoïdes* (French), Comment. Math. Helv. **32** (1958), 248–329, DOI 10.1007/BF02564582. MR100269

[7] I. B. Kim, J. T. Cho, and K. K. Hwang, *On almost everywhere warped product manifolds with harmonic curvatures*, J. Korean Math. Soc. **28** (1991), no. 1, 23–35. MR1107209

[8] S. Kobayashi, *Fixed points of isometries*, Nagoya Math. J. **13** (1958), 63–68. MR103508

[9] S. Kobayashi and K. Nomizu, *Foundations of differential geometry. Vol I*, Interscience Publishers, a division of John Wiley & Sons, New York-London, 1963. MR0152974

[10] G. Maschler, *Conformally Kähler Ricci solitons and base metrics for warped product Ricci solitons*, Pacific J. Math. **286** (2017), no. 2, 361–384, DOI 10.2140/pjm.2017.286.361. MR3599864

Department of Mathematics, The Ohio State University, 231 W. 18th Avenue, Columbus, OH 43210, USA
Email address: andrzej@math.ohio-state.edu

Departamento de Matemática, Instituto de Matemática e Estatística, Universidade de São Paulo, Rua do Matão 1010, CEP 05508-900, São Paulo, SP, Brazil
Email address: piccione@ime.usp.br

Contemporary Mathematics
Volume **756**, 2020
https://doi.org/10.1090/conm/756/15199

CR-submanifolds of Chen-type two in non-flat complex space forms

Ivko Dimitrić and Mirjana Djorić

Dedicated to Professor B.-Y. Chen on the occasion of his 75th birthday

Abstract. We prove some nonexistence results for certain families of CR-submanifolds of Chen-type two in complex space forms. For example, there exist no holomorphic submanifolds of the complex hyperbolic space which are of 2-type via the standard embedding by projectors. This is in contrast to the situation in complex projective space, which is known to contain Einstein-Kähler submanifolds of 2-type. We further show that there are no mass-symmetric ruled real hypersurfaces in $\mathbb{C}Q^m(4c)$ of Chen-type 2. Additionally, we characterize mass-symmetric totally real submanifolds of 2-type in terms of detailed intrinsic and extrinsic conditions and derive some corollaries for Lagrangian submanifolds.

1. Introduction

According to B.-Y. Chen [**2**], a Riemannian submanifold $x : M^n \to E^N_{(K)}$ of Euclidean or pseudo-Euclidean space is said to be of finite type in $E^N_{(K)}$ if the position vector x allows a decomposition into a finite sum of vector eigenfunctions of the Laplacian Δ_M on M, viz.

$$x = x_0 + x_1 + \cdots + x_k,$$

where $x_0 = \text{const}$, $\quad x_i \neq \text{const}$, $\quad$ and $\quad \Delta x_i = \lambda_i x_i$, $\; i = 1, ..., k$. If all eigenvalues λ_i are different, the submanifold is said to be of Chen-type k, or simply of $k-$type. For a compact submanifold, x_0 is its center of mass. This notion can be extended to submanifolds $x : M^n \to \overline{M}$ of a more general manifold $\overline{M}$, as long as there is a suitable equivariant embedding $\Phi : \overline{M} \to E^N_{(K)}$ of the ambient manifold $\overline{M}$ into an appropriate (pseudo) Euclidean space, in which case M is said to be of Chen-type k (via Φ) if the composite immersion $\Phi \circ x$ is of Chen-type k.

Next to spheres and real projective and hyperbolic spaces, the complex projective space $\mathbb{C}P^m(4)$ and the complex hyperbolic space $\mathbb{C}H^m(-4)$ are the symmetric spaces which allow one of the simplest equivariant embeddings into a certain

2010 *Mathematics Subject Classification.* Primary 53C40, 53C42.
Key words and phrases. Complex space form, Laplacian, CR-submanifold, finite Chen-type.
The first author partially supported by the Eberly Science Fund, PSU Fayette.
The second author is partially supported by Ministry of Education, Science and Technological Development, Republic of Serbia, project 174012.

(pseudo) Euclidean space $E^N_{(K)}$ of suitable Hermitian matrices by the projection operators (see, for example [**16**], [**12**], [**8**], [**5**]). We use the symbol $\mathbb{C}Q^m(4c)$, $c = \pm 1$, to denote jointly either of the two simply-connected model space forms $\mathbb{C}P^m(4)$ or $\mathbb{C}H^m(-4)$ and $\Phi : \mathbb{C}Q^m(4c) \to E^N_{(K)}$ for the embedding that associates to a complex line in $\mathbb{C}^{m+1}_{(1)}$ the operator (i.e. its matrix) of the orthogonal projection onto it. The study of finite-type submanifolds $x : M^n \to \mathbb{C}Q^m(4c)$ is then the study of the spectral behavior of the associated immersion $\tilde{x} = \Phi \circ x$ of M^n into $E^N_{(K)}$, i.e. of the possibility of decomposing $\tilde{x}$ into finitely many eigenfunctions of Δ_M.

It is not difficult to see that a $k-$type immersion x satisfies a polynomial equation in the Laplacian, $P(\Delta)(x - x_0) = 0$, where $P(t)$ is the monic polynomial of degree k with coefficients that are the elementary symmetric functions of the eigenvalues λ_i, $i = 1, \cdots, k$ [**2**]. Thus, in the study of $k-$type submanifolds one necessarily deals with the k-th iterated Laplacian $\Delta^k x$ of the immersion, which makes the investigation computationally very complex for higher values of k. Hence, the most promising study is that of submanifolds of low type: 1, 2, or 3. The study of 1-type submanifolds of $\mathbb{C}P^m$ was begun in [**12**] and a parallel investigation for hypersurfaces of $\mathbb{C}H^m$ was carried out in [**8**]. The complete classification of 1-type submanifolds of a non-flat complex space form $\mathbb{C}Q^m(4c)$ was achieved in [**5**]. Submanifolds of Chen-type 2 have been studied by several authors. Compact Kähler submanifolds of $\mathbb{C}P^m$ were characterized by Ros [**13**] and subsequently classified by Udagawa [**18**] as compact Einstein-Kähler parallel submanifolds, an example of which is the complex hyperquadric. The classification of real hypersurfaces of 2-type was undertaken in [**11**], [**20**] for hypersurfaces of $\mathbb{C}P^m$ with constant mean curvature and in [**6**] for Hopf hypersurfaces of $\mathbb{C}Q^m$. Moreover, 2-type totally-real minimal submanifolds of $\mathbb{C}P^m$ were studied in [**14**]. An investigation of 3-type submanifolds can be found in [**17**], [**18**], and [**6**].

Since real hypersurfaces, Kähler submanifolds, and totally real submanifolds are all examples of CR-submanifolds, it is natural to try to extend the study of 2-type submanifolds of $\mathbb{C}Q^m$ to general CR-submanifolds. We recall that a CR-submanifold, according to Bejancu [**1**] (see also [**7**]), is a submanifold M of an almost Hermitian manifold $(\overline{M}, g, J)$ whose tangent space splits at each point into a direct sum of two complementary orthogonal distributions (subspaces) $\mathcal{D}$ and $\mathcal{D}^\perp$ of constant dimensions so that at each point $p \in M$,

$$T_pM = \mathcal{D}_p \oplus \mathcal{D}^\perp_p, \quad \text{with} \quad J\mathcal{D}_p \subset \mathcal{D}_p \quad \text{and} \quad J\mathcal{D}^\perp_p \subset T^\perp_p M.$$

When dim $\mathcal{D}^\perp_p = 0$ at every $p \in M$, a submanifold is said to be *holomorphic* (the terms *complex*, or *J-invariant* are also used), and when dim $\mathcal{D}_p = 0$, it is said to be *totally real*. At each point of a CR-submanifold M, the tangent space of the ambient manifold $\overline{M}$ is decomposed into an orthogonal direct sum of subspaces as follows

$$T_p\overline{M} = \mathcal{D}_p \oplus \mathcal{D}^\perp_p \oplus J\mathcal{D}^\perp_p \oplus \mathcal{L}_p,$$

where the first two subspaces span T_pM, the last two span $T^\perp_p M$, and $\mathcal{D}$ and $\mathcal{L}$ are $J-$invariant (complex) subspaces.

In this paper, we provide various characterizations and classification results for 2-type holomorphic and totally-real submanifolds of $\mathbb{C}Q^m$ and examine the case of ruled hypersurfaces of 2-type.

2. The notation and relevant formulas

In this section we compute the second iterated Laplacian $\Delta^2\tilde{x}$ for an arbitrary submanifold of $\mathbb{C}Q^m(4c)$ of dimension ≥ 2. For a more detailed explanation of the notation and the background, in particular as it regards the embedding Φ, see [**5**], [**8**],[**12**]. Here, we only give the essentials.

Let $x : M^n \to \mathbb{C}Q^m(4c)$ be an isometric immersion of a Riemannian $n-$manifold into a non-flat model space form $\mathbb{C}Q^m(4c)$, $c = \pm 1$ in the projective, resp. hyperbolic case. Let $\Phi : \mathbb{C}Q^m \to H^{(1)}(m+1)$ be the standard embedding of $\mathbb{C}Q^m$ into the set of Hermitian matrices of order $m+1$ obtained by identifying a complex line with the projection operator onto it. The composite immersion $\tilde{x} = \Phi \circ x$ realizes M as a submanifold of a (pseudo) Euclidean space $E^N_{(K)} = H^{(1)}(m+1)$, equipped with the usual trace metric $\langle A, B\rangle = \frac{c}{2}\mathrm{tr}\,(AB)$. Let $\overline{\nabla}, \overline{A}, \overline{D}$, denote respectively the Levi-Civita connection, the Weingarten endomorphism, and the metric connection in the normal bundle, related to $\mathbb{C}Q^m$ and the embedding Φ. Let the same letters without bar denote the respective objects for a submanifold M and the immersion x, whereas the same symbols with tilde will denote the corresponding objects related to the composite immersion $\tilde{x} = \Phi \circ x$ of M into the (pseudo) Euclidean space $H^{(1)}(m+1)$. As usual, we use σ for the second fundamental form of $\mathbb{C}Q^m$ in $E^N_{(K)}$ via Φ and symbol h for the second fundamental form of a submanifold in $\mathbb{C}Q^m$. An orthonormal basis of the tangent space T_pM at a general point will be denoted by $\{e_i\}$, $i = 1, 2, \cdots, n$, and a basis of the normal space $T_p^\perp M$ will be represented by $\{e_r\}$. In general, indices i, j, k will range from 1 to n and indices r, s from $n+1$ to $2m$. The tangent vector fields to M are usually denoted by letters $X, Y, \cdots$, and vectors normal to M in $\mathbb{C}Q^m$ by $\xi, \eta, \cdots$. For a normal basis vector e_r the Weingarten map A_{e_r} is abreviated to A_r and is related to the second fundamental form by

$$\langle h(X, Y), e_r\rangle = \langle A_r X, Y\rangle.$$

The mean curvature vector H is defined by $H := \frac{1}{n}\sum_i h(e_i, e_i) = \frac{1}{n}\sum_r (\mathrm{tr}\, A_r)e_r$, and the squared norm of the second fundamental form by

$$||h||^2 := \sum_{i,j}\langle h(e_i, e_j), h(e_i, e_j)\rangle = \sum_r \mathrm{tr}\, A_r^2.$$

For a normal vector $\xi \in T^\perp M$ we define its *ancillary normal vector field* $\hat{\mathfrak{a}}(\xi)$ by

$$\hat{\mathfrak{a}}(\xi) := \sum_i h(e_i, A_\xi e_i) = \sum_r \mathrm{tr}\,(A_\xi A_r)\, e_r, \tag{2.1}$$

which is related to the allied vector field $\mathfrak{a}(\xi)$ of Chen [**2**], the latter being the component of $\hat{\mathfrak{a}}(\xi)$ in the direction perpendicular to ξ. Then $\hat{\mathfrak{a}} : \Gamma(T^\perp M) \to \Gamma(T^\perp M)$ is a symmetric endomorphism of the normal bundle satisfying

$$\langle \hat{\mathfrak{a}}(\xi), \eta\rangle = \langle \xi, \hat{\mathfrak{a}}(\eta)\rangle = \mathrm{tr}\,(A_\xi A_\eta), \qquad \mathrm{tr}\,\hat{\mathfrak{a}} = ||h||^2. \tag{2.2}$$

We call $\hat{\mathfrak{a}}$ *the ancillary shape operator* (also known as *Simons operator*, [**15**]) of the immersion. We give next some important formulas which will be repeatedly used throughout this paper. For a general submanifold M, local tangent fields $X, Y \in \Gamma(TM)$ and a local normal field $\xi \in \Gamma(T^\perp M)$, the formulas of Gauss and Weingarten are

$$\overline{\nabla}_X Y = \nabla_X Y + h(X, Y), \qquad \overline{\nabla}_X \xi = -A_\xi X + D_X \xi. \tag{2.3}$$

Let J be the standard Kähler almost complex structure of $\mathbb{C}Q^m$ and M an arbitrary submanifold. Define an endomorphism S of the tangent space and a normal bundle valued 1-form F by $SX = (JX)_T$, $FX = (JX)_N$, i.e. for $X \in \Gamma(TM)$, $JX = SX + FX$ is the decomposition of JX into tangential and normal parts. Similarly, for a normal vector ξ we set $J\xi = (J\xi)_T + (J\xi)_N = L\xi + K\xi$, K defining an endomorphism of the normal space and $L : T^\perp M \to TM$ a linear map from the normal to the tangent space. Then (see [**22**], p. 215)

(2.4) $(\nabla_X S)Y = L(h(X,Y)) + A_{FY}X$, where $(\nabla_X S)Y := \nabla_X(SY) - S(\nabla_X Y)$,

The curvature tensor of $\mathbb{C}Q^m(4c)$ is given by

(2.5) $\overline{R}(X,Y)Z = c\,[\langle Y,Z\rangle X - \langle X,Z\rangle Y + \langle JY,Z\rangle JX - \langle JX,Z\rangle JY - 2\langle JX,Y\rangle JZ\,]$,

and the equations of Gauss and Codazzi for a submanifold of $\mathbb{C}Q^m(4c)$ are respectively given by

(2.6) $R(X,Y,Z,W) = \overline{R}(X,Y,Z,W) + \langle h(X,W), h(Y,Z)\rangle - \langle h(X,Z), h(Y,W)\rangle$,

(2.7) $(\overline{\nabla}_X h)(Y,Z) - (\overline{\nabla}_Y h)(X,Z) = c\,[\langle SY,Z\rangle FX - \langle SX,Z\rangle FY - 2\langle SX,Y\rangle FZ]$.

An alternative form of the Codazzi equation is

$$(\nabla_Y A)_\xi X - (\nabla_X A)_\xi Y = (\overline{R}(X,Y)\xi)_T, \tag{2.8}$$

where $(\nabla_Y A)_\xi X := \nabla_Y(A_\xi X) - A_{D_Y\xi}X - A_\xi(\nabla_Y X)$, for any normal vector ξ. The Ricci (1,1)-tensor Q of M as an endomorphism of the tangent space of M is defined by $Q(X) = \sum_i R(X,e_i)\,e_i$. From (2.5) and (2.6) one gets

$$Q(X) = c(n-1)X - 3cS^2X + nA_HX - \sum_r A_r^2 X. \tag{2.9}$$

The following formula of A. Ros for the shape operator of Φ in the direction of $\sigma(X,Y)$ is also well known (see, for example, [**12**] and [**5**])

(2.10) $\bar{A}_{\sigma(X,Y)}V = c\,[2\langle X,Y\rangle V + \langle Y,V\rangle X + \langle X,V\rangle Y + \langle JY,V\rangle JX + \langle JX,V\rangle JY]$.

One also verifies

(2.11) $\sigma(JX,JY) = \sigma(X,Y)$, $\quad\langle \sigma(X,Y), \tilde{x}\rangle = -\langle X,Y\rangle$, $\quad\langle \sigma(X,Y),\, I\rangle = 0$.

The gradient of a smooth function f is a vector field given by $\nabla f = \sum_i (e_i f)e_i$. The Hessian of f is a symmetric tensor field defined by

$$\text{Hess}_f(X,Y) = XYf - (\nabla_X Y)f,$$

and the Laplacian acting on smooth functions is defined as $\Delta = -\text{tr Hess}_f$. The Laplace operator is then extended to act on a vector field V along $\tilde{x}(M)$ by

$$\Delta V = \sum_i [\widetilde{\nabla}_{\nabla_{e_i}e_i} V - \widetilde{\nabla}_{e_i}\widetilde{\nabla}_{e_i} V].$$

The product formula for the Laplacian, which is used in the ensuing computations, is

$$\Delta(f\,g) = (\Delta f)\,g + f(\Delta g) - 2\sum_i (e_i f)(e_i g), \tag{2.12}$$

for smooth functions $f,\, g \in C^\infty(M)$ and it can be then extended to the scalar product of vector valued functions, so that the formula holds true also for matrix products. For an endomorphism B of the tangent space of M we define $\operatorname{tr}(\nabla B) := \sum_{i=1}^{n}(\nabla_{e_i}B)e_i$.

From the Beltrami's formula $\Delta\tilde{x} = -n\widetilde{H}$ we get

$$\Delta\tilde{x} = -n\widetilde{H} = -nH - \sum_{i=1}^{n}\sigma(e_i, e_i), \tag{2.13}$$

where, here and in the ensuing computations, we understand the Laplacian to be applied to vector fields along M (viewed as $E^N_{(K)}$−valued functions, i.e. matrices) componentwise, and thus

$$\Delta^2\tilde{x} := \Delta(\Delta\tilde{x}) = -n\Delta H - \sum_{i=1}^{n}\Delta(\sigma(e_i, e_i)). \tag{2.14}$$

The Laplacian of the mean curvature vector equals

$$\Delta H = \sum_i[\widetilde{\nabla}_{\nabla_{e_i}e_i}H - \widetilde{\nabla}_{e_i}\widetilde{\nabla}_{e_i}H].$$

To facilitate the following computations we may assume that $\nabla_{e_i}e_j = 0$ for every $i, j = 1, 2, \dots n$ at a point considered. Thus, by a repeated use of Gauss and Weingarten formulas (2.3) we have

$$\begin{aligned} \Delta H &= -\sum_i \widetilde{\nabla}_{e_i}(-A_He_i + D_{e_i}H + \sigma(e_i, H)) \\ &= \sum_i [\nabla_{e_i}(A_He_i) + h(e_i, A_He_i) + \sigma(e_i, A_He_i) - \sigma(e_i, D_{e_i}H) \\ &\qquad + A_{D_{e_i}H}e_i - D_{e_i}D_{e_i}H + \overline{A}_{\sigma(e_i,H)}e_i - \overline{D}_{e_i}(\sigma(e_i, H))]. \end{aligned}$$

Using formula (2.10) to evaluate $\sum_i \overline{A}_{\sigma(e_i,H)}e_i$, the parallelism of σ, and the expression

$$\sum_i \overline{D}_{e_i}(\sigma(e_i, H)) = \sum_i[\sigma(h(e_i, e_i), H) - \sigma(e_i, A_He_i) + \sigma(e_i, D_{e_i}H)]$$

we get

$$\begin{aligned} \Delta H &= \operatorname{tr}(\nabla A_H) + \operatorname{tr} A_{DH} + \Delta^{\perp}H + cnH + \hat{\mathfrak{a}}(H) + cJ(JH)_T \\ &\quad -n\sigma(H, H) + 2\sum_i \sigma(e_i, A_He_i) - 2\sum_i \sigma(e_i, D_{e_i}H), \end{aligned} \tag{2.15}$$

where

$$\operatorname{tr}(\nabla A_H) := \sum_i(\nabla_{e_i}A_H)e_i, \qquad \operatorname{tr} A_{DH} := \sum_i A_{D_{e_i}H}e_i,$$

and $\Delta^{\perp}$ stands for the Laplacian in the normal bundle $T^{\perp}M$ in $\mathbb{C}Q^m$, i.e. with respect to the normal connection D.

Noting that $\sum_i(\nabla_{e_i}S)e_i = n(JH)_T - \sum_{i,j}\langle Jh(e_j, e_i), e_j\rangle e_i$, by (2.4), and using (2.7) and (2.11) one obtains after a somewhat long computation

$$
\begin{aligned}
\sum_i \Delta(\sigma(e_i,e_i)) &= 2cn(n+2)H - 2cnJ(JH)_T + 4c\sum_{i,j}\langle Jh(e_j,e_i),e_j\rangle Je_i \\
&\quad + 2c(n+1)\sum_i \sigma(e_i,e_i) + 2c\sum_i \sigma(Se_i,Je_i) \\
&\quad + 2\sum_{i,r}\sigma(A_re_i,A_re_i) - 2\sum_{r,s}\mathrm{tr}\,(A_rA_s)\sigma(e_r,e_s) \\
&\quad - 2n\sum_i \sigma(D_{e_i}H,e_i) - 6c\sum_i \sigma(FSe_i,e_i).
\end{aligned} \tag{2.16}
$$

Combining (2.14), (2.15), and (2.16) we obtain

$$
\begin{aligned}
\Delta^2\tilde{x} &= -n\,\mathrm{tr}\,(\nabla A_H) - n\,\mathrm{tr}\,A_{DH} - n\Delta^{\perp}H - cn(3n+4)H \\
&\quad + cnJ(JH)_T - n\hat{\mathfrak{a}}(H) - 4c\sum_{i,j}\langle Jh(e_j,e_i),e_j\rangle Je_i \\
&\quad + n^2\sigma(H,H) - 2c(n+1)\sum_i \sigma(e_i,e_i) + 6c\sum_i \sigma(FSe_i,e_i) \\
&\quad - 2n\sum_i \sigma(e_i,A_He_i) + 4n\sum_i \sigma(e_i,D_{e_i}H) - 2c\sum_i \sigma(Se_i,Je_i) \\
&\quad - 2\sum_{i,r}\sigma(A_re_i,A_re_i) + 2\sum_{r,s}\mathrm{tr}\,(A_rA_s)\sigma(e_r,e_s),
\end{aligned} \tag{2.17}
$$

which holds for any submanifold of $\mathbb{C}Q^m(4c)$. Additionally, by applying the equation of Codazzi (2.8) we have

$$
\mathrm{tr}\,(\nabla A_H) = \mathrm{tr}\,A_{DH} + \frac{n}{2}\nabla\alpha^2 + 3c\,S(LH), \tag{2.18}
$$

where $\alpha^2 = \langle H,H\rangle$ is the squared mean curvature. Consequently, when $DH = 0$ (this condition is weaker than minimality, stronger than CMC), formula (2.17) simplifies to

$$
\begin{aligned}
\Delta^2\tilde{x} &= -cn(3n+4)H + cnJ(LH) - 3cn\,S(LH) - n\hat{\mathfrak{a}}(H) \\
&\quad - 4c\sum_{i,j}\langle Jh(e_j,e_i),e_j\rangle Je_i + n^2\sigma(H,H) - 2c(n+1)\sum_i \sigma(e_i,e_i) \\
&\quad + 6c\sum_i \sigma(FSe_i,e_i) - 2n\sum_i \sigma(e_i,A_He_i) - 2c\sum_i \sigma(Se_i,Je_i) \\
&\quad - 2\sum_{i,r}\sigma(A_re_i,A_re_i) + 2\sum_r \sigma(e_r,\hat{\mathfrak{a}}(e_r)).
\end{aligned} \tag{2.19}
$$

Note that for CR-submanifolds $S\circ L = 0$ and $F\circ S = 0$, so that the above expressions simplify accordingly.

3. The ancillary shape operator and the ancillary curvatures

The ancillary shape operator $\hat{\mathfrak{a}}:\ \xi \to \sum_r \mathrm{tr}\,(A_\xi A_r)e_r$ of a submanifold M is a symmetric endomorphism of $T^{\perp}M$. It is an endomorphism version of the tensor $T(\xi,\eta) = \mathrm{tr}\,(A_\xi A_\eta)$ considered by Ros [**13**], see also [**22**]. It plays a kind of a role dual to the Casorati operator $\sum_r A_r^2$ acting on the tangent space - both have trace equal to $\|h\|^2$. Its matrix in an orthonormal frame $\{e_r\}$ of normal vectors is given by $[\mathrm{tr}(A_rA_s)]_{n+1\le r,s\le m}$. Since $\langle\hat{\mathfrak{a}}(\xi),\xi\rangle = \mathrm{tr}\,A_\xi^2 \ge 0$, the operator $\hat{\mathfrak{a}}$ has non-negative eigenvalues. These eigenvalues ρ_r we call *the ancillary curvatures* of M. They are solutions of the polynomial equation $\det(\hat{\mathfrak{a}}-\rho I) = \det[\operatorname{tr}(A_rA_s)-\delta_{rs}\rho] = 0$. As a matter of fact, for $\xi=\sum_r x_re_r$ when the quadratic form $\mathfrak{Q}(\xi) = \operatorname{tr}A_\xi^2 = \sum_i\langle A_\xi e_i, A_\xi e_i\rangle = \sum_{r,s}\operatorname{tr}(A_rA_s)x_rx_s$ is diagonalized for a certain basis $\{e_r\}$, then $\operatorname{tr}(A_rA_s)=0$, $r\neq s$, and $\operatorname{tr}A_r^2$ are precisely the ancillary curvatures of M. For the operator norm $\|\hat{\mathfrak{a}}\|^2=\operatorname{tr}\hat{\mathfrak{a}}^2$ one has

$$\frac{1}{2m-n}\,\|h\|^4 \le \|\hat{\mathfrak{a}}\|^2 \le \|h\|^4. \tag{3.1}$$

The equality in the left inequality holds if and only if $\hat{\mathfrak{a}}$ is homothetic. Indeed, for any symmetric endomorphism A on a $p-$dimensional vector space (in our case $\hat{\mathfrak{a}}$ operating on the space $T^\perp M$ of real dimension $p=2m-n$) one has

$$p\operatorname{tr}A^2 = (\operatorname{tr}A)^2 + \sum_{i<j}(k_i-k_j)^2 \ge (\operatorname{tr}A)^2,$$

where k_i are the eigenvalues of A and the equality holds when these eigenvalues are all equal. Moreover, for any two symmetric endomorphisms $A=A_r$, $B=A_s$ we have $[\operatorname{tr}(AB)]^2\le(\operatorname{tr}A^2)(\operatorname{tr}B^2)$ (proof by diagonalizing A and using Cauchy-Schwarz inequality). The equality holds if and only if one of the two matrices is a scalar multiple of the other. This gives the right-hand inequality. For holomorphic submanifolds, the right-hand inequality in (3.1) can be replaced by a better estimate $\|\hat{\mathfrak{a}}\|^2\le\frac{1}{2}\|h\|^4$, see **[13]**, **[22]**. For a complex n-dimensional Kähler submanifold of a complex space form $\overline{M}^m(c)$ of constant holomorphic sectional curvature c one has

$$\|\hat{\mathfrak{a}}\|^2 = \frac{1}{2}(n+2)c\,\|h\|^2 + g(\Delta h, h) - \sum_{r,s}\|[A_r,A_s]\|^2$$

(**[22]**, p. 185).

Of particular interest are those submanifolds for which $\hat{\mathfrak{a}}$ is homothetic, i.e. proportional to the identity map. Let us look at some examples.

1. For a hypersurface of any Riemannian manifold, $\hat{\mathfrak{a}}(\xi) = (\operatorname{tr}A^2)\xi = ||h||^2\xi$ holds, so $\hat{\mathfrak{a}} = ||h||^2 I$ is homothetic and has all ancillary curvatures equal to $||h||^2$. They are constant when the second fundamental form has constant length.

2. On a complex hypersurface in a Kähler manifold, if ξ is an arbitrary unit normal vector then $\{\xi, J\xi\}$ form an orthonormal basis of $T^\perp M$. Then, since $A_{J\xi}=JA_\xi$, we get $\operatorname{tr}(A_\xi A_{J\xi}) = -\operatorname{tr}(A_{J\xi}A_\xi)=0$ and hence $\hat{\mathfrak{a}}(\xi) = \operatorname{tr}(A_\xi A_\xi)\xi + \operatorname{tr}(A_\xi A_{J\xi})J\xi = (\operatorname{tr}A_\xi^2)\xi$. Thus $\hat{\mathfrak{a}}$ is homothetic with only one ancillary curvature $\rho=\frac{1}{2}\|h\|^2$. For a general codimension-2 submanifold of a Riemannian manifold with normal basis $\{\xi,\eta\}$, the ancillary operator is homothetic if and only if $\operatorname{tr}(A_\xi A_\eta)=0$ and $\operatorname{tr}A_\xi^2=\operatorname{tr}A_\eta^2=\frac{1}{2}\|h\|^2$.

3. Let M be a Lagrangian submanifold of a Kähler manifold. Then for any two normal vectors $\xi=JX$, $\eta=JY$, $X,Y\in\Gamma(TM)$ we have by (2.9)

$$\begin{aligned}\operatorname{tr}(A_\xi A_\eta) &= \sum_i\langle A_{JX}e_i, A_{JY}e_i\rangle = \sum_i\langle A_{Je_i}X, A_{Je_i}Y\rangle = \sum_i\langle A_{Je_i}^2X, Y\rangle\\ &= \langle\sum_r A_r^2X, Y\rangle = \langle c(n-1)X + nA_HX - QX, Y\rangle \qquad (3.2)\\ &= c(n-1)\langle\xi,\eta\rangle + n\langle h(J\xi,J\eta),H\rangle - Ric(J\xi,J\eta).\end{aligned}$$

Thus, if M is pseudoumbilical (in particular minimal) Lagrangian Einstein submanifold of a complex space form then $\operatorname{tr}(A_\xi A_\eta)=0$ when $\xi\perp\eta$ and $\operatorname{tr}A_\xi^2=$

$[c(n-1)+nb-k]$ for every unit vector $\xi \in T^{\perp}M$, where k is the Ricci curvature (constant when $n \geq 3$) and b the eigenvalue of A_H. So, in this case $\hat{\mathfrak{a}}$ is also homothetic. Moreover, in the set of three conditions for a Kähler submanifold - M is pseudoumbilical, M is Einstein, and $\hat{\mathfrak{a}}$ is homothetic - any two imply the third. Note that minimal Lagrangian submanifolds of $\mathbb{C}Q^m(4c)$ are of 1-type [**12**].

4. Holomorphic submanifolds of $\mathbb{C}Q^m(4c)$ of type 2

Let us assume that M is a holomorphic (i. e. $J-$invariant) submanifold fully immersed in a non-flat complex space form. Since $\mathbb{C}Q^m(4c)$ is a Kähler manifold we have

$$h(JX,Y)=h(X,JY)=Jh(X,Y), \qquad (\nabla_X J)Y=0, \quad \text{and} \tag{4.1}$$

$$A_{J\xi}=JA_\xi=-A_\xi J, \qquad D_X(J\xi)=JD_X\xi, \tag{4.2}$$

which hold for every tangent vectors $X, Y \in \Gamma(TM)$ and a normal vector $\xi \in \Gamma(T^{\perp}M)$. Condition (4.1) implies that M is minimal, i.e. $H=0$, and thus (2.13) and (2.17) reduce to

$$\Delta\tilde{x} = -\sum_i \sigma(e_i,e_i), \tag{4.3}$$

$$\Delta^2\tilde{x} = 2\sum_r \sigma(e_r,\hat{\mathfrak{a}}(e_r)) - 2\sum_{i,r}\sigma(A_re_i,A_re_i) - 2c(n+2)\sum_i \sigma(e_i,e_i). \tag{4.4}$$

We assume that M is of 2-type in $E^N_{(K)}=H^{(1)}(m+1)$, i.e.

$$\tilde{x}=\tilde{x}_0+\tilde{x}_u+\tilde{x}_v, \quad \Delta\tilde{x}_u=\lambda_u\tilde{x}_u, \quad \Delta\tilde{x}_v=\lambda_v\tilde{x}_v\,,$$

where $\tilde{x}_0=$ const and λ_u, λ_v are two distinct eigenvalues of the Laplacian on M. By eliminating $\tilde{x}_p$, $\tilde{x}_q$ we get

$$\Delta^2\tilde{x}-p\,\Delta\tilde{x}+q(\tilde{x}-\tilde{x}_0)=0, \quad \text{with} \quad p:=\lambda_u+\lambda_v, \quad q:=\lambda_u\lambda_v. \tag{4.5}$$

This is a necessary condition for M to be of 2-type. When M is compact or when the polynomial $P(t)=t^2-pt+q$ has simple real roots, this condition implies that M is of type ≤ 2, [**3**], [**2**], p. 164. In order to eliminate constant vector $\tilde{x}_0$ we differentiate with respect to an arbitrary tangent field X to get

$$\widetilde{\nabla}_X(\Delta^2\tilde{x})-p\,\widetilde{\nabla}_X(\Delta\tilde{x})+q\,X=0. \tag{4.6}$$

We now find covariant derivative of every term appearing in $\Delta\tilde{x}$ and $\Delta^2\tilde{x}$. By a repeated use of Gauss and Weingarten formulas (2.3) and (4.2), we find

$$\widetilde{\nabla}_X\sum_i\sigma(e_i,e_i)=-2c(n+2)X+2\sum_r\sigma(e_r,A_rX), \tag{4.7}$$

$$\begin{aligned}\widetilde{\nabla}_X\sum_{i,r}\sigma(A_re_i,A_re_i) &= -2c\,||h||^2X-4c\sum_r A_r^2X\\ &\quad +2\sum_{i,r}\sigma((\nabla_XA_r)e_i,A_re_i)+2\sum_{i,r}\sigma(h(X,A_re_i),A_re_i)\end{aligned} \tag{4.8}$$

$$\widetilde{\nabla}_X \sum_r \sigma(e_r, \hat{\mathfrak{a}}(e_r)) = -2c\|h\|^2 X + \sum_r \sigma(e_r, (D_X \hat{\mathfrak{a}})\, e_r) - 2\sum_r \sigma(A_r X, \hat{\mathfrak{a}}(e_r)). \tag{4.9}$$

Using (4.3), (4.4) and substituting (4.7)-(4.9) into (4.6) we then separate tangential and normal components of (4.6) to get, respectively,

$$8c\Big(\sum_r A_r^2\Big)X + 4(n+2)^2 X - 2cp(n+2)X + qX = 0, \qquad \text{i.e.} \tag{4.10}$$

$$\sum_r A_r^2 = \mu I, \qquad \mu = \frac{c}{8}\,[2cp(n+2) - 4(n+2)^2 - q], \quad \text{and} \tag{4.11}$$

$$\sum_r \sigma(e_r, (D_X \hat{\mathfrak{a}})\, e_r) - 2\sum_r \sigma(A_r X, \hat{\mathfrak{a}}(e_r)) + [p - 2c(n+2)] \sum_r \sigma(e_r, A_r X) - 2\sum_{i,r} \sigma((\nabla_X A_r)e_i, A_r e_i) - 2\sum_{i,r} \sigma(h(X, A_r e_i), A_r e_i) = 0. \tag{4.12}$$

Taking trace of (4.11) we get

$$\|h\|^2 = \frac{1}{8}[2n(n+2)p - cnq - 4cn(n+2)^2], \tag{4.13}$$

that is, the squared norm of the second fundamental form is constant. Since M is minimal, from the Gauss equation (2.6) this condition is equivalent to the scalar curvature τ of M being constant. Moreover, from (2.9) and (4.11) we have

$$Q(X) = \frac{1}{8}[4c(n+2)(n+4) - 2p(n+2) + cq]X, \tag{4.14}$$

which shows that M must be an Einstein manifold.

We analyze next the normal-part equation (4.12) by looking at the metric product with various $\sigma(U, V)$ where U, V could be tangent or normal to M. More precisely, denote by Ψ the left hand side of the equation (4.12). Then Ψ is a linear combination of vectors of the form $\sigma(U, V)$. Since in that case $\Psi = 0$ if and only if $A_\Psi = 0$ (cf. [**4**]), we compute $\bar{A}_\Psi Y$ and $\bar{A}_\Psi \xi$ for arbitrary vectors $Y \subset \Gamma(TM)$ and $\xi \in \Gamma(T^\perp M)$. Using (2.10), (4.1), (4.11) and $\|h\|^2 = \sum_r \operatorname{tr} A_r^2$ being constant, the condition $\bar{A}_\Psi Y = 0$ yields

$$J\hat{\mathfrak{a}}(Jh(X,Y)) - \hat{\mathfrak{a}}(h(X,Y)) + [p - 2c(n+2) - 2\mu]\, h(X,Y) = 0. \tag{4.15}$$

By (2.1), (4.1) and (4.2), for a holomorphic submanifold we have $\hat{\mathfrak{a}}(J\xi) = J\hat{\mathfrak{a}}(\xi)$, for $\xi \in \Gamma(TM)$, so that

$$\hat{\mathfrak{a}}(h(X,Y)) = k\, h(X,Y), \qquad k = \frac{p}{2} - c(n+2) - \mu. \tag{4.16}$$

This means that $\hat{\mathfrak{a}}(\xi) - k\xi$ for ξ in the first normal space $\mathcal{N}^1$ of the immersion. In a similar manner, we analyze the equation $\bar{A}_\Psi \xi = 0$ to get

$$(D_X \hat{\mathfrak{a}})\xi - A_{\hat{\mathfrak{a}}(\xi)} X + k A_\xi X = 0, \tag{4.17}$$

for every $\xi \in \Gamma(T^\perp M)$, i.e. $D_X \hat{\mathfrak{a}} = 0$ and $A_{\hat{\mathfrak{a}}(\xi)} = kA_\xi$. Let $\mathcal{H}$ be a subbundle of the normal space of M on which $\hat{\mathfrak{a}}$ is homothetic, i.e. $\hat{\mathfrak{a}}(\xi) = k\xi$, for some constant k as in (4.16). From the parallelism of $\hat{\mathfrak{a}}$ it follows that $\mathcal{H}$ is a parallel subbundle containing the first normal space $\mathcal{N}^1$. Under the assumption that the immersion is full, $\mathcal{H}$ must be the entire normal space, or else there would be reduction of codimension, which would contradict fullness. Thus, $\hat{\mathfrak{a}}$ is constant homothetic on the entire normal space, or equivalently $\mathrm{tr}\,(A_\xi A_\eta) = k\langle \xi, \eta\rangle$. This condition and Einstein property of M give necessary and sufficient conditions for a full holomorphic immersion into a space form $\mathbb{C}Q^m$ to be of 2-type via Φ, since (4.11), (4.12), (4.15), (4.17) would hold, returning back to the 2-type equation (4.5). For holomorphic submanifolds of $\mathbb{C}P^m$ this was first proved, using a different approach, by Ros [**13**], who showed first that such submanifolds are necessarily mass-symmetric, i.e. $\tilde{x}_0 = I/(m+1)$. Subsequently, Udagawa proved that these submanifolds in $\mathbb{C}P^m$ are exactly the Einstein-Kähler parallel submanifolds and classified them as one of the following [**18**], [**19**]: $\mathbb{C}P^n(1/2)$, the complex quadric Q^n, $\mathbb{C}P^n \times \mathbb{C}P^n$, $U(s+2)/U(2)\times U(s)$, for $s \geq 3$, $SO(10)/U(5)$, $E_6/Spin(10)\times T$.

However, the situation in $\mathbb{C}H^m(-4)$ looks markedly different. Having proved above that M must be an Einstein manifold, by the result of Umehara ([**21**]), stating that the only Kähler-Einstein submanifolds of $\mathbb{C}H^m$ are the totally geodesic ones, which are of Chen 1-type [**5**], we have

THEOREM 4.1. *There exists no holomorphic isometric immersion of a Riemannian (even-dimensional) manifold into $\mathbb{C}H^m(-4)$ which is of Chen 2-type via the embedding Φ by projectors.*

5. Totally-real submanifolds of $\mathbb{C}Q^m(4c)$ of Chen 2-type

On the opposite end from holomorphic submanifolds among all CR-submanifolds are totally real (or anti-invariant) submanifolds, for which the tangent space of the submanifold is sent into the normal space by the almost complex structure of $\mathbb{C}Q^m(4c)$. From $\overline{\nabla}_X(JY) = J\overline{\nabla}_X Y$ we get

$$D_X(JY) = J(\nabla_X Y) + K(h(X,Y)), \tag{5.1}$$

$$A_{JY}X = A_{JX}Y = -[Jh(X,Y)]_T = -L(h(X,Y)). \tag{5.2}$$

Thus, the cubic form $\theta(X,Y,Z) := \langle h(X,Y), JZ\rangle$ is totally symmetric in $X, Y, Z \in \Gamma(TM)$. Then we have

$$\sum_{i,j}\langle Jh(e_j,e_i), e_j\rangle Je_i = -\sum_j JA_{Je_j}e_j = nJ(LH). \tag{5.3}$$

By definition, M is mass-symmetric via Φ if the center of mass $\tilde{x}_0$ coincides with the center $I/(m+1)$ of the hyperquadric that $\Phi(\mathbb{C}Q^n)$ lies in. We examine first totally real submanifolds of $\mathbb{C}Q^m(4c)$ which are mass-symmetric and of 2-type. They satisfy the 2-type equation

$$\Delta^2\tilde{x} - p\Delta\tilde{x} + q(\tilde{x} - I/(m+1)) = 0. \tag{5.4}$$

When we separate in this equation the parts tangent to M, the parts normal to M but tangent to $\mathbb{C}Q^m$, and the parts normal to $\mathbb{C}Q^m$ using formulas (2.13), (2.17) and (2.18), observing that $\tilde{x}$ and I are normal to $\mathbb{C}Q^m$. we get respectively,

$$f\nabla f + 2n\mathrm{tr}\,A_{DH} = 0, \quad \text{where} \quad f := n\alpha, \tag{5.5}$$

$$\Delta^\perp H + \hat{\mathfrak{a}}(H) + [c(3n+4) - p]\,H + 3cJ(LH) = 0, \tag{5.6}$$

$$(5.7)\quad \Psi := n^2\sigma(H,H) + [p - 2c(n+1)]\sum_i \sigma(e_i,e_i) - 2n\sum_i \sigma(e_i, A_H e_i) + 4n\sum_i \sigma(e_i, D_{e_i}H) - 2\sum_{i,r}\sigma(A_r e_i, A_r e_i) + 2\sum_r \sigma(e_r, \hat{\mathfrak{a}}(e_r)) + q(\tilde{x} - I/(m+1)) = 0.$$

Taking the inner product of (5.7) with $\tilde{x}$ and using $\langle\sigma(V,W),\tilde{x}\rangle = -\langle V,W\rangle$ and $\langle\tilde{x},\tilde{x}\rangle = \frac{c}{2}\mathrm{tr}\tilde{x} = \frac{c}{2}$, we get $f^2 + cqm/2(m+1) = n[p-2c(n+1)]$ and thus $f = \mathrm{const}$ and from (5.5) also $\mathrm{tr}\, A_{DH} = 0$. The other conditions will come from examining the equations $\bar{A}_\Psi X = 0$ for arbitrary tangent vector $X \in \Gamma(TM)$ and $\bar{A}_\Psi \xi = 0$ for a normal vector $\xi \in \Gamma(T^\perp M)$, which are computed by using (2.10). Note that, by (2.9), for totally real submanifolds we have

$$(5.8)\qquad \sum_r A_r^2 X = c(n-1)X + nA_H X - Q(X).$$

Then, using $\overline{A}_{\tilde{x}} = -I$, $\overline{A}_I = 0$ and separating $\bar{A}_\Psi X$ into parts which are tangent to M and normal to M (but tangent to $\mathbb{C}Q^m(4c)$), we get respectively

$$(5.9)\qquad \begin{aligned}&[2pc(n+1) - 4n(n+3) - 2cf^2 - q]\, X + 2cn^2\langle JH, X\rangle LH\\ &-8cnA_H X + 4c\,Q(X) - 4cL(\hat{\mathfrak{a}}(JX)) = 0\end{aligned}$$

$$(5.10)\qquad 2n^2\langle JH, X\rangle KH + 4nD_X H + 4n\sum_i \langle JD_{e_i}H, X\rangle Je_i - 4K(\hat{\mathfrak{a}}(JX)) = 0.$$

In a similar fashion, when we separate parts of the equation $\bar{A}_\Psi \xi = 0$ which are tangent to M and normal to it we get

$$(5.11)\qquad 2n^2\langle JH,\xi\rangle LH - 4nL(D_{L\xi}H) + 4n\sum_i \langle D_{e_i H}, \xi\rangle e_i - 4L(\hat{\mathfrak{a}}(K\xi)) = 0.$$

$$(5.12)\qquad \begin{aligned}&\{2cn[p - 2c(n+1)] - 2cf^2 - q\}\,\xi + 2cn^2\langle H,\xi\rangle H + 2cn^2\langle KH,\xi\rangle KH\\ &+2(4n - cp)J(L\xi) + 8cnJA_H(L\xi) - 4cnK(D_{L\xi}H) - 4c\,JQ(L\xi)\\ &+4c\,\hat{\mathfrak{a}}(\xi) - 4c\,K(\hat{\mathfrak{a}}(K\xi)) + 4cn\sum_i \langle JD_{e_i}H,\xi\rangle Je_i = 0.\end{aligned}$$

THEOREM 5.1. *Let $x : M^n \to \mathbb{C}Q^m(4c)$ be a totally real isometric immersion of a Riemannian $n-$manifold into a non-flat complex space form of constant holomorphic sectional curvature $4c$, $c = \pm 1$ and let $T^\perp M = \mathcal{V} \oplus \mathcal{W}$ be the orthogonal decomposition of the normal bundle into the totally real and holomorphic subbundles. Then if $\tilde{x} = \Phi \circ x : M^n \to H^{(1)}(m+1)$ is 2-type mass-symmetric immersion satisfying*

$$\Delta^2\tilde{x} - p\Delta\tilde{x} + q(\tilde{x} - I/(m+1)) = 0,$$

we have

(i) *The mean curvature α is constant.*
(ii) *$\mathrm{tr}\, A_{DH} = 0$.*
(iii) *$\Delta^\perp H + \hat{\mathfrak{a}}(H) + [c(3n+4) - p]\, H + 3cJ(LH) = 0$.*
(iv) *$Q(X) - L(\hat{\mathfrak{a}}(JX)) - 2nA_H X - kX + \frac{n^2}{2}\langle LH, X\rangle LH = 0$, for every $X \in \Gamma(TM)$, where $k = \frac{c}{4}[q + 2cf^2 + 4n(n+3) - 2cp(n+1)]$ is constant.*
(v) *$\langle JD_X H, Y\rangle = \langle JD_Y H, X\rangle$, for every $X, Y \in \Gamma(TM)$.*

(vi) $\frac{n^2}{2}\langle LH,X\rangle KH - nK^2(D_XH) - K(\hat{\mathfrak{a}}(JX)) = 0$, *for all* $X \in \Gamma(TM)$.
(vii) $\frac{n^2}{2}\langle KH,\xi\rangle LH + n\sum_i\langle D_{e_i}H,\xi\rangle e_i - L(\hat{\mathfrak{a}}(J\xi)) = 0$, *for* $\xi \in \mathcal{W}$.
(viii)

$$\frac{n^2}{2}\langle H,\xi\rangle H + \frac{n^2}{2}\langle KH,\xi\rangle KH + k'\xi \\ - n\sum_i\langle D_{e_i}H,K\xi\rangle Je_i + \hat{\mathfrak{a}}(\xi) - K(\hat{\mathfrak{a}}(K\xi)) = 0,$$

for any $\xi \in \mathcal{W}$ *where* $k' = \frac{c}{4}[2cnp - 4n(n+1) - 2cf^2 - q]$.

Conversely, if $(i)-(viii)$ *hold, then the immersion is mass-symmetric and of type* ≤ 2, *provided that the polynomial* $t^2 - pt + q = 0$ *has simple real roots or* M *is compact.*

PROOF. These conditions follow from the preceeding computations and the fact that the components in all tangential and normal directions were considered. Parts (v) and (vi) are obtained from (5.10) by separating into $\mathcal{V}$- and $\mathcal{W}$-components. Note that, because of (v), the $\mathcal{V}$-component of (5.11) reduces to zero and the $\mathcal{W}$-component results in (vii). On the other hand, the $\mathcal{V}$-component of (5.12) resolves into (iv) and (vi), so that the $\mathcal{W}$-component results in $(viii)$. □

For a Lagrangian immersion these conditions reduce to

COROLLARY 5.2. *If* $x : M^n \to \mathbb{C}Q^n(4c)$ *is a Lagrangian immersion for which* $\tilde{x}$ *is mass-symmetric and of 2-type, then*

(i) $f := n\alpha = const \neq 0$;
(ii) $tr\, A_{DH} = 0$;
(iii) $\Delta^{\perp}H + \hat{\mathfrak{a}}(H) + [c(3n+1) - p]\, H = 0$;
(iv) $Q(X) - J(\hat{\mathfrak{a}}(JX)) - 2nA_HX - kX + \frac{n^2}{2}\langle JH,X\rangle JH = 0$, *for every* $X \in \Gamma(TM)$;
(v) $\langle JD_XH,Y\rangle = \langle JD_YH,X\rangle$, *for every* $X,Y \in \Gamma(TM)$.

As is known, minimal Lagrangian submanifolds in $\mathbb{C}Q^m(4c)$ are actually of 1-type (they are also mass-symmetric) and the only totally real submanifolds M^n of $\mathbb{C}Q^m$ which are of 1-type are minimal Lagrangian submanifolds in a canonically embedded $\mathbb{C}Q^n \subset \mathbb{C}Q^m$, [**12**] , [**5**]. It would be interesting to find out which totally real minimal submanifolds (that are not Lagrangian in some smaller-dimensional $\mathbb{C}Q^n$) are of 2-type. Towards that goal, we have

COROLLARY 5.3. *Let* M *be a totally real minimal submanifold of* $\mathbb{C}Q^m(4c)$ *which is mass-symmetric and of 2-type via* Φ. *Then*

(i) $tr(A_\xi A_\eta) = 0$ *for any pair of normal vectors* $\xi \in \Gamma(\mathcal{V})$ *and* $\eta \in \Gamma(\mathcal{W})$, *i.e the subspaces* $\mathcal{V}$ *and* $\mathcal{W}$ *are invariant under the action of* $\hat{\mathfrak{a}}$;
(ii) $\hat{\mathfrak{a}}(JX) + JQ(X) = \frac{c}{4}[-2pc(n+1) + 4n(n+3) + 2cf^2 + q]\, JX$, *for every* $X \in \Gamma(TM)$;
(iii) $\hat{\mathfrak{a}}(\xi) - J(\hat{\mathfrak{a}}(J\xi)) = \frac{c}{4}[-2cnp + 4n(n+1) + 2cf^2 + q]\,\xi$, *for every* $\xi \in \Gamma(\mathcal{W})$.

Conversely if $(i)-(iii)$ *hold and* M^n *is compact or the equation* $t^2 - pt + q = 0$ *has simple real roots, then* M^n *is of type* ≤ 2.

Namely, parts $(i)-(iii)$ and (v) of Theorem 5.1 are trivially satisfied, parts (vi) and (vii) of the theorem reduce to (i), part (iv) of the theorem and the invariance of

$\mathcal{V}$ under $\hat{\mathfrak{a}}$ give (*ii*) and part (*iii*) follows from (*viii*) of Theorem 5.1 and invariance of $\mathcal{W}$ under $\hat{\mathfrak{a}}$.

COROLLARY 5.4. *Lagrangian submanifold M^n of $\mathbb{C}Q^n(4c)$ which is mass-symmetric and of 2-type via $\tilde{x}$ has the following properties*

(*i*) $D_{JH}H = 0, \quad \nabla_{JH}(JH) = 0, \quad$ *and* $\quad A_H(JH) = J(h(JH, JH))$.
In particular, the integral curves of JH are geodesics.

(*ii*) *If, in addition, $DH = 0$ then M is an $\mathfrak{a}-$submanifold, i.e. $\mathfrak{a}(H) = 0$.*

(*iii*) *If M is a Maslov submanifold (i.e. JH is proper for shape operator A_H) that has parallel mean curvature vector, then $Q(JH) = qJH$, for a suitable function q, i.e. JH is a principal vector of the Ricci endomorphism Q at each point of M, and, moreover, $\mathfrak{a}(H) = 0$.*

Indeed, (*i*) follows from (*i*), (*v*) of Corollary 5.2 by putting $Y = JH$, (*ii*) follows from part (*iii*) of Corollary 5.2, and (*iii*) follows from parts (*iv*) and (*iii*) of Corollary 5.2.

The classification of 2-type Hopf hypersurfaces in $\mathbb{C}Q^m$ (both those that are mass-symmetric and in general) was done in [**6**]. Another interesting class of hypersurfaces is the class of ruled real hypersurfaces [**9**], [**10**]. They are never Hopf hypersurfaces, as AU has components in both U and $U^{\perp}$ directions where $U := -J\xi$ is the structure vector field (the Reeb vector field). Thus, there is a unit tangent vector $W \perp U$ in the holomorphic subspace so that $AW = \nu U$, $\nu \neq 0$. This is the unit vector of $(AU)|_{\mathcal{D}}$. Then, $AU = \mu U + \nu W$, where $\mu = \langle AU, U\rangle$. Now we prove

THEOREM 5.5. *There exist no real ruled hypersurfaces of $\mathbb{C}Q^m$ which are mass-symmetric and of 2-type in $E^N_{(K)}$.*

PROOF. Assume a ruled hypersurface M^{2m-1} is mass-symmetric and of 2-type, so that

$$\Delta^2\tilde{x} - p\Delta\tilde{x} + q\Big(\tilde{x} - \frac{I}{m+1}\Big) = 0. \tag{5.13}$$

The expressions for $\Delta\tilde{x}$ and $\Delta^2\tilde{x}$ for real hypersurfaces are obtained from (2.13) and (2.17) (cf. [**6**], p. 208). Since $\langle\Delta\tilde{x}, \tilde{x}\rangle = n$ and $\langle\Delta^2\tilde{x}, \tilde{x}\rangle = f^2 + 2c(n^2 + 2n - 1)$, where $f = \mathrm{tr}A = n\alpha$ and $f_2 = \mathrm{tr}A^2$, the $\tilde{x}$-component of (5.13) gives us $f = \mathrm{const}$, hence $\nabla f = 0$ and $\Delta^2\tilde{x}$ reduces to

$$\begin{aligned}\Delta^2\tilde{x} = &-[f(f_2 + c(3n+5)) - 4c\mu]\,\xi + 4cSAU\\ &+ (2c + 2f_2 + f^2)\,\sigma(\xi,\xi) - 2c(n+2)\sum_i \sigma(e_i, e_i)\\ &- 2f\sum_i \sigma(e_i, Ae_i) - 2\sum_{i,r} \sigma(Ae_i, Ae_i).\end{aligned}$$

But $4cSAU = 4c(\mu U + \nu W) = 4cSW$ is the only term in equation (5.13) which is tangent to M, so it must be equal to zero. However, since $\nu \neq 0$ and $SW \neq 0$, that is a contradiction. $\square$

References

[1] A. Bejancu, *Geometry of CR-submanifolds*, Mathematics and its Applications (East European Series), vol. 23, D. Reidel Publishing Co., Dordrecht, 1986. MR861408

[2] B.-Y. Chen, *Total mean curvature and submanifolds of finite type*, Series in Pure Mathematics, vol. 1, World Scientific Publishing Co., Singapore, 1984. MR749575

[3] B.-Y. Chen and M. Petrovic, *On spectral decomposition of immersions of finite type*, Bull. Austral. Math. Soc. **44** (1991), no. 1, 117–129, DOI 10.1017/S0004972700029518. MR1120399

[4] I. Dimitrić, 1*-type submanifolds of the complex projective space*, Kodai Math. J. **14** (1991), no. 2, 281–295, DOI 10.2996/kmj/1138039399. MR1123421

[5] I. Dimitrić, 1*-type submanifolds of non-Euclidean complex space forms*, Bull. Belg. Math. Soc. Simon Stevin **4** (1997), no. 5, 673–684. MR1600245

[6] I. Dimitrić, *Hopf hypersurfaces of low type in non-flat complex space forms*, Kodai Math. J. **34** (2011), no. 2, 202–243, DOI 10.2996/kmj/1309829547. MR2811641

[7] M. Djorić and M. Okumura, *CR submanifolds of complex projective space*, Developments in Mathematics, vol. 19, Springer, New York, 2010. MR2566776

[8] O. J. Garay and A. Romero, *An isometric embedding of the complex hyperbolic space in a pseudo-Euclidean space and its application to the study of real hypersurfaces*, Tsukuba J. Math. **14** (1990), no. 2, 293–313, DOI 10.21099/tkbjm/1496161454. MR1085200

[9] M. Kimura, *Real hypersurfaces and complex submanifolds in complex projective space*, Trans. Amer. Math. Soc. **296** (1986), no. 1, 137–149, DOI 10.2307/2000565. MR837803

[10] Y. Maeda, *On real hypersurfaces of a complex projective space*, J. Math. Soc. Japan **28** (1976), no. 3, 529–540, DOI 10.2969/jmsj/02830529. MR407772

[11] A. Martínez and A. Ros, *On real hypersurfaces of finite type of* $\mathbf{C}P^m$, Kodai Math. J. **7** (1984), no. 3, 304–316, DOI 10.2996/kmj/1138036953. MR760040

[12] A. Ros, *Spectral geometry of CR-minimal submanifolds in the complex projective space*, Kodai Math. J. **6** (1983), no. 1, 88–99. MR698330

[13] A. Ros, *On spectral geometry of Kaehler submanifolds*, J. Math. Soc. Japan **36** (1984), no. 3, 433–448, DOI 10.2969/jmsj/03630433. MR746704

[14] Y. B. Shen, *Spectral geometry of totally real minimal submanifolds* (Chinese), Chinese Ann. Math. Ser. A **12** (1991), no. 6, 745–753. MR1154155

[15] J. Simons, *Minimal varieties in riemannian manifolds*, Ann. of Math. (2) **88** (1968), 62–105, DOI 10.2307/1970556. MR233295

[16] S.-s. Tai, *Minimum imbeddings of compact symmetric spaces of rank one*, J. Differential Geometry **2** (1968), 55–66. MR231395

[17] S. Udagawa, *Spectral geometry of compact Hermitian symmetric submanifolds*, Math. Z. **192** (1986), no. 1, 57–72, DOI 10.1007/BF01162020. MR835391

[18] S. Udagawa, *Spectral geometry of Kaehler submanifolds of a complex projective space*, J. Math. Soc. Japan **38** (1986), no. 3, 453–472, DOI 10.2969/jmsj/03830453. MR845713

[19] S. Udagawa, *Einstein parallel Kaehler submanifolds in a complex projective space*, Tokyo J. Math. **9** (1986), no. 2, 335–340, DOI 10.3836/tjm/1270150722. MR875192

[20] S. Udagawa, *Bi-order real hypersurfaces in a complex projective space*, Kodai Math. J. **10** (1987), no. 2, 182–196, DOI 10.2996/kmj/1138037413. MR897253

[21] M. Umehara, *Einstein Kaehler submanifolds of a complex linear or hyperbolic space*, Tohoku Math. J. (2) **39** (1987), no. 3, 385–389, DOI 10.2748/tmj/1178228285. MR902577

[22] K. Yano and M. Kon, *Structures on manifolds*, Series in Pure Mathematics, vol. 3, World Scientific Publishing Co., Singapore, 1984. MR794310

Pennsylvania State University Fayette, Lemont Furnace, Pennsylvania 15456
Email address: ivko@psu.edu

University of Belgrade, Faculty of Mathematics, Studentski trg 16, pp. 550, 11000 Belgrade, Serbia
Email address: mdjoric@matf.bg.ac.rs

Contemporary Mathematics
Volume **756**, 2020
https://doi.org/10.1090/conm/756/15200

On certain contact CR-submanifolds in $\mathbb{S}^7$

Mirjana Djorić and Marian Ioan Munteanu

Dedicated to Professor B.-Y. Chen on the occasion of his 75th birthday.

ABSTRACT. The odd-dimensional analogue of CR-submanifolds in Kählerian manifolds is the concept of contact CR-submanifolds in Sasakian manifolds. We present here several examples of four-dimensional and five-dimensional contact CR-submanifolds of product and warped product type of seven-dimensional unit sphere, which are nearly totally geodesic, minimal and which satisfy the equality sign in some Chen type inequalities.

1. Introduction

CR-submanifolds of complex manifolds have been an active field of research for years and many authors have described a lot of interesting geometric properties of these submanifolds. We refer the reader to [**1**] for initial studies and also [**11**] for more details and further references, especially when the ambient space is a complex projective space. When the ambient space is not a Kähler manifold, fewer results are known but may be expected, see for example [**2**] and [**12**], where CR-submanifolds of the nearly Kähler six-dimensional sphere have been studied.

The analogue of CR-submanifolds in Kählerian manifolds is the concept of contact CR-submanifolds in Sasakian manifolds. In this paper we consider submanifolds M of the seven-dimensional unit sphere $\mathbb{S}^7$ whose structure vector field ξ is tangent to M and whose tangent bundle of M can be decomposed as $T(M) = H(M) \oplus E(M) \oplus \mathbb{R}\xi$, where $H(M)$ is invariant and $E(M)$ is anti-invariant with respect to φ, i.e. contact CR-submanifolds. We are especially interested in those submanifolds which are as close as possible to totally geodesic, namely in those whose restrictions of the second fundamental form both to $H(M)$ and $E(M)$ vanish identically, while the dimensions of both $H(M)$ and $E(M)$ are different from zero.

We also pay particular attention to the so-called Chen type inequalities, involving the length of the second fundamental form, or linking the squared mean

2010 *Mathematics Subject Classification.* Primary 53B25; Secondary 53C15, 53C25.

Key words and phrases. contact CR-submanifold, nearly totally geodesic submanifold, seven-dimensional unit sphere, warped product, Chen's invariants.

The first author is partially supported by Ministry of Education, Science and Technological Development, Republic of Serbia, project 174012.

The last author is supported by the project funded by the Ministry of Research and Innovation within Program 1 – Development of the national RD system, Subprogram 1.2 – Institutional Performance – RDI excellence funding projects, Contract no. 34PFE/19.10.2018.

curvature with the intrinsic δ-curvature of Chen. At this place we limit our attention to just mentioning one of the most simple such curvatures, called the $\delta(2)$-invariant, which is defined by $\delta(2)(p) = \tau(p) - (\inf K)(p)$, where K is the sectional curvature function on M and τ denotes the scalar curvature of M.

We provide here the detailed construction of several examples of four-dimensional and five-dimensional contact CR-submanifolds of product and warped product type of seven-dimensional unit sphere, which are nearly totally geodesic, minimal and which satisfy the equality sign in some Chen type inequalities.

2. Preliminaries

In this section we review some standard facts on Sasakian manifolds and contact CR-submanifolds, which will be used later.

Sasakian manifolds and Sasakian structure on $\mathbb{S}^{2m+1}(1)$. A differentiable manifold $\widetilde{M}$ is said to have an almost contact structure if it admits a vector field ξ (called *the characteristic vector field* or *the Reeb vector field*), a 1-form η and a (1, 1)-tensor field φ satisfying

$$\eta(\xi) = 1, \quad \varphi^2 = -I + \eta \otimes \xi,$$

where I denotes the identity endomorphism. As a consequence, the dimension of $\widetilde{M}$ is odd and $\eta \circ \varphi = 0$, $\varphi\xi = 0$. A Riemannian metric $\tilde{g}$ on $\widetilde{M}$ is called *compatible* with the almost contact structure if it satisfies $\tilde{g}(\varphi X, \varphi Y) = \tilde{g}(X, Y) - \eta(X)\eta(Y)$, for all vector fields X and Y. The manifold $\widetilde{M}$ together with $(\varphi, \xi, \eta, \tilde{g})$ is called an *almost contact metric manifold.* Then, we immediately conclude: $\eta(X) = \tilde{g}(X, \xi)$, for all $X \in \chi(\widetilde{M})$, φ is skew-symmetric with respect to $\tilde{g}$ and ξ is unitary. The structure $(\varphi, \xi, \eta, \tilde{g})$ is called *contact metric* if the differential of η and the metric $\tilde{g}$ are related by $d\eta(X, Y) = \tilde{g}(\varphi X, Y)$, for all $X, Y \in \chi(\widetilde{M})$, where we considered $d\eta(X, Y) = \frac{1}{2}\big(X(\eta(Y) - Y(\eta(X) - \eta([X, Y])\big)$. If the relation

$$(\widetilde{\nabla}_X \varphi)Y = -\tilde{g}(X, Y)\xi + \eta(Y)X,$$

holds on $\widetilde{M}$ for all $X, Y \in \chi(\widetilde{M})$, where $\widetilde{\nabla}$ denotes the Levi-Civita connection on $\widetilde{M}$, it characterizes Sasakian manifolds among almost contact Riemannian metric manifolds. Note that we follow the convention used in [**13**].

A φ-section is a plane section in $T_p\widetilde{M}$ spanned by $\{X, \varphi X\}$ for some $X \in T_p\widetilde{M}$ orthogonal to ξ. Its sectional curvature is called *φ-sectional curvature* and a *Sasakian space form* is a Sasakian manifold of constant φ-sectional curvature.

Let $\mathbb{S}^{2m+1}(1) = \{\mathbf{p} \in \mathbb{R}^{2m+2} : \langle \mathbf{p}, \mathbf{p} \rangle = 1\}$ be the $(2m+1)$-dimensional unit sphere, where $\langle\ ,\ \rangle$ is the usual scalar product in $\mathbb{R}^{2m+2}$. It is well-known that $\mathbb{S}^{2m+1}(1)$ carries a natural Sasakian structure induced from the canonical complex structure of $\mathbb{R}^{2m+2}$ (see [**17**]). Namely, if $\mathbb{R}^{2m+2}$ is identified with $\mathbb{C}^{m+1}$, let J denote the multiplication with the imaginary unit $i = \sqrt{-1}$, on $\mathbb{R}^{2m+2}$. Since at any point $\mathbf{p} \in \mathbb{S}^{2m+1}(1)$, the outward unit normal to sphere coincides with the position vector p, putting $\xi = J\mathbf{p}$ to be the characteristic vector field, for X tangent to $\mathbb{S}^{2m+1}$, JX fails in general to be tangent and we decompose it into the tangent and the normal part, respectively $JX = \varphi X - \eta(X)\mathbf{p}$. Thus, $\mathbb{S}^{2m+1}(1)$ is equipped with an almost contact structure (φ, ξ, η). Together with the induced metric, this structure is Sasakian (see [**4**]).

Contact CR-submanifolds. Let (M, g) be a submanifold in the Riemannian manifold $(\widetilde{M}, \tilde{g})$, where g is the induced metric and let ∇ and $\widetilde{\nabla}$ be the Levi Civita connections on M and $\widetilde{M}$, respectively. Recall the formulae of Gauss and Weingarten

(G) $\widetilde{\nabla}_X Y = \nabla_X Y + h(X, Y)$,

(W) $\widetilde{\nabla}_X N = -A_N X + \nabla^{\perp}_X N$,

for $X, Y \in T(M)$, $N \in T^{\perp}M$, where $\nabla^{\perp}$ is a connection in the normal bundle of M, h and A_N are the second fundamental form and the shape operator corresponding to N, respectively, related by $\tilde{g}(h(X,Y), N) = g(A_N X, Y)$.

A submanifold M in the Sasakian manifold $(\widetilde{M}, \varphi, \xi, \eta, \tilde{g})$ carrying a φ-invariant distribution $\mathcal{D}$, i.e. $\varphi_p \mathcal{D}_p \subseteq \mathcal{D}_p$, for any $p \in M$, such that the orthogonal complement $\mathcal{D}^{\perp}$ of $\mathcal{D}$ in $T(M)$ is φ-anti-invariant, i.e. $\varphi_p \mathcal{D}^{\perp}_p \subseteq T^{\perp}_p M$, for any $p \in M$, is called a *contact CR-submanifold*. This notion was used by Bejancu and Papaghiuc in [**3**] using the terminology of *semi-invariant submanifold*. It is customary to require that ξ is tangent to M rather than normal, which is too restrictive (by Prop. 1.1 in [**18**], p. 43, M must be φ-anti-invariant, i.e. $\varphi T_p M \subseteq T^{\perp}_p M$, for all $p \in M$), or oblique which leads to highly complicated embedding equations.

For a contact CR-submanifold M of a Sasakian manifold $\widetilde{M}$, with ξ tangent to M, the tangent space at each point decomposes orthogonally as

$$T(M) = H(M) \oplus E(M) \oplus \mathbb{R}\xi,$$

where $\varphi H(M) = H(M)$, $\varphi^2 = -I$ along $H(M)$ and $\varphi E(M) \subseteq T^{\perp}M$. The distribution $H(M)$ is known as the *Levi distribution* or the maximal complex distribution on M. It should be remarked that $H(M)$ is never integrable (see [**5**]), while $H(M) \oplus \mathbb{R}\xi$ can be (as in the case of contact CR-products, see [**15**]). On the other hand, the normal bundle of M can be decomposed as $T^{\perp}M = \varphi E(M) \oplus \nu(M)$, where $\nu(M)$ is the orthogonal complement of $\varphi E(M)$ in $T^{\perp}M$, invariant under the action of φ. We set $2s = \dim\ H(M)$, $p = \dim\ E(M)$ and $2r = \dim\ \nu(M)$. The contact CR-submanifold is called *proper* if both distributions $H(M)$ and $E(M)$ are non-trivial distributions. As $H(M)$ is always even dimensional the lowest possible dimension for a proper contact CR-submanifold M is 4.

When the second fundamental form is identically zero, the submanifold M is called *totally geodesic*. Yet, it is straightforward to show that a proper contact CR-submanifold can never be totally geodesic. A contact CR-submanifold is called *mixed totally geodesic* if $h(H(M), E(M)) = 0$. On the other hand, if $\mathcal{D}$ is a certain distribution on M and if $h(\mathcal{D}, \mathcal{D}) = 0$, then M is called *$\mathcal{D}$-totally geodesic*. We say that a contact CR-submanifold M of a Sasakian manifold is *nearly totally geodesic* if M is simultaneously $H(M)$-totally geodesic and $E(M)$-totally geodesic. In [**9**] and [**10**] the authors obtain a complete classification of four-dimensional and five-dimensional contact nearly totally geodesic CR-submanifolds in $\mathbb{S}^7(1)$.

3. Four and five-dimensional contact CR-submanifolds in $\mathbb{S}^7(1)$

We start this section presenting an example of a four-dimensional nearly totally geodesic contact CR-submanifold in $\mathbb{S}^7(1)$.

Considering the map

$$\tilde{F} : \mathbb{R}^4 \times \mathbb{R}^2 \longrightarrow \mathbb{R}^8$$
$$\tilde{F}(x_1, y_1, x_2, y_2; u, v) = (x_1 u, y_1 u, x_2 u, y_2 u, x_1 v, y_1 v, x_2 v, y_2 v), \tag{3.1}$$

we conclude that $\tilde{F}$ is not an immersion. However, its restriction to $\mathbb{S}^3 \times \mathbb{S}^1$ takes values in $\mathbb{S}^7$ and is an immersion. Namely, denoting it by $F : M = \mathbb{S}^3 \times \mathbb{S}^1 \longrightarrow \mathbb{S}^7$ and using the standard metrics, we check that F is an isometric immersion. Moreover, the characteristic vector field $\xi = (-y_1, x_1, -y_2, x_2) = Jp$ for $p = (x_1, y_1, x_2, y_2)$ of $\mathbb{S}^3$ is mapped to the characteristic vector field of $\mathbb{S}^7$, and hence we continue to use the same letter, for simplicity of notation.

Choosing the following vector fields on $M = \mathbb{S}^3 \times \mathbb{S}^1$

$$\zeta = (-x_2, y_2, x_1, -y_1) \ , \ \mu = (-y_2, -x_2, y_1, x_1) \text{ and } \psi = (-v, u), \tag{3.2}$$

which satisfy

$$F_*\mu = \varphi F_*\zeta \quad \text{and} \quad \varphi F_*\psi \perp F_*(T(M))$$

and setting $H(M) = \text{span}\{\zeta, \mu\}$ and $E(M) = \text{span}\{\psi\}$, we can regard M as a contact CR-submanifold of $\mathbb{S}^7$.

For the second fundamental form h of the immersion F, one can prove that $h(H(M), H(M)) = 0$ and $h(E(M), E(M)) = 0$, but $h(H(M), E(M)) \neq 0$, which implies that M is nearly totally geodesic, but not mixed totally geodesic.

We continue recalling, from [**15**], some Chen's type inequalities for contact CR-submanifolds in Sasakian manifolds of product and warped product type, respectively, concerning the norm of the second fundamental form. Let us emphasize three of them.

Theorem A ([**15**], Theorem 2.5). *Let $\widetilde{M}^{2m+1}(c)$ be a Sasakian manifold of constant φ-sectional curvature c and let $M = N_1 \times N_2$ be a contact CR product, that is a contact CR-submanifold in $\widetilde{M}$, such that N_1 is φ-invariant and tangent to ξ, while N_2 is φ-anti-invariant. Then the norm of the second fundamental form of M satisfies the inequality*

$$||h||^2 \geq p\big((c+3)s + 2\big). \tag{3.3}$$

The equality sign holds if and only if both N_1 and N_2 are totally geodesic in $\widetilde{M}$, meaning that the contact CR-submanifold M is nearly totally geodesic.

Let us remark that $\dim(N_1) = 2s + 1$ and $\dim(N_2) = p$.

Theorem B ([**15**], Theorem 3.3). *Let $M = N_1 \times_f N_2$ be a contact CR warped product of a Sasakian space form $\widetilde{M}^{2m+1}(c)$, that is a contact CR-submanifold in $\widetilde{M}$, such that N_1 is φ-invariant and tangent to ξ, while N_2 is φ-anti-invariant. Then the second fundamental form of M satisfies the following inequality*

$$||h||^2 \geq 2p\left[||\nabla \ln f||^2 - \Delta \ln f + \frac{c+3}{2}\ s + 1\right]. \tag{3.4}$$

Here f is the warping function which has to satisfy $\xi(f) = 0$ and Δf is the Laplacian of f defined by

$$\Delta f = -\text{div}\nabla f = \sum_{j=1}^{k} \left\{(\nabla_{e_j} e_j) f - e_j e_j(f)\right\}, \tag{3.5}$$

where ∇f is the gradient of f and $\{e_1, \ldots, e_k\}$ is an orthonormal frame on M.

Note that **Theorem A** is a special case of **Theorem B** when the warping function f is a constant (in particular 1). The next result is a general inequality satisfied by the norm of the second fundamental form h of a contact CR warped product in Sasakian manifolds (which are not necessary Sasakian space forms).

Theorem C ([**15**], Theorem 3.4). *Let $M = N_1 \times_f N_2$ be a contact CR warped product in a Sasakian manifold $\widetilde{M}$. We have*

(1) *The norm of the second fundamental form of M satisfies*

$$\|h\|^2 \geq 2p\left(\|\nabla \ln f\|^2 + 1\right). \tag{3.6}$$

(2) *If the equality sign in (3.6) holds identically, then N_1 is a totally geodesic submanifold and N_2 is a totally umbilical submanifold of $\widetilde{M}$. The product manifold M is a minimal submanifold in $\widetilde{M}$.*

(3) *The case $TM^\perp = \varphi\mathcal{D}^\perp$. If $p > 1$ then the equality sign holds identically if and only if N_2 is a totally umbilical submanifold of $\widetilde{M}$.*

(4) *If, moreover, $p = 1$ then the equality sign holds identically if and only if the characteristic vector field $\varphi\omega$ of M satisfies $A_\omega\varphi\omega = -\varphi\nabla \ln f - \xi$.* (Notice that in this case, M is a hypersurface in $\widetilde{M}$ with the unitary normal vector field denoted by ω).

Remark that, either in **Theorem B**, or in **Theorem C**, the equality sign implies that the contact CR-submanifold is nearly totally geodesic.

We end this part recalling also the Theorem 2.8 in [**15**], which is a (kind of) converse of **Theorem A**.

Theorem D. *Let $M = N_1 \times N_2$ be a strictly proper contact CR product in S^7 whose second fundamental form has the norm $\sqrt{6}$. Then M is, locally, the Riemannian product between S^3 and S^1. Moreover, up to a rigid transformation of $\mathbf{R}^8$, the embedding of $\mathbb{S}^3 \times \mathbb{S}^1$ in $\mathbb{S}^7$ is given by*

$$\begin{gathered} F : S^3 \times S^1 \longrightarrow S^7 \\ F(x_1, y_1, x_2, y_2; u, v) = (x_1u, y_1u, x_2u, y_2u, x_1v, y_1v, x_2v, y_2v). \end{gathered} \tag{3.7}$$

Let us observe that the immersion given by (3.7) is exactly the immersion obtained as the restriction to $\mathbb{S}^3 \times \mathbb{S}^1$ of the map given by (3.1). Also, remark that the version presented in this note is a slightly modification of the original Theorem 2.8 in [**15**] and this is caused by changing the initial conditions satisfied by the immersion F. Moreover, the $\sqrt{6}$ is the minimum value for $\|h\|$ obtained in **Theorem A** with $c = 1$, $s = 1$ and $p = 1$.

The following result expresses several properties of the example, given by the immersion (3.7), of the four-dimensional contact CR-submanifold.

THEOREM 3.1. *Let M be the contact CR-submanifold in $\mathbb{S}^7$ (of product type) defined by the isometric immersion F given by (3.7). Then,*

(i) *M is nearly totally geodesic;*
(ii) *M is a minimal submanifold in $\mathbb{S}^7$;*
(iii) *M satisfies the equality in the Chen type inequality (3.3) in* **Theorem A**.

PROOF. Let us give some details for (iii). To do this, we need to compute the square of the length of the second fundamental form, that is

$$\|h\|^2 = \sum_{a,b=1}^{4} \|h(E_a, E_b)\|^2,$$

where $\{E_a\}_{a=1,\dots,4}$ is an orthonormal basis on M. More precisely, we can write

$$\|h\|^2 = \|h(\mathcal{D}, \mathcal{D})\|^2 + 2\|h(\mathcal{D}, E(M))\|^2 + \|h(E(M), E(M))\|^2,$$

where $\mathcal{D} = H(M) \oplus [\xi]$. Due to (i) and using some properties of the Reeb vector field ξ, we obtain

$$||h||^2 = 2\big(||h(\xi,\psi)||^2 + ||h(\zeta,\psi)||^2 + ||h(\mu,\psi)||^2\big).$$

It follows that $||h|| = \sqrt{6}$. □

We continue with an example of five-dimensional contact CR-submanifold in $\mathbb{S}^7(1)$ considering the immersion

$$\begin{aligned} &F : M = \mathbb{S}^3 \times \mathbb{S}^2 \longrightarrow \mathbb{S}^7 \\ &F(x_1, y_1, x_2, y_2; u, v, w) = (x_1u, y_1u, x_1v, y_1v, x_1w, y_1w, x_2, y_2). \end{aligned} \tag{3.8}$$

In order to have an isometric immersion we need to consider the following warped metric on M

$$g_M = g_{\mathbb{S}^3} + f^2 g_{\mathbb{S}^2}, \text{ where } f : D \subset \mathbb{S}^3 \to \mathbb{R}, \quad f(x_1, y_1, x_2, y_2) = \sqrt{x_1^2 + y_1^2}. \tag{3.9}$$

We need to remove the circle $x_2^2 + y_2^2 = 1$ in order to have a non vanishing warping function on the domain. So, let D be the open set of $\mathbb{S}^3$ on which f is not zero.

On the 3-sphere we consider the three vector fields ξ, ζ and μ, defined by (3.2), which give the parallelizability of $\mathbb{S}^3$. It can be proved that if U is a vector field tangent to $\mathbb{S}^2$, then F_*U is mapped by φ to the normal bundle. Moreover, after computing $\varphi F_*\zeta = F_*\mu$ and $\varphi F_*\mu = -F_*\zeta$, we conclude that $\text{span}\{\zeta, \mu\}$ is φ-invariant. Thus, since M is tangent to the Reeb vector field ξ of $\mathbb{S}^7$, we can regard M as a contact CR-submanifold of $\mathbb{S}^7$ of warped product type by setting $H(M) = \text{span}\{\zeta, \mu\}$ and $E(M) = T\mathbb{S}^2$.

For later use we first compute the gradient of $\log(f)$ and its norm, respectively:

$$\nabla \log(f) = \frac{1}{f^2}\Big((1-f^2)x_1, (1-f^2)y_1, -f^2x_2, -f^2y_2\Big),$$

$$||\nabla \log(f)||^2 = \frac{1-f^2}{f^2}.$$

Now, using (3.5), we compute the Laplacian of $\log(f)$

$$\Delta \log(f) = 2. \tag{3.10}$$

The following theorem yields the information about the example whose immersion and warped metric are given by (3.8) and (3.9), respectively.

THEOREM 3.2. *Let $M = \mathbb{S}^3 \times \mathbb{S}^2$ be the contact CR-submanifold (of warped product type) in $\mathbb{S}^7$ defined by the isometric immersion F given by* (3.8). *Then*

(i) *M is nearly totally geodesic;*
(ii) *M is minimal;*
(iii) *M satisfies the equality in the Chen type inequality* (3.4) *in* **Theorem B***;*
(iv) *M satisfies the equality in the Chen type inequality* (3.6) *in* **Theorem C**.

PROOF. Let us sketch the proof for (i) and (ii). It is easy to prove that $h|_{H(M)\times H(M)} = 0$. In order to show that also $h|_{E(M)\times E(M)} = 0$ we recall the following formula. If U and V are vector fields on $\mathbb{S}^2$ and ∇ denotes the Levi-Civita connection on M, then

$$\nabla_U V = \nabla_U^{\mathbb{S}^2} V - g(U,V)\nabla \log(f),$$

where $\nabla^{\mathbb{S}^2}$ stands for the usual connection on the 2-sphere. For more details on the geometry of warped products we suggest for example [**16**, page 204]. A detailed study on the geometry of warped product manifolds and of submanifolds of warped product type is given in Chen's recent book **[8]**. Thus, we find that M is simultaneously $H(M)$-totally geodesic and $E(M)$-totally geodesic, namely, M is nearly totally geodesic. The minimality is immediately obtained.

To prove (iii) and (iv) we need to compute the square of the length of the second fundamental form. As in the proof of Theorem 3.1, using (i), we obtain

$$||h||^2 = 2\sum_{i=1,2}\left\{||h(\xi,U_i)||^2 + ||h(\zeta,U_i)||^2 + ||h(\mu,U_i)||^2\right\},$$

where $\{U_1, U_2\}$ is a system of tangent vectors to $\mathbb{S}^2$, orthonormal with respect to g_M.

Straightforward computations lead to

$$||h(\xi,U)||^2 = (x_1^2+y_1^2)||U||^2, \quad ||h(\zeta,U)||^2 = \frac{(x_1y_2+x_2y_1)^2}{f^2}||U||^2,$$
$$||h(\mu,U)||^2 = \frac{(-x_1x_2+y_1y_2)^2}{f^2}||U||^2,$$

where U is tangent to $\mathbb{S}^2$ and $||U||$ denotes the norm of U on $\mathbb{S}^2$. In order to compute $||h||^2$, we consider the orthonormal basis on $\mathbb{S}^2$, with respect to the metric g_M, namely $f^2||U||^2 = 1$. Consequently,

$$||h(\xi,U)||^2 + ||h(\zeta,U)||^2 + ||h(\mu,U)||^2 = \frac{1}{f^2},$$

and we compute that $||h||^2 = \frac{4}{f^2}$. Since $2s = \dim\ H(M) = 2$, $p = \dim\ E(M) = 2$, using (3) and (3.10), we have actually proved the statements (iii) and (iv). $\square$

Finally, let us consider the immersion

$$\begin{gathered} F : M = \mathbb{S}^3 \times \mathbb{S}^1 \longrightarrow \mathbb{S}^7 \\ F(x_1,y_1,x_2,y_2;u,v) = (x_1u, y_1u, x_1u, y_1v, x_2, y_2, 0, 0), \end{gathered} \tag{3.11}$$

with the warped metric as in (3.9). Namely, considering on M the warped metric

$$g_M = g_{\mathbb{S}^3} + f^2 g_{\mathbb{S}^1}, \text{ where } f : D \subset \mathbb{S}^3 \to \mathbb{R}, \quad f(x_1,y_1,x_2,y_2) = \sqrt{x_1^2+y_1^2}, \tag{3.12}$$

F is an isometric immersion. Here D is, as previously, an open subset of the 3-sphere where f does not vanish.

Since M is tangent to the Reeb vector field ξ of $\mathbb{S}^7$, defining the four vector fields ξ, ζ, μ and ψ as in (3.2), one can consider M as a contact CR-submanifold of $\mathbb{S}^7$ by setting $H(M) = \text{span}\{\zeta,\mu\}$ and $E(M) = \text{span}\{\psi\}$.

The following statement presents the features of the example given by the immersion (3.11) and the warping function defined by (3.12).

THEOREM 3.3. *Let $M = \mathbb{S}^3 \times \mathbb{S}^1$ be the contact CR-submanifold (of warped product type) in $\mathbb{S}^7$ defined by the isometric immersion F given by* (3.11). *Then*

(i) *M is nearly totally geodesic;*
(ii) *M is minimal;*
(iii) *M satisfies the equality in the two Chen type inequalities* (3.4) *in* **Theorem B** *and* (3.6) *in* **Theorem C**.

PROOF. Since the computations are similar to those presented in the proof of Theorem 3.2, the proof may be handled in much the same way and we will not be presenting it in much detail here.

However, let us remark that the length of the second fundamental form is given by

$$||h||^2 = \frac{2}{f^2},$$

since

$$||h(\xi,\psi)||^2 = f^2,\ ||h(\zeta,\psi)||^2 = \frac{(x_2y_1 + x_1y_2)^2}{f^2},\ ||h(\mu,\psi)||^2 = \frac{(x_1x_2 - y_1y_2)^2}{f^2}.$$

The two equalities follow immediately. □

In [**6**], Chen introduced a new invariant, nowadays called $\delta(2)$, for a Riemannian manifold M. More precisely, this invariant is given by:

$$\delta(2)(p) = \tau(p) - (\inf K)(p), \tag{3.13}$$

where $(\inf K)(p) = \inf\{K(\pi)\,|\,\pi \text{ is a 2-dimensional subspace of } T_pM\}$. Here $K(\pi)$ is the sectional curvature of π and $\tau(p) = \sum_{i<j} K(e_i \wedge e_j)$ denotes the scalar curvature defined in terms of an orthonormal basis $\{e_1, \ldots, e_n\}$ of the tangent space T_pM of M at p. Later, in [**7**], Chen introduced many other curvature invariants.

One of the aims of introducing these invariants is to use them to obtain a lower bound for the length of the mean curvature vector for an immersion in a real space form $\widetilde{M}(c)$. A submanifold is called an *ideal submanifold*, for that curvature invariant, if it realizes equality at every point.

For a submanifold M^n in a Riemannian manifold $\widetilde{M}(c)$ of constant sectional curvature c, the following basic inequality involving the intrinsic invariant $\delta(2)$ and the length of the mean curvature vector $H = \frac{1}{n}\mathrm{trace}\, h$ was first established in [**6**]:

$$\delta(2) \leq \frac{n^2(n-2)}{2(n-1)}||H||^2 + \frac{1}{2}(n-2)(n+1)c. \tag{3.14}$$

Our next goal is to determine the invariant $\delta(2)$ on M. Firstly, we have $K(\xi,\zeta) = K(\xi,\mu) = K(\zeta,\mu) = 1$. Secondly, we can calculate

$$K(\xi,\psi) = 0, \quad K(\zeta,\psi) = 1 - \frac{(x_1y_2 + x_2y_1)^2}{f^4}, \quad K(\mu,\psi) = 1 - \frac{(-x_1x_2 + y_1y_2)^2}{f^4}$$

and therefore evaluate the scalar curvature of M:

$$\tau(M) = 5 - \frac{1 - f^2}{f^2}. \tag{3.15}$$

At this point we want to determine the infimum for all sectional curvatures of the 2-planes in M. If $A = (A_1, A_2, A_3, A_4)$ is an arbitrary unitary vector field tangent to $\mathbb{S}^3 \subset \mathbb{R}^4$ then one can directly compute

$$K(A,\psi) = 1 - \frac{(-y_1A_1 + x_1A_2)^2}{f^4}. \tag{3.16}$$

Note that, if one takes $A = f\xi + \frac{x_1y_2+x_2y_1}{f}\zeta + \frac{-x_1x_2+y_1y_2}{f}\mu$ one obtains

$$K(A,\psi) = \frac{f^2 - 1}{f^2}. \tag{3.17}$$

The most general situation is when π is spanned by the vectors $A + a\psi$ and $B + b\psi$, where a, b are smooth functions on M and A and B are tangent to $\mathbb{S}^3$. If $a \neq 0$, it follows that $B - \frac{b}{a}A$ lies in π, too. Consequently, we can always find in π a vector field tangent to $\mathbb{S}^3$. Therefore, we now need to compute the sectional curvature of the 2-plane spanned by A and $B+b\psi$. Since

$$K(A,B+b\psi)=\frac{g(R(A,B+b\psi)(B+b\psi),A)}{g(A,A)g(B+b\psi,B+b\psi)-g(A,B+b\psi)^2},$$

doing all the computations and taking into account that the 3-sphere is of constant sectional curvature 1, together with the formula (3.16), we get

$$K(A,B+b\psi)=1-\frac{1}{f^4}\frac{b^2f^2|A|^2(-y_1A_1+x_1A_2)^2}{|A|^2|B|^2+b^2f^2|A|^2-\langle A,B\rangle^2}.$$

It is easily inferred that

$$\frac{b^2f^2|A|^2(-y_1A_1+x_2A_2)^2}{|A|^2|B|^2+b^2f^2|A|^2-\langle A,B\rangle^2}\leq(-y_1A_1+x_1A_2)^2,$$

which implies

$$K(A,B+b\psi)\geq K(A,\psi),\ \text{for any } b\in C^\infty(M) \text{ and } B\in T\mathbb{S}^3. \tag{3.18}$$

Finally, since $(-y_1A_1+x_1A_2)^2\leq(x_1^2+y_1^2)(A_1^2+A_2^2)\leq f^2$, we get

$$K(A,\psi)\geq 1-\frac{1}{f^2}. \tag{3.19}$$

Combining (3.17), (3.18), (3.19), we deduce that

$$\inf K=1-\frac{1}{f^2}, \tag{3.20}$$

which is attained for $A=f\xi+\frac{x_1y_2+x_2y_1}{f}\zeta+\frac{-x_1x_2+y_1y_2}{f}\mu=\frac{1}{f}(-y_1,x_1,0,0)$.

Plugging (3.15) and (3.20) into (3.13) we obtain

$$\delta(2)(M)=5.$$

As a consequence, we can state the following.

THEOREM 3.4. *Let $M=\mathbb{S}^3\times_f\mathbb{S}^1$ be the warped product contact CR-submanifold in $\mathbb{S}^7$ defined by the isometric immersion F, given by* (3.11). *Then M is a $\delta(2)$-ideal in $\mathbb{S}^7$.*

PROOF. The inequality (3.14) is verified with the equality sign because M is minimal, $n=4$ and $c=1$. □

Note that the image of M through F lies on the 5-dimensional sphere $\mathbb{S}^5$ and therefore, we end this Section recalling the following result obtained in [**14**]:

Theorem E. *Let M^n be a proper, minimal, contact CR-submanifold of $\mathbb{S}^{2m+1}$ which is a $\delta(2)$-ideal. Then n is even and there exists a totally geodesic Sasakian $\mathbb{S}^{n+1}$ in $\mathbb{S}^{2m+1}$ containing M as a hypersurface.*

References

[1] A. Bejancu, *Geometry of CR-submanifolds*, Mathematics and its Applications (East European Series), vol. 23, D. Reidel Publishing Co., Dordrecht, 1986. MR861408

[2] M. Antić, M. Djorić, and L. Vrancken, *4-dimensional minimal CR submanifolds of the sphere S^6 satisfying Chen's equality*, Differential Geom. Appl. **25** (2007), no. 3, 290–298, DOI 10.1016/j.difgeo.2006.11.007. MR2330458

[3] A. Bejancu and N. Papaghiuc, *Semi-invariant submanifolds of a Sasakian manifold*, An. Ştiinţ. Univ. "Al. I. Cuza" Iaşi Secţ. I a Mat. (N.S.) **27** (1981), no. 1, 163–170. MR618723

[4] D. E. Blair, *Riemannian geometry of contact and symplectic manifolds*, Progress in Mathematics, vol. 203, Birkhäuser Boston, Inc., Boston, MA, 2002. MR1874240

[5] M. Capursi and S. Dragomir, *Contact Cauchy-Riemann submanifolds of odd-dimensional spheres* (English, with Serbo-Croatian summary), Glas. Mat. Ser. III **25(45)** (1990), no. 1, 167–172. MR1108961

[6] B.-Y. Chen, *Some pinching and classification theorems for minimal submanifolds*, Arch. Math. (Basel) **60** (1993), no. 6, 568–578, DOI 10.1007/BF01236084. MR1216703

[7] B.-Y. Chen, *Pseudo-Riemannian geometry, δ-invariants and applications*, World Scientific Publishing Co. Pte. Ltd., Hackensack, NJ, 2011. With a foreword by Leopold Verstraelen. MR2799371

[8] B.-Y. Chen, *Differential geometry of warped product manifolds and submanifolds*, World Scientific Publishing Co. Pte. Ltd., Hackensack, NJ, 2017. With a foreword by Leopold Verstraelen. MR3699316

[9] M. Djorić, M.I. Munteanu, *Five-dimensional contact CR-submanifolds in* $S^7(1)$, to appear in Mathematics, Special Issue "Riemannian Geometry of Submanifolds" (2020), **8**, 1278; DOI:10.3390/math8081278.

[10] M. Djorić, M. I. Munteanu, and L. Vrancken, *Four-dimensional contact CR-submanifolds in* $\mathbb{S}^7(1)$, Math. Nachr. **290** (2017), no. 16, 2585–2596, DOI 10.1002/mana.201600437. MR3722495

[11] M. Djorić and M. Okumura, *CR submanifolds of complex projective space*, Developments in Mathematics, vol. 19, Springer, New York, 2010. MR2566776

[12] M. Djorić and L. Vrancken, *Three-dimensional minimal CR submanifolds in* S^6 *satisfying Chen's equality*, J. Geom. Phys. **56** (2006), no. 11, 2279–2288, DOI 10.1016/j.geomphys.2005.12.002. MR2248875

[13] M. Harada, *On Sasakian submanifolds*, Tohoku Math. J. (2) **25** (1973), 103–109, DOI 10.2748/tmj/1178241369. MR324590

[14] M. I. Munteanu and L. Vrancken, *Minimal contact CR submanifolds in* $\mathbb{S}^{2n+1}$ *satisfying the* $\delta(2)$*-Chen equality*, J. Geom. Phys. **75** (2014), 92–97, DOI 10.1016/j.geomphys.2013.09.003. MR3126936

[15] M.-I. Munteanu, *Warped product contact CR-submanifolds of Sasakian space forms*, Publ. Math. Debrecen **66** (2005), no. 1-2, 75–120. MR2128686

[16] B. O'Neill, *Semi-Riemannian geometry with applications to relativity*, Pure and Applied Mathematics, vol. 103, Academic Press, Inc. [Harcourt Brace Jovanovich, Publishers], New York, 1983. MR719023

[17] Y. Tashiro, *On contact structure of hypersurfaces in complex manifolds. I*, Tohoku Math. J. (2) **15** (1963), 62–78, DOI 10.2748/tmj/1178243870. MR146760

[18] K. Yano and M. Kon, *CR submanifolds of Kaehlerian and Sasakian manifolds*, Progress in Mathematics, vol. 30, Birkhäuser, Boston, Mass., 1983. MR688816

University of Belgrade, Faculty of Mathematics, Studentski trg 16, pp. 550, 11000, Belgrade, Serbia

Email address: mdjoric@matf.bg.ac.rs

Alexandru Ioan Cuza University of Iasi, Faculty of Mathematics, Bd. Carol I, no. 11, 700506 Iasi, Romania

Email address: marian.ioan.munteanu@gmail.com

Contemporary Mathematics
Volume **756**, 2020
https://doi.org/10.1090/conm/756/15201

A class of strictly convex hypersurfaces satisfying Weingarten-type inequalities, I

Leonard M. Giugiuc and Bogdan D. Suceavă

Dedicated to Dr. Bang-Yen Chen on the occasion of his 75th birthday.

ABSTRACT. Linear Weingarten surfaces in three-dimensional ambient space satisfy a relation between mean curvature and Gaussian curvature: $aH^2+bK = c$. We investigate whether for coreciprocal points on smooth strictly convex hypersurfaces of dimensions 3 and 4 there are any curvature inequalities inspired by the classical Weingarten condition. We also consider the globalization of these pointwise inequalities. This question is suggested by the investigations of Bang-Yen Chen's fundamental inequalities, as we reflect on the geometric interpretations of these relations. The results can be regarded as a response to a question posed by B.-Y. Chen in 1975: Let (M,g) be a closed Riemannian manifold. What are the relationships between the total mean curvature of an isometric immersion $\Phi : M \to \mathbb{R}^m$ and the Riemannian structure of (M,g)?

1. Introduction and geometric motivation

In the theory of surfaces in the three dimensional real space, the two important curvature invariants are the Gaussian curvature and the mean curvature. Starting with the second half of the 19th century, various classes of relations involving K and H have been extensively investigated. The history of the original idea of the study of the relationships involving K and H^2 can be traced back to Weingarten's original papers [**18, 19**], as some authors suggest, e.g. [**15**], with later important developments in [**11**], and S.-S. Chern's classical paper [**9**].

It is natural to inquire whether there are further relationships involving K and H^2 in various contexts, and there are works dedicated fully to the study of the relationship between H and K, e.g. [**17**]. There are many substantive recent works focused on the study of linear Weingarten surfaces, i.e. of the surfaces in $\mathbb{R}^3$ satisfying the relation $aH^2 + bK = c$, where a, b, c are real constants; see e.g. [**14**]. Of interest is whether some of these relations admit a global expression, and this is the question that motivates the present study.

Our perspective is influenced by the consequences of B.-Y. Chen's curvature invariants [**7**] and their topological consequences. As a matter of fact, the investigation we pursue here is entirely inspired by B.-Y. Chen's fundamental inequalities

2010 *Mathematics Subject Classification.* Primary 53B20, 53B25 ; Secondary 26D15, 51M16, 53A07, 53A55, 53C20, 53C21.

Key words and phrases. principal curvatures, Gauss-Kronecker curvature, total mean curvature, shape operator, inequalities with real numbers.

e.g. [**3–6**], investigations that were in their turn inspired by a question of S.-S. Chern in [**10**] about the intrinsic control on minimality of submanifolds.

To better state our idea, we describe first the context. Consider (M^n, g) a Riemannian smooth manifold of dimension n. Let p be an arbitrary point on M. Denote by *sec* the sectional curvature corresponding to the metric g. For an orthonormal basis $e_1, ..., e_n$ of the tangent space T_pM of a Riemannian n-manifold M, the scalar curvature at p is:

$$scal(p) = \sum_{i<j} sec(e_i \wedge e_j)$$

For any r-dimensional subspace of T_pM denoted by L, with orthonormal basis $e_1, ..., e_r$, one defines

$$scal(L) = \sum_{1 \le i < j \le r} sec(e_i \wedge e_j).$$

B.-Y. Chen considered the finite set $S(n)$ of k-tuples $(n_1, ..., n_k)$ with $k \ge 0$ which satisfy the conditions: $n_1 < n$, $n_i \ge 2$ $(i = 1, ..., k)$, and $n_1 + ... + n_k \le n$. For each $(n_1, ..., n_k) \in S(n)$ B.-Y. Chen introduced the following Riemannian invariants:

$$\delta(n_1, ..., n_k)(p) = scal(p) - \inf\{scal(L_1) + ... + scal(L_k)\}(p),$$

where infimum is taken for all possible choices of mutually orthogonal subspaces $L_1, ..., L_k$, satisfying $n_j = \dim L_j$, $(j = 1, ..., k)$.

These invariants are now known as Chen invariants. When $k = 0$, the Chen invariant is nothing but the scalar curvature.

By denoting

$$c(n_1, ..., n_k) = \frac{n^2(n + k - 1 - \sum n_j)}{2(n + k - \sum n_j)},$$

$$b(n_1, ..., n_k) = \frac{1}{2}\{(n(n-1) - \sum_{j=1}^{k} n_j(n_j - 1)\},$$

Chen's fundamental inequalities are stated as follows (see [**7**] Thm 13.3/page 262).

THEOREM 1.1 (Bang-Yen Chen, 2005). *For any n-dimensional submanifold M of an arbitrary Riemannian space $\bar{M}^{n+m}$ and for any k-tuple $(n_1, ..., n_k) \in S(n)$, we have:*

$$\delta(n_1, ..., n_k) \le c(n_1, ..., n_k)|H|^2 + b(n_1, ..., n_k) \max \bar{sec}(p),$$

where max $\bar{sec}(p)$ *denotes the maximum of the sectional curvature function of $\bar{M}^{n+m}$ restricted to 2-plane sections of T_pM.*

The equality case is fully described in the indicated source. In consequence, this is the perspective from which we regard Weingarten-type relations, combining them with B.-Y. Chen's fundamental inequalities stated above, originally introduced in e.g. [**3–6**].

We consider the following problems.

QUESTION 1. Are there any other pointwise curvature inequalities involving mean curvature and Gaussian curvature (or, for hypersurfaces, Gauss-Kronecker curvature)? Namely, we are looking for geometric inequalities at some point p on the hypersurface so that, for a well-defined class of geometric objects, an inequality of the following type holds:

$$aH^\alpha + bK^\beta \ge c,$$

for constants $a, b, c, \alpha, \beta \in \mathbb{R}$. We call such a relation a Weingarten-type inequality.

In 1975, B.-Y.Chen posed the following problem.

PROBLEM. Let (M, g) be a closed Riemannian manifold. What are the relationships between the total mean curvature of an isometric immersion $\phi : M \to \mathbb{R}^m$ and the Riemannian structure of (M, g)?

By applying finite type theory, in **[2]** B.-Y. Chen obtained the following result (see **[8]**, p. 306 et al.).

THEOREM 1.2. *Let $\phi : M \to \mathbb{R}^m$ be an isometric immersion of a closed Riemannian $n-$manifold M into the Euclidean space of dimension m. Then, for $k \geq 2$, we have*

$$\int_M H^k dV \geq \left(\frac{\lambda_p}{n}\right)^{k/2} vol(M),$$

where p is the lower order of M in the Euclidean space $\mathbb{R}^m$. The equality sign in this inequality holds for some real number $k \geq 2$ if and only if M is of 1-type.

This theorem generalizes a result of Reilly, who obtained in **[16]** that

$$\int_M H^2 dV \geq \frac{\lambda_1}{n} vol(M),$$

with the equality holding if and only if M is a minimal submanifold of a hypersphere.

For further geometric inequalities involving total mean curvature and volume of a closed Riemannian manifold, see **[8]**, Chapter 9.

Assuming we find a positive answer to the first question, the discussion leads to the following.

QUESTION 2. Are there any classes of compact hypersurfaces on which we can derive global consequences from the answers to the first Question?

The Corollaries 3.3, 4.3 we obtain below respond to the aforementioned problem posed by B. Y. Chen in **[1]**, in the view of Question 2.

In all these situations we are much interested to insure the fact the inequalities we provide are optimal, i.e. the constants we derive are the best in the given context.

2. Notations in the geometry of hypersurfaces

To establish our notations, let $\sigma : U \subset \mathbb{R}^n \to \mathbb{R}^{n+1}$ be a hypersurface given by the smooth map σ. Let p be a point on the hypersurface. Denote $\sigma_k(p) = \frac{\partial \sigma}{\partial x_k}$, for all k from 1 to n. Consider $\{\sigma_1(p), \sigma_2(p), ..., \sigma_n(p), N(p)\}$, the Gauss frame of the hypersurface, where N denotes the normal vector field. We denote by $g_{ij}(p)$ the coefficients of the first fundamental form and by $h_{ij}(p)$ the coefficients of the second fundamental form. Then we have

$$g_{ij}(p) = \langle \sigma_i(p), \sigma_j(p) \rangle, \qquad h_{ij}(p) = \langle N(p), \sigma_{ij}(p) \rangle.$$

The Weingarten map $L_p = -dN_p \circ d\sigma_p^{-1} : T_{\sigma(p)}\sigma \to T_{\sigma(p)}\sigma$ is linear. Denote by $(h^i_j(p))_{1 \leq i,j \leq n}$ the matrix associated to Weingarten map, that is:

$$L_p(\sigma_i(p)) = h^k_i(p)\sigma_k(p),$$

where the repeated index and upper script above indicates Einstein's summation convention. Weingarten's operator is self-adjoint, which implies that the roots of the algebraic equation

$$\det(h_j^i(p) - \lambda(p)\delta_j^i) = 0$$

are real. The eigenvalues of Weingarten's linear map are called principal curvatures of the hypersurface. They are the roots $k_1(p), k_2(p), ..., k_n(p)$ of this algebraic equation. The mean curvature at the point p is

$$H(p) = \frac{1}{n}[k_1(p) + ... + k_n(p)],$$

and the Gauss-Kronecker curvature is

$$K(p) = k_1(p)k_2(p)...k_n(p).$$

DEFINITION 2.1. A hypersurface $\sigma : U \subset \mathbb{R}^n \to \mathbb{R}^{n+1}$ is *convex* if at every point p all the principal curvatures are non-negative: $k_1(p) \geq 0$, $k_2(p) \geq 0$, ..., $k_n(p) \geq 0$. If the principal curvatures are strictly positive, we say the hypersurface is strictly convex.

DEFINITION 2.2. A point on a smooth strictly convex hypersurface $\sigma : U \subset \mathbb{R}^n \to \mathbb{R}^{n+1}$ is called *coreciprocal* if the principal curvatures satisfy the condition

$$k_1(p) + ... + k_n(p) = \frac{1}{k_1(p)} + \frac{1}{k_2(p)} + ... + \frac{1}{k_n(p)}.$$

A smooth strictly convex hypersurface $\sigma : U \subset \mathbb{R}^n \to \mathbb{R}^{n+1}$ is called *coreciprocal* if every point p is coreciprocal.

We focus the present investigation on pointwise relations taking place at coreciprocal points on smooth strictly convex hypersurfaces.

3. The case of three-dimensional hypersurfaces

We start by inquiring what is the best real constant $k > 0$ such that for any three real numbers $a, b, c > 0$ subject to the constraint $a + b + c = \frac{1}{a} + \frac{1}{b} + \frac{1}{c}$ the following inequality holds true:

$$(a + b + c)^2 + ka^2b^2c^2 \geq 9 + k. \tag{3.1}$$

We start by homogenizing the inequality. Since

$$abc = \frac{ab + bc + ca}{a + b + c},$$

and

$$1 = \frac{abc(a + b + c)}{ab + bc + ca},$$

we homogenize and obtain

$$(a + b + c)^2 + k\left(\frac{ab + bc + ca}{a + b + c}\right)^2 \geq (9 + k)\left(\frac{abc(a + b + c)}{ab + bc + ca}\right). \tag{3.2}$$

Hence, the statement of our investigation is: Determine k_{max} for which the inequality (3.2) holds true. Since this last inequality is homogeneous, without any loss of generality we may assume that $a + b + c = 3$. Since $3(ab + bc + ca) \leq (a + b + c)^2$, the function $f : [0, 1) \to (0, 3]$, $f(t) = 3(1 - t^2)$ is strictly decreasing and continuous, therefore by the Intermediate Value Theorem there exists $t \in [0, 1)$ such that $ab + bc + ca = 3(1 - t^2)$.

By an argument from [20], the maximum of abc is reached at $(a, b, c) = (1 - t, 1 - t, 1 + 2t)$ (or circular permutations of these values) and this maximum is $(1 - t)^2(1 + 2t)$.

Thus, we now look for the largest positive real number k_{max} satisfying

$$9 + k(1 - t^2)^2 \geq (9 + k)\left(\frac{(1 - t)^2(1 + 2t)}{1 - t^2}\right).$$

This inequality is equivalent to

$$\frac{18t^2}{1 + t} \geq k \cdot \frac{t^3(1 - t)(t + 2)}{1 + t}, \quad \forall t \in [0, 1).$$

This last inequality is true for all k at $t = 0$, so we pursue our investigation on the interval $t \in (0, 1)$. This means we obtain

$$\frac{18}{t(1 - t)(t + 2)} \geq k \quad \forall t \in (0, 1),$$

hence

$$k_{max} = \min_{t \in (0,1)} \left(\frac{18}{t(1 - t)(t + 2)}\right).$$

Since $[t(1 - t)(t + 2)]' = -3t^2 - 2t + 2$, we have that

$$\min_{t \in (0,1)} \left(\frac{18}{t(1 - t)(t + 2)}\right) = \frac{18}{\left(\frac{\sqrt{7}-1}{3}\right)\left(1 - \frac{\sqrt{7}-1}{3}\right)\left(\frac{\sqrt{7}-1}{3} + 2\right)} = 10 + 7\sqrt{7}.$$

This minimum is attained at $t = \frac{\sqrt{7}-1}{3}$. Considering this value in the assertion that the maximum of abc is reached at $(a, b, c) = (1 - t, 1 - t, 1 + 2t)$ (or circular permutations of these values) when we obtained that the maximum is $(1-t)^2(1+2t)$, we see that k_{max} is attained at either (u, u, u) or at $(u(4 - \sqrt{7}), u(4 - \sqrt{7}), u(1 + 2\sqrt{7}))$ and their circular permutations, where u is a positive parameter. Hence, the equality in (3.1) holds for either $(1, 1, 1)$ or for $\left(\frac{1}{\sqrt{3}}, \frac{1}{\sqrt{3}}, \frac{2+\sqrt{7}}{\sqrt{3}}\right)$ or their circular permutations.

Therefore, we proved the following.

LEMMA 3.1. *Let $a, b, c > 0$ such that $a + b + c = \frac{1}{a} + \frac{1}{b} + \frac{1}{c}$. Then*

$$(a + b + c)^2 + (10 + 7\sqrt{7})a^2b^2c^2 \geq 19 + 7\sqrt{7}.$$

Equality holds for either $(1, 1, 1)$ or for $\left(\frac{1}{\sqrt{3}}, \frac{1}{\sqrt{3}}, \frac{2+\sqrt{7}}{\sqrt{3}}\right)$ or their circular permutations.

Based on this result, we have a Weingarten-type inequality which holds for a class of points on strictly convex smooth hypersurfaces in $\mathbb{R}^4$. The result which responds to Question 1 from our introduction is the following.

THEOREM 3.2. *Let $M^3 \subset \mathbb{R}^4$ be a strictly convex smooth hypersurface in the Euclidean four dimensional space. If $p \in M$ is a coreciprocal point on the hypersurface, i.e. at $p \in M$ the principal curvatures a, b, c satisfy $a + b + c = \frac{1}{a} + \frac{1}{b} + \frac{1}{c}$, then the mean curvature H and the Gauss-Kronecker curvature K satisfy*

$$9H^2 + (10 + 7\sqrt{7})K^2 \geq 19 + 7\sqrt{7}. \tag{3.3}$$

The equality case holds for either umbilical points where the principal curvatures are $(1,1,1)$ *or for points where the principal curvatures take the values* $\left(\frac{1}{\sqrt{3}}, \frac{1}{\sqrt{3}}, \frac{2+\sqrt{7}}{\sqrt{3}}\right)$ *or their circular permutations.*

The global consequence of this result, which responds to Question 2 in our introduction, is the following.

COROLLARY 3.3. Let $M^3 \subset \mathbb{R}^4$ be a compact strictly convex smooth hypersurface in the Euclidean four dimensional space. If M is coreciprical, i.e. at every point the principal curvatures satisfy $a+b+c = \frac{1}{a}+\frac{1}{b}+\frac{1}{c}$. Then

$$9\int_M H^2 dV + (10+7\sqrt{7})\int_M K^2 dV \geq (19+7\sqrt{7}) vol(M).$$

4. The case of four-dimensional hypersurfaces

In the previous section, when we worked with hypersurfaces of dimension 3, we started our exploration on the argument of what the best constant can be. For hypersurfaces of dimension 4, we prove first the following.

LEMMA 4.1. *Let a, b, c, d be positive real numbers such that*

$$a+b+c+d = \frac{1}{a}+\frac{1}{b}+\frac{1}{c}+\frac{1}{d}.$$

Then

$$(a+b+c+d)^2 + 48abcd \geq 64.$$

The equality holds when either $a=b=c=d=1$ or when

$$(a,b,c,d) = \left(\sqrt{10}, \frac{2}{\sqrt{10}}, \frac{2}{\sqrt{10}}, \frac{2}{\sqrt{10}}\right),$$

or the permutations of these values.

Proof: From hypothesis, we have immediately that

$$1 = \frac{abcd(a+b+c+d)}{abc+abd+acd+bcd}$$

and

$$abcd = \frac{abc+abd+acd+bcd}{a+b+c+d}.$$

The inequality we need to prove is

$$(a+b+c+d)^2 + 48\left(\frac{abc+abd+acd+bcd}{a+b+c+d}\right) \geq \frac{64abcd(a+b+c+d)}{abc+abd+acd+bcd} \tag{4.1}$$

We may assume without the loss of generality in (4.1) that $a+b+c+d=4$. By MacLaurin's inequality we have that

$$\frac{abc+abd+acd+bcd}{4} \leq \left(\frac{a+b+c+d}{4}\right)^3,$$

which yields immediately

$$abc+abd+acd+bcd \leq 4. \tag{4.2}$$

We now state and prove the following.

CLAIM 1. The function $f:[0,1) \to (0,4]$, given by $f(t) = 4(1-t)^2(1+2t)$, for all $t \in [0,1)$, is bijective.

To prove this claim, note that $f'(t) = -8(1-t)(1+2t)+8(1-t)^2 = -24t(1-t) \leq 0$, $\forall t \in [0,1)$. This means the function f is strictly decreasing. We have $f(0) = 4$, and $\lim_{x\to 1^-} f(t) = 0$. Thus f is well defined and bijective. This concludes the proof of Claim 1.

From Claim 1, the Intermediate Value Theorem, and relation (4.2) it follows that there exists a value $t \in [0,1)$ such that

$$abc + abd + acd + bcd = 4(1-t)^2(1+2t) \tag{4.3}$$

We now state and prove the following.

CLAIM 2.

$$abcd \leq (1-t)^3(1+3t). \tag{4.4}$$

For the proof of this claim, denote $ab+bc+cd+da+ac+bd = 6q$, and $abcd = p$. It follows that both $p, q > 0$. Define the polynomial $P : (0,\infty) \to \mathbb{R}$ as

$$P(x) = (x-a)(x-b)(x-c)(x-d), \quad \forall x > 0.$$

By combining Viète's relations with (4.3) we obtain for this polynomial

$$P(x) = x^4 - 4x^3 + 6qx^2 - 4(1-t)^2(1+2t)x + p, \quad \forall x > 0.$$

Define the function $g : (0,\infty) \to \mathbb{R}$ as $g(x) = \frac{P(x)}{x^2}$, $\forall x > 0$. Since the zeroes of the function g are a, b, c, and d, by Rolle's Theorem it follows that the first derivative of g has at least three roots in the interval $(0,\infty)$. Calculate

$$g'(x) = 2 \cdot \frac{x^4 - 2x^3 + 2(1-t)^2(1+2t)x - p}{x^3}, \qquad \forall x > 0.$$

We see that $g'(x) = 0$ if and only if $h(x) = 0$, where

$$h(x) = x^4 - 2x^3 + 2(1-t)^2(1+2t)x - p, \qquad \forall x > 0.$$

By computing the derivative we obtain that $h'(x)$ equals

$$4[x-(1-t)]\left[x - \frac{1+2t+\sqrt{3(1+2t)(3-2t)}}{4}\right]\left[x - \frac{1+2t-\sqrt{3(1+2t)(3-2t)}}{4}\right],$$

for all $x > 0$. Hence h' has two zeroes for $x > 0$ and the zeroes of h' are

$$1 - t \leq \frac{1+2t+\sqrt{3(1+2t)(3-2t)}}{4}.$$

The polynomial h admits on the interval $(0,\infty)$ the Rolle sequence

$$0_+ < 1 - t \leq \frac{1+2t+\sqrt{3(1+2t)(3-2t)}}{4} < \infty.$$

On the other hand, $h(0_+) = -p < 0$, and $\lim_{t\to\infty} h(t) = \infty > 0$, and h admits at least three roots in the interval $(0,\infty)$. As a consequence of Rolle's theorem, we obtain $h(1-t) \geq 0$, and

$$h\left(\frac{1+2t+\sqrt{3(1+2t)(3-2t)}}{4}\right) \leq 0.$$

From $h(1-t) \geq 0$ we obtain right away $p \leq (1-t)^3(1+3t)$, which yields

$$abcd \leq (1-t)^3(1+3t).$$

This concludes the argument for Claim 2. The elementary details of the calculus argument in Claim 2 have also been used in [**12**].

We remark that if $a = b = c = 1 - t$, and $d = 1 + 3t$, then

$$a + b + c + d = 4, \quad abc + abd + acd + bcd = 4(1 - t)^2(1 + 2t),$$

and $abcd = (1 - t)^3(1 + 3t)$.

With this preparation, we are ready to approach now the proof of the lemma. Claim 2 states that $abcd \le (1-t)^3(1+3t)$. Hence, it suffices to prove the following single variable inequality:

$$1 + 3(1 - t)^2(1 + 2t) \ge \frac{4(1 - t)^3(1 + 3t)}{(1 - t)^2(1 + 2t)},$$

which is equivalent to $t^2(2t - 1)^2 \ge 0$. This concludes the proof of the inequality.

We now turn our attention to the equality case. The equality holds at $t = 0$ and $t = \frac{1}{2}$, which means that the quadruples $(a, b, c, d) = (k, k, k, k)$, or $(a, b, c, d) = (k, k, k, 5k)$, where $k > 0$ is a real parameter. Plugging in the hypothesis of the original statement, we derive that the equality holds when either $a = b = c = d = 1$ or when

$$(a, b, c, d) = \left(\sqrt{10}, \frac{2}{\sqrt{10}}, \frac{2}{\sqrt{10}}, \frac{2}{\sqrt{10}}\right),$$

or the corresponding permutations of these values.

The proof of Lemma 4.1 is concluded, and now we investigate why the real number $k = 3$ is the largest positive number k such that the following inequality holds:

$$(a + b + c + d)^2 + 16kabcd \ge 16(1 + k).$$

The whole idea of constructing a Weingarten-type inequality resides in determining whether the work is done with the best possible constant.

If $k = 3$, Lemma 4.1 establishes the stated inequality.

If $0 < k < 3$, we claim that $(a+b+c+d)^2 + 16kabcd - 16(1+k) \ge 0$. To prove this assertion, we start from $a + b + c + d \ge 4$, which yields

$$(3 - k)(a + b + c + d)^2 \ge 16(3 - k),$$

which is equivalent to

$$3(a + b + c + d)^2 + 48kabcd - 48(1 + k) \ge k(a + b + c + d)^2 + 48kabcd - 64k.$$

On the other hand,

$$k(a + b + c + d)^2 + 48kabcd - 64k \ge 0,$$

therefore

$$(a + b + c + d)^2 + 16kabcd - 16(1 + k) \ge 0.$$

This concludes the discussion in the case $0 < k < 3$.

For $k > 3$, a counterexample can be reached for the quadruple

$$\left(\frac{\sqrt{10}}{5}, \frac{\sqrt{10}}{5}, \frac{\sqrt{10}}{5}, \sqrt{10}\right),$$

which will get $\frac{2+k}{5} < \frac{1+k}{4}$.

We therefore proved that $k = 3$ is the best possible constant. □

With this fact established, we are ready now to state the Weingarten-type inequality for a class of hypersurfaces in $\mathbb{R}^5$. The result which responds to Question 1 is the following.

THEOREM 4.2. *Let $p \in M^4 \subset \mathbb{R}^5$ be a coreciprocal point on a smooth strictly convex hypersurface. Then the following holds:*

$$H^2 + 3K \geq 4. \tag{4.5}$$

The equality holds at the coreciprocal points where the principal curvatures satisfy either $a = b = c = d = 1$ or when

$$(a, b, c, d) = \left(\sqrt{10}, \frac{2}{\sqrt{10}}, \frac{2}{\sqrt{10}}, \frac{2}{\sqrt{10}}\right),$$

or the corresponding permutations of these values.

Additionally, if the hypersurface is compact and has all the points coreciprocal, we obtain the following fact, which responds to Question 2 stated in our Introduction.

COROLLARY 4.3. Let $M^4 \subset \mathbb{R}^5$ be a compact coreciprocal strictly convex hypersurface in $\mathbb{R}^5$. Then

$$\frac{1}{4}\int_M |H|^2 dV + \frac{3}{4}\int_M K dV \geq vol(M).$$

The case when the hypersurface has dimension 4 is somewhat special, in the following sense. For dimension 3, Corollary 3.3 does not respond completely to the question posed by B.-Y. Chen in 1975, while the situation for $n = 4$ is different. This happens because the inequalities we state also involve the Gauss-Kronecker curvature, which is originally an extrinsic object. When $n = 4$, it follows from the Gauss equation that the Gauss-Kronecker curvature only depends on the intrinsic structure of M^4, since

$$k_1 k_2 k_3 k_4 = sec(e_1 \wedge e_2) sec(e_3 \wedge e_4), \tag{4.6}$$

where $sec(e_i \wedge e_j)$ denotes the sectional curvature on the plane generated by e_i and e_j.

5. Higher dimensional cases

We presented in this work the technical details of the cases for dimensions 3 and 4. For five variables, we have the following [**13**].

LEMMA 5.1. *Let a, b, c, d, e be five strictly positive real numbers such that $a + b + c + d + e = \frac{1}{a} + \frac{1}{b} + \frac{1}{c} + \frac{1}{d} + \frac{1}{e}$. Then*

$$a + b + c + d + e + 4\sqrt[3]{4} - 5 \geq 4\sqrt[3]{\sqrt{4abcde}}. \tag{5.1}$$

Equality holds only for $a = b = c = d = e = 1$.

The proof will be presented in a second paper on this topic, [**13**]. This yields the following geometric consequence.

THEOREM 5.2. *Let $p \in M^5 \subset \mathbb{R}^6$ be a coreciprocal point on a smooth strictly convex hypersurface. Then at p the following holds :*

$$5H + 4\sqrt[3]{4} - 5 \geq 4\sqrt[3]{\sqrt{4}K^{\frac{1}{3}}}. \tag{5.2}$$

The equality case occurs only for hyperspheres $S^5(1) \subset \mathbb{R}^6$.

The general case of the lemma is the following [**13**].

LEMMA 5.3. *Let n be a natural number $n \geq 5$, and $x_i > 0$, for $i = 1, 2, ..., n$. Suppose the coreciprocal condition holds, i.e. $\sum_{i=1}^{n} x_i = \sum_{i=1}^{n} \frac{1}{x_i}$. Then*

$$\left(\sum_{i=1}^{n} x_i\right) + (n-1)\sqrt[n-2]{n-1} - n \geq (n-1)\sqrt[n-2]{(n-1)\prod_{i=1}^{n} x_i}.$$

Equality holds when all $x_i = 1$, $i = 1, 2, ..., n$.

6. Further investigations

The results we proved in the present paper raise a more general thought. It would be interesting if we obtain inequalities similar to (3.3) and (4.5) for other classes of geometric objects, which would correspond to other natural algebraic constraints in the hypothesis of the inequalities.

The authors thank both referees for the very careful and insightful array of suggestions which improved both the content and the presentation of the present work. The important remark at the end of Section 4 is due to one of the referees.

References

[1] B.-Y. Chen, *Mean curvature vector of a submanifold*, Differential geometry (Proc. Sympos. Pure Math., Vol. XXVII, Stanford Univ., Stanford, Calif., 1973), Amer. Math. Soc., Providence, R. I., 1975, pp. 119–123. MR0377769

[2] B.-Y. Chen, *Conformal mappings and first eigenvalue of Laplacian on surfaces*, Bull. Inst. Math. Acad. Sinica **7** (1979), no. 4, 395–400. MR558248

[3] B.-Y. Chen, *Some pinching and classification theorems for minimal submanifolds*, Arch. Math. (Basel) **60** (1993), no. 6, 568–578, DOI 10.1007/BF01236084. MR1216703

[4] B.-Y. Chen, *A Riemannian invariant and its applications to submanifold theory*, Results Math. **27** (1995), no. 1-2, 17–26, DOI 10.1007/BF03322265. Festschrift dedicated to Katsumi Nomizu on his 70th birthday (Leuven, 1994; Brussels, 1994). MR1317820

[5] B.-Y. Chen, *Strings of Riemannian invariants, inequalities, ideal immersions and their applications*, The Third Pacific Rim Geometry Conference (Seoul, 1996), Monogr. Geom. Topology, vol. 25, Int. Press, Cambridge, MA, 1998, pp. 7–60. MR1751063

[6] B.-Y. Chen, *Some new obstructions to minimal and Lagrangian isometric immersions*, Japan. J. Math. (N.S.) **26** (2000), no. 1, 105–127, DOI 10.4099/math1924.26.105. MR1771434

[7] B.-Y. Chen, *Pseudo-Riemannian geometry, δ-invariants and applications*, World Scientific Publishing Co. Pte. Ltd., Hackensack, NJ, 2011. With a foreword by Leopold Verstraelen. MR2799371

[8] B.-Y. Chen, *Total mean curvature and submanifolds of finite type*, 2nd ed., Series in Pure Mathematics, vol. 27, World Scientific Publishing Co. Pte. Ltd., Hackensack, NJ, 2015. With a foreword by Leopold Verstraelen. MR3362186

[9] S.-S. Chern, *On special W-surfaces*, Proc. Amer. Math. Soc. **6** (1955), 783–786, DOI 10.2307/2032934. MR74857

[10] S.-S. Chern, *Minimal submanifolds in a Riemannian manifold*, University of Kansas, Department of Mathematics Technical Report 19 (New Series), Univ. of Kansas, Lawrence, Kan., 1968. MR0248648

[11] P. Hartman and A. Wintner, *Umbilical points and W-surfaces*, Amer. J. Math. **76** (1954), 502–508, DOI 10.2307/2372698. MR63082

[12] L. M. Giugiuc, D. T. Oai, and K. Altintas, *An inequality related to the lengths and area of a convex quadrilateral*, Int. J. Geom. **7** (2018), no. 1, 81–86. MR3796259

[13] L.M. Giugiuc, B.D. Suceavă, A Class of Strictly Convex Hypersurfaces Satisfying Weingarten-Type Inequalities, II (to appear).

[14] W. Kühnel and M. Steller, *On closed Weingarten surfaces*, Monatsh. Math. **146** (2005), no. 2, 113–126, DOI 10.1007/s00605-005-0313-4. MR2176338

[15] R. López, *Rotational linear Weingarten surfaces of hyperbolic type*, Israel J. Math. **167** (2008), 283–301, DOI 10.1007/s11856-008-1049-3. MR2448026

[16] R. C. Reilly, *On the first eigenvalue of the Laplacian for compact submanifolds of Euclidean space*, Comment. Math. Helv. **52** (1977), no. 4, 525–533, DOI 10.1007/BF02567385. MR482597

[17] H. Rosenberg and R. Sa Earp, *The geometry of properly embedded special surfaces in* $\mathbf{R}^3$, *e.g., surfaces satisfying* $aH + bK = 1$, *where* a *and* b *are positive*, Duke Math. J. **73** (1994), no. 2, 291–306, DOI 10.1215/S0012-7094-94-07314-6. MR1262209

[18] J. Weingarten, *Ueber eine Klasse auf einander abwickelbarer Flächen* (German), J. Reine Angew. Math. **59** (1861), 382–393, DOI 10.1515/crll.1861.59.382. MR1579186

[19] J. Weingarten, *Ueber die Flächen deren Normalen eine gegebene Fläche berühren* (German), J. Reine Angew. Math. **62** (1863), 61–63, DOI 10.1515/crll.1863.62.61. MR1579229

[20] Vo Quoc Ba Can, On a class of three-variable inequalities, *Mathematical Reflections*, 2(2007), 1-7. https://www.cut-the-knot.org/arithmetic/algebra/3variable.pdf

Bd. Carol I No. 6, National College *Trajan*, Drobeta-Turnu Severin, Romania
Email address: leonardgiugiuc@yahoo.com

Department of Mathematics, California State University, Fullerton, 800 N. State College Blvd., Fullerton, CA 92834-6850, U.S.A.
Email address: bsuceava@fullerton.edu

Contemporary Mathematics
Volume **756**, 2020
https://doi.org/10.1090/conm/756/15202

Bounding the invariant spectrum when the scalar curvature is non-negative

Stuart J. Hall and Thomas Murphy

Abstract. On compact Riemannian manifolds with a large isometry group we investigate the invariant spectrum of the ordinary Laplacian. For either a toric Kähler metric, or a rotationally-symmetric metric on the sphere, we produce upper bounds for all eigenvalues of the invariant spectrum assuming non-negative scalar curvature.

1. Introduction

Motivation for finding universal bounds on the spectrum of metrics on the sphere comes from a famous result due to Hersch [**12**]: any metric g on the sphere $\mathbb{S}^2$, normalised to have volume 4π, satisfies

$$\lambda_1(g) \leq 2.$$

In fact, Hersch also proved that $\lambda_1(g) = 2$ if and only if the metric g is isometric to the canonical round metric.

If $G \leqslant \mathrm{Iso}(g)$ is a subgroup of the isometry group of a metric g, the G-invariant spectrum is defined by restricting the Laplacian to G-invariant functions. The kth G invariant eigenvalue of the ordinary Laplacian is denoted $\lambda_k^G(g)$. In this setting, the analogue of Hersch's Theorem for metrics invariant under the standard action of $\mathbb{S}^1$ on $\mathbb{S}^2$ was studied by Abreu and Freitas. They proved that no counterpart for Hersch's Theorem holds for the $\mathbb{S}^1$-invariant spectrum. More specifically, for any $c \in (0, \infty)$, they were able to construct metrics g with volume 4π and first invariant eigenvalue $\lambda_1^{\mathbb{S}^1}(g) = c$.

However, under additional assumptions on the geometry defined by g, they also proved the following theorem which is our primary inspiration in this paper. A related set of results was also proved independently by Engman [**9**] [**10**]. To state the result, denote by ξ_k the $\frac{1}{2}(k+1)^{\mathrm{th}}$ positive zero of the Bessel function J_0 if k is odd, and the $\frac{1}{2}k^{\mathrm{th}}$ positive zero of J_0' if k is even.

Theorem (Abreu–Freitas, [**3**]). *Let g be a metric on $\mathbb{S}^2$ which is invariant under the standard action of $\mathbb{S}^1$, with non-negative Gaussian curvature and volume 4π. Then*

$$\lambda_k^{\mathbb{S}^1}(g) < \frac{1}{2}\xi_k^2.$$

In the present note we are concerned with two natural generalisations of this theorem to higher dimensions. In both of them the analogue of non-negative Gaussian curvature is to assume the metrics have non-negative scalar curvature. The two generalisations correspond to two different ways of viewing $\mathbb{S}^1$-invariant metrics on $\mathbb{S}^2$. The first is to realize $\mathbb{S}^2$ as a complex manifold (with complex dimension one), g as a Kähler metric, and to view $\mathbb{S}^1$ as a real, one-dimensional torus $\mathbb{T}$. In higher dimensions, a well-developed theory of Kähler metrics which are invariant under the Hamiltonian action of a real n-torus $\mathbb{T}^n$ exists, namely *toric Kähler manifolds.* In this setting we prove the following result.

THEOREM A. *Let (M, ω) be a compact toric Kähler manifold with non-negative scalar curvature. Then there exist constants $0 < C_k([\omega])$, depending only upon the cohomology class $[\omega] \in H^2(M; \mathbb{R})$, such that*

$$\lambda_k^{\mathbb{T}^n} \leq C_k.$$

The $k = 1$ bound in Theorem A was established by the authors with a slightly different proof in [**11**]; here a refined method yields all higher eigenvalue bounds.

The second viewpoint we take is to identify $\mathbb{S}^1 \cong SO(2)$ and $SO(2) \leqslant SO(3)$ via the embedding

$$g \in SO(2) \to \begin{pmatrix} 1 & 0 \\ 0 & g \end{pmatrix} \in SO(3).$$

The action is inherited from the natural action of $SO(3)$ on $\mathbb{R}^3$ by regarding the sphere as a hypersurface $\mathbb{S}^2 \subset \mathbb{R}^3$. In higher dimensions this leads to the action of the group $SO(n)$ on the n-dimensional sphere $\mathbb{S}^n \subset \mathbb{R}^{n+1}$. This example is part of another well-established theory, that of *cohomogeneity one metrics.* We prove the following result.

THEOREM B. *There exist constants $D_k > 0$ such that for any unit-volume, $SO(n)$-invariant metric g on $\mathbb{S}^n$ with non-negative scalar curvature, we have the bounds*

$$\lambda_k^{SO(n)}(g) \leq D_k.$$

We should say immediately that Theorem B is not really new and is essentially contained in the work of Colbois, Dryden and El Soufi (c.f. Theorem 1.7 in [**6**]). However, in our proof one can see very directly how the assumption on the non-negativity of the scalar curvature is used.

2. Background

2.1. Previous results in the literature. In all the results we mention here, M is assumed to be compact and without boundary. Korevaar [**14**], building on work by Yang and Yau [**16**], gave a far-reaching generalisation of the Hersch result to higher eigenvalues proving that, for any oriented Riemannian surface (S, g) of genus γ, there exists a universal constant $C > 0$ such that

$$\lambda_k(g) \leq \frac{C(\gamma+1)k}{\mathrm{Vol}(S, g)},$$

where $\lambda_k(g)$ is the kth eigenvalue in the spectrum of the Laplacian.

In the case that the manifold has dimension 3 or higher, Colbois and Dodziuk [**5**] proved that there is a unit volume metric g with arbitrarily large $\lambda_1(g)$; hence there is no hope of a general analogue of the Hersch and Korevaar results for arbitrary manifolds. However, in the case that the manifold M is a compact, projective, complex manifold, Bourguignon, Li and Yau [**4**] demonstrated an upper bound for $\lambda_1(g)$ depending upon some of the data associated with embedding M into complex projective space $\mathbb{CP}^N$. This result has been recently extended to higher eigenvalues by Kokarev [**13**].

The study of the invariant spectrum for the (effective and non-transitive) action for a general compact Lie group G was taken up by Colbois, Dryden and El Soufi in [**6**]. They proved that, if the dimension of G is at least one, then

$$\sup\left\{\lambda_1^G(g)\mathrm{Vol}(g)^{2/n}\right\} = \infty,$$

where the supremum is taken over all G-invariant metrics conformal to a fixed reference metric g_0.

Finally, in the case of n-complex-dimensional toric Kähler manifolds, we mention that Legendre and Sena-Dias [**15**] have generalised the results of Abreu and Freitas to show that the first $\mathbb{T}^n$-invariant eigenvalue, $\lambda_1^{\mathbb{T}^n}(g)$, can take any value in $(0, \infty)$ when the metric g is allowed to vary over the set of toric Kahler metrics in a fixed cohomology class. The Colbois–Dryden–El Soufi and Legendre–Sena-Dias results make no assumption about the curvature of the metrics and so do not contradict Theorems A and B.

2.2. Toric Kähler manifolds. In order to prove Theorem A we recall some facts about toric Kähler metrics. We cannot hope to give a comprehensive introduction to this subject but refer readers to the works of Abreu [**2**] and Donaldson [**8**] which both contain detailed discussions of the theory. In a nutshell, associated to any Hamiltonian action of an n-torus $\mathbb{T}^n$ on an n-complex-dimensional Kähler manifold (M, ω, J) is a convex polytope $P \subset \mathbb{R}^n$. The polytope P, often referred to in the literature as the *moment polytope*, depends only upon the cohomology class of the Kähler metric $[\omega] \in H^2(M;\mathbb{R})$. There is also a dense open set $M^\circ \subset M$ such that $M^\circ \cong P^\circ \times (0, 2\pi)^n$, where P° is the interior of P. In the natural coordinates, the metric $g(\cdot,\cdot) := \omega(J\cdot,\cdot)$ is given by

$$g = u_{ij}dx_idx_j + u^{ij}d\theta_id\theta_j,$$

where u_{ij} is the Euclidean Hessian of a convex function $u : P^\circ \to \mathbb{R}$ and u^{ij} is the inverse of this matrix. The function u is referred to as the *symplectic potential* of the metric g. Clearly, in these coordinates, the volume form of the metric is the standard Euclidean form

$$dV = dx_1 \wedge dx_2 \wedge ... \wedge dx_n \wedge d\theta_1 \wedge d\theta_2 \wedge ... \wedge d\theta_n.$$

Crucial to our proof is the formula for the scalar curvature in the (x_i, θ_j) coordinates which was first given by Abreu [**1**] (often called the Abreu equation in the literature):

$$\mathrm{Scal}(g) = -\sum_{i,j}\frac{\partial^2 u^{ij}}{\partial x_i \partial x_j}.$$

This expression yields the following useful integration-by-parts formula due to Donaldson [**7**].

LEMMA 1 (Donaldson's Integration-by-parts formula, Lemma 3.3.5 in [**7**]). *Let $u : P^\circ \to \mathbb{R}$ be a symplectic potential of a toric Kähler metric g and let $F \in C^\infty(P)$.*

Then

$$\int_P u^{ij}F_{ij}d\mu = \int_{\partial P} 2F d\sigma - \int_P \mathrm{Scal}(g)F d\mu, \tag{2.1}$$

where μ is the usual Lesbegue measure on $\mathbb{R}^n$ and σ is the integral Lesbegue measure on the boundary ∂P.

Here the subscripts on F denote partial differentiation. Equation (2.1) differs slightly from the expression in Lemma 3.3.5 in [**7**] since we adopt a different convention to describe the singular part of the symplectic potential. We refer the reader to [**11**] for futher discussion of our conventions and the precise definition of the integral Lesbegue measure σ.

3. Toric Kähler eigenvalue bounds

We now give the proof of Theorem A. Central to the proof is the fact that σ does not depend upon the metric g. Rather it depends only on the cohomology class of the associated Kähler form ω.

PROOF. Fix the moment polytope $P \subset \mathbb{R}^n$ and pick a coordinate $x \in [x_{\min}, x_{\max}]$. Throughout the proof we shall abuse notation by denoting by f functions in this single variable $f : [x_{\min}, x_{\max}] \to \mathbb{R}$ and the extension to the whole polytope P given by $(f \circ \pi)$ where $\pi : P \to [x_{\min}, x_{\max}]$ is the projection to this coordinate.

Let $V_k \subset W^{1,2}$ be the subspace given by

$$V_k = \mathrm{span}\{1, x, x^2, \ldots, x^k\}.$$

Hence the min-max characterisation of $\lambda_k^{\mathbb{T}^n}$ yields

$$\lambda_k^{\mathbb{T}^n} \leq \sup_{\varphi\in V_k, \varphi\neq 0} \left\{ \frac{\int_P \|\nabla\varphi\|^2 d\mu}{\int_P \varphi^2 d\mu} \right\}.$$

We denote by $\varrho : [x_{\min}, x_{\max}] \to \mathbb{R}$ the unique function satisfying

$$\varrho_{xx} = (\varphi_x)^2 \qquad \text{and} \qquad \varrho(x^*) = \varrho_x(x^*) = 0,$$

where $x^* = \dfrac{x_{\min} + x_{\max}}{2}$. As ϱ is manifestly convex, the point x^* is a global minimum and so ϱ is non-negative on $[x_{\min}, x_{\max}]$.

The Raleigh quotient thus furnishes us with a map $R : V_k\backslash\{0\} \to \mathbb{R}$

$$R(\varphi) = \frac{\int_P \|\nabla\varphi\|^2 d\mu}{\int_P \varphi^2 d\mu}.$$

The numerator can be evaluated using Equation (2.1)

$$\int_P \|\nabla\varphi\|^2 d\mu = \int_P u^{xx}(\varphi_x)^2 d\mu = \int_P u^{xx}\varrho_{xx}d\mu = \int_{\partial P} 2\varrho d\sigma - \int_P \mathrm{Scal}(g)\varrho d\mu.$$

As the final term in the previous equation is non-negative, we obtain the bound

$$\int_P \|\nabla\varphi\|^2 d\mu \leq \int_{\partial P} 2\varrho d\sigma.$$

Hence

$$R(\varphi) \leq \frac{\int_{\partial P} 2\varrho d\sigma}{\int_P \varphi^2 d\mu},$$

where the bound is independent of the metric. The function $B : V_k\backslash\{0\} \to \mathbb{R}$ given by

$$B(\varphi) = \frac{\int_{\partial P} 2\varrho d\sigma}{\int_P \varphi^2 d\mu},$$

is continuous. We demonstrate this by considering the numerator and denominator separately. We endow the space V_k with the standard Euclidean topology. The map $\mathcal{F}_1 : V_k \to V_{2(k-1)}$ given by $\mathcal{F}_1(\varphi) = (\varphi_x)^2$ is continuous (this could be seen by writing it out in coordinates). The map $\mathcal{F}_2 : V_{2(k-1)} \to V_{2k}$ given by $\mathcal{F}_2(f) = \varrho$, where ϱ solves the equation

$$\varrho_{xx} = f \qquad \text{and} \qquad \varrho(x^*) = \varrho_x(x^*) = 0,$$

is linear and so continuous (note the domain is finite dimensional). The map $\mathcal{F}_3 : V_{2k} \to \mathbb{R}$ given by

$$\mathcal{F}_3(f) = \int_{\partial P} 2f d\sigma,$$

is also linear and so continuous. The numerator of the function B is given by the composition $\mathcal{F}_3 \circ \mathcal{F}_2 \circ \mathcal{F}_1(\varphi)$ and thus is continuous. A very similar argument yields the continuity of the denominator as a map from $V_k\backslash\{0\}$ to $\mathbb{R}$. Thus we obtain a continuous function $B : V_k\backslash\{0\} \to \mathbb{R}$.

If we scale $\varphi \to \alpha\varphi$ for $\alpha \in \mathbb{R}$, then $\varrho \to \alpha^2\varrho$. Hence B is determined by the values it takes on the functions $\varphi \in \mathbb{S}^k \subset V_k$ (where we take the obvious coordinates on V_k). Restricting B to the sphere yields a continuous function, so the constant C_k can be determined by taking the maximum of B as φ varies over the sphere $\mathbb{S}^k$. □

4. Cohomogeneity one metrics on the sphere

Let g be a unit-volume metric on $\mathbb{S}^n$ given by

$$g = dt^2 + \rho^2(t)d\mathbb{S}^2_{n-1},$$

where $d\mathbb{S}^2_{n-1}$ is the unit-volume round metric on the sphere $\mathbb{S}^{n-1}$ and $\rho : [0, b] \to \mathbb{R}$ is such that

$$\rho(0) = \rho(b) = 0, \qquad \dot\rho(0) = -\dot\rho(b) = 1 \qquad \text{and} \quad \rho > 0 \quad \text{on} \quad (0, b),$$

where dots over a function denotes derivatives with respect to t. In these coordinates the scalar curvature is given by

$$\text{Scal}(t) = -2(n-1)\frac{\ddot\rho}{\rho} + (n-1)(n-2)\frac{1-\dot\rho^2}{\rho^2}. \tag{4.1}$$

This immediately yields the following result.

LEMMA 2. *Let ρ be as above and suppose the metric g has non-negative scalar curvature. Then*

$$|\dot\rho| \le 1.$$

PROOF. Suppose the bound does not hold, then the boundary conditions imply that $\dot\rho$ has a global maximum or minimum, t^* say, where $t^* \in (0, b)$. From Equation (4.1), at such a point the scalar curvature is given by

$$\text{Scal}(t^*) = (n-1)(n-2)\frac{1-\dot\rho^2(t^*)}{\rho^2(t^*)}.$$

If $|\dot{\rho}(t^*)| > 1$ there is now an immediate contradiction as the scalar curvature would be strictly negative at this point. □

We will now use a new coordinate s defined by a map $s(t) : [0, b] \to [0, 1]$ given by

$$s(t) := \int_0^t \rho^{n-1}(\tau)d\tau.$$

As the integrand is positive, $s(t)$ is invertible and we denote its inverse as $t(s)$. In these s–coordinates the metric becomes

$$g = \frac{ds^2}{\Phi^{2n-2}} + \Phi^2 d\mathbb{S}^2_{n-1}, \tag{4.2}$$

where $\Phi(s) := \rho(t(s))$.

Straightforward calculation, with the convention that derivatives with respect to s are denoted with a prime, yields

$$\dot{\rho}(t(s)) = \Phi'(s)(\Phi(s))^{n-1} = \frac{1}{n}(\Phi^n)',$$

and so

$$\Phi^n(0) = \Phi^n(1) = 0, \qquad (\Phi^n)'(0) = -(\Phi^n)'(1) = n.$$

Combining this with Lemma 2 gives

LEMMA 3. *Let $\Phi : [0,1] \to \mathbb{R}$ be as above. Then*

$$\Phi(s) \leq \Phi_{\max}(s),$$

where

$$\Phi_{\max}(s) = \begin{cases} \sqrt[n]{ns} & \text{if } \ 0 \leq s \leq \frac{1}{2}, \\ \sqrt[n]{\frac{n}{2} - n(s - \frac{1}{2})} & \text{if } \ \frac{1}{2} \leq s \leq 1. \end{cases}$$

We can now prove Theorem B. It essentially follows from the fact that Φ_{max} is independent of metric.

PROOF. Working in the 's' coordinate, we consider the subspace $V_k \subset W^{1,2}$ given by

$$V_k := \text{span}\left\{1, s, s^2, ..., s^k\right\}.$$

We take $\varphi \in \mathbb{S}^k \subset V_k$ (where we use the obvious coordinates on V_k to define the sphere) and consider

$$R(\varphi) := \frac{\int_{\mathbb{S}^n} |\nabla\varphi|^2 dV_g}{\int_{\mathbb{S}^n} \varphi^2 dV_g}.$$

Using the specific form of the metric g in Equation (4.2) we obtain

$$R(\varphi) = \frac{\int_0^1 \Phi^{2n-2}(\varphi')^2 ds}{\int_0^1 \varphi^2 ds} \leq \frac{\int_0^1 \Phi_{\max}^{2n-2}(\varphi')^2 ds}{\int_0^1 \varphi^2 ds}$$

where the last inequality uses Lemma 3. Thus the Raleigh quotient is clearly bounded above by a quantity that does not depend upon the metric g. The result follows from the min-max characterisation of the kth eigenvalue. □

References

[1] M. Abreu, *Kähler geometry of toric varieties and extremal metrics*, Internat. J. Math. **9** (1998), no. 6, 641–651, DOI 10.1142/S0129167X98000282. MR1644291

[2] M. Abreu, *Kähler geometry of toric manifolds in symplectic coordinates*, Symplectic and contact topology: interactions and perspectives (Toronto, ON/Montreal, QC, 2001), Fields Inst. Commun., vol. 35, Amer. Math. Soc., Providence, RI, 2003, pp. 1–24. MR1969265

[3] M. Abreu and P. Freitas, *On the invariant spectrum of S^1-invariant metrics on S^2*, Proc. London Math. Soc. (3) **84** (2002), no. 1, 213–230, DOI 10.1112/plms/84.1.213. MR1863401

[4] J.-P. Bourguignon, P. Li, and S.-T. Yau, *Upper bound for the first eigenvalue of algebraic submanifolds*, Comment. Math. Helv. **69** (1994), no. 2, 199–207, DOI 10.1007/BF02564482. MR1282367

[5] B. Colbois and J. Dodziuk, *Riemannian metrics with large λ_1*, Proc. Amer. Math. Soc. **122** (1994), no. 3, 905–906, DOI 10.2307/2160770. MR1213857

[6] B. Colbois, E. B. Dryden, and A. El Soufi, *Extremal G-invariant eigenvalues of the Laplacian of G-invariant metrics*, Math. Z. **258** (2008), no. 1, 29–41, DOI 10.1007/s00209-007-0154-z. MR2350031

[7] S. K. Donaldson, *Scalar curvature and stability of toric varieties*, J. Differential Geom. **62** (2002), no. 2, 289–349. MR1988506

[8] S. K. Donaldson, *Kähler geometry on toric manifolds, and some other manifolds with large symmetry*, Handbook of geometric analysis. No. 1, Adv. Lect. Math. (ALM), vol. 7, Int. Press, Somerville, MA, 2008, pp. 29–75. MR2483362

[9] M. Engman, *Trace formulae for S^1 invariant Green's operators on S^2*, Manuscripta Math. **93** (1997), no. 3, 357–368, DOI 10.1007/BF02677477. MR1457734

[10] M. Engman, *The spectrum and isometric embeddings of surfaces of revolution*, Canad. Math. Bull. **49** (2006), no. 2, 226–236, DOI 10.4153/CMB-2006-023-7. MR2226245

[11] S. J. Hall and T. Murphy, *Bounding the first invariant eigenvalue of toric Kähler manifolds*, Math. Res. Lett. **24** (2017), no. 1, 67–81, DOI 10.4310/MRL.2017.v24.n1.a4. MR3662275

[12] J. Hersch, *Quatre propriétés isopérimétriques de membranes sphériques homogènes* (French), C. R. Acad. Sci. Paris Sér. A-B **270** (1970), A1645–A1648. MR292357

[13] G. Kokarev, *Bounds for Laplace eigenvalues of Kähler metrics*, Adv. Math. **365** (2020), 107061, DOI 10.1016/j.aim.2020.107061. MR4065716

[14] N. Korevaar, *Upper bounds for eigenvalues of conformal metrics*, J. Differential Geom. **37** (1993), no. 1, 73–93. MR1198600

[15] E. Legendre and R. Sena-Dias, *Toric aspects of the first eigenvalue*, J. Geom. Anal. **28** (2018), no. 3, 2395–2421, DOI 10.1007/s12220-017-9908-y. MR3833798

[16] P. C. Yang and S. T. Yau, *Eigenvalues of the Laplacian of compact Riemann surfaces and minimal submanifolds*, Ann. Scuola Norm. Sup. Pisa Cl. Sci. (4) **7** (1980), no. 1, 55–63. MR577325

School of Mathematics, Statistics and Physics, Herschel Building, Newcastle University, Newcastle upon Tyne, NE1 7RU, UK

Email address: stuart.hall@newcastle.ac.uk

Department of Mathematics, California State University Fullerton, 800 N. State College Blvd., Fullerton, California 92831

Email address: tmurphy@fullerton.edu

Contemporary Mathematics
Volume **756**, 2020
https://doi.org/10.1090/conm/756/15203

Isometric immersions of surfaces: classical approaches and integrability

Thomas A. Ivey

Abstract. The problem of finding a surface in 3-dimensional Euclidean space with a given first fundamental form was classically known as Bour's problem, and is closely related to the problem of determining which surfaces in space are 'applicable' (i.e., isometric without being congruent) to a given surface. We review classical approaches to these problems, including coordinate-based approaches exposited by Darboux and Eisenhart, as well as the moving-frames based approach employed by Cartan. In particular, one approach involves reducing Bour's problem to solving a single PDE; settling the question of when this PDE is integrable by the method of Darboux is surprisingly easy, once we answer the analogous question for the isometric immersion system arising from the moving frames approach. The latter question was settled in recent joint work with Clelland, Tehseen and Vassiliou.

1. Introduction

The problem of finding a surface in 3-dimensional Euclidean space with a given first fundamental form was known in the 19th century as *Bour's problem* (see, e.g., [**11**]). With the introduction of the concept of an abstract manifold, we can reformulate the problem more precisely as follows: given an abstract two-dimensional manifold M with a smooth metric g, find an immersion $\mathsf{f} : M \to \mathbb{R}^3$ that is isometric, i.e., such that the pullback via f of the Euclidean metric coincides with g. In terms of local coordinates x^1, x^2 on M, and the components g_{ij} of the metric in these coordinates, the equations we are trying to solve are

$$\frac{\partial \mathsf{f}}{\partial x^i} \cdot \frac{\partial \mathsf{f}}{\partial x^j} = g_{ij}. \tag{1.1}$$

This is a nonlinear system of first-order partial differential equations with a large symmetry group – namely, the 'rigid motions' or isometries of Euclidean space. By differentiating these equations, we can arrive at some necessary conditions for the existence of an isometric immersion – namely, the Gauss-Codazzi equations stated below; these conditions are invariant under the symmetries and are also sufficient for the existence of a local isometric immersion. This result, due to the 19th-century geometer Bonnet, is sometimes known as the 'fundamental theorem of surface theory' (see for example [**10**], Vol. 3, Ch. 2). We state it here in a more concise form, in terms of tensors:

2010 *Mathematics Subject Classification.* Primary 37K35; Secondary 35H30, 35L60, 37K10, 58A15.

THEOREM 1.1 (Bonnet). *Let (M, g) be an abstract Riemannian surface, and let h be a symmetric 2-tensor on M satisfying the conditions*

$$\det(g^{-1}h) = K, \tag{1.2}$$

$$\nabla_i h_{jk} - \nabla_j h_{ik} = 0, \tag{1.3}$$

where K is the Gauss curvature of g and $\nabla_i h_{jk}$ denotes the components in local coordinates of the covariant derivative of h with respect to the Levi-Civita connection of g. Then given any point $p \in M$, there exists an open set $\mathcal{U} \subset M$ containing p, and an isometric immersion $\mathsf{f} : \mathcal{U} \to \mathbb{R}^3$ with second fundamental form h. Moreover, f is unique up to rigid motions.

Individually, (1.2) and (1.3) are respectively known as the Gauss and Codazzi equations, and the latter can be more succinctly formulated as saying ∇h is a totally symmetric 3-tensor. One way to prove the existence portion of Bonnet's theorem is to re-formulate (1.1) as an overdetermined system of linear differential equations for the mapping f and an adapted moving frame, where the Gauss and Codazzi equations are the solvability conditions for the linear system (see [**10**] *loc. cit.*).

Despite the elegance of the formulation of the Gauss-Codazzi equations, they present a combination of algebraic and differential conditions for multiple unknowns that would be difficult to solve using methods available in Bonnet's time. In hope of writing down solutions of (1.1) in at least some particular cases, 19th-century geometers such as Darboux and Weingarten sought to reduce the solvability conditions to a single scalar PDE. Two approaches to this are described in [**4**]; we discuss the first one here, following the exposition in Chapter 3 of [**8**]. (The second approach is the method of Weingarten, which we hope to discuss in a future paper.)

Let $\mathsf{f} : M \to \mathbb{R}^3$ be an isometric immersion and let $\mathsf{n} : M \to \mathbb{R}^3$ be a unit normal vector field along the image. Regarding f and n as vector-valued functions on M, differentiating $\mathsf{n} \cdot d\mathsf{f} = 0$ shows that the second fundamental form of the image surface is given by

$$h = \mathsf{n} \cdot \nabla^2 \mathsf{f} = -d\mathsf{n} \cdot d\mathsf{f}. \tag{1.4}$$

Fix a unit vector $\hat{\mathsf{v}}$ in $\mathbb{R}^3$, and on M define the function $u = \hat{\mathsf{v}} \cdot \mathsf{f}$. Differentiating this gives

$$du = \hat{\mathsf{v}} \cdot d\mathsf{f} \tag{1.5}$$

and $\nabla^2 u = \hat{\mathsf{v}} \cdot \nabla^2 \mathsf{f}$, and then from (1.4) we have

$$\nabla^2 u = (\hat{\mathsf{v}} \cdot \mathsf{n})h$$

at points where $\hat{\mathsf{v}} \cdot \mathsf{n} \neq 0$. This in turn, along with the Gauss equation (1.2), implies that

$$\det(g^{-1}\nabla^2 u) = (\hat{\mathsf{v}} \cdot \mathsf{n})^2 K.$$

Moreover, resolving $\hat{\mathsf{v}}$ in terms of its tangential and normal components and using (1.5) shows that $(\hat{\mathsf{v}} \cdot \mathsf{n})^2 = 1 - |du|^2$, where the norm on the right is taken with respect to g. As a result, $u = \hat{\mathsf{v}} \cdot \mathsf{f}$ must satisfy the scalar PDE

$$\det(g^{-1}\nabla^2 u) = (1 - |du|^2)K, \tag{1.6}$$

which has come to be known as the *Darboux equation* (see, e.g., Chapter 3 in [**8**]). Notice that form of (1.6) does not depend on the choice of unit vector $\hat{\mathsf{v}}$. In fact, by letting $\hat{\mathsf{v}}$ run over the standard orthonormal basis $\{\hat{\imath}, \hat{\jmath}, \hat{k}\}$ for $\mathbb{R}^3$, we see that every component of f must be a solution of (1.6). Not only are solutions of this PDE necessary to an isometric immersion, but they are also sufficient (at least locally). For, given a single solution u of (1.6) such that $|du|^2 < 1$, the metric $g - du^2$ is nondegenerate and flat, and thus there exist local coordinates x, y such that $g - du^2 = dx^2 + dy^2$ (see, e.g., Lemma 3.1.1 in [**8**] for details), and so $\mathsf{f} = (x, y, u)$ constitutes a local isometric immersion .

The main object of this article is determining the set of metrics of non-vanishing Gauss curvature for which the partial differential equation (1.6) is integrable in the sense of the *method of Darboux*; see Theorem 4.3 below for the precise statement of the main result. The article is organized as follows. In §2 we review some basic notions from the theory of exterior differential systems (EDS), including the concepts of prolongation and integrable extension, and the definition of Darboux integrability. In §3 we use moving frames to formulate the isometric immersion problem as an EDS $\mathcal{I}_0$, show how it is related to the Gauss-Codazzi equations, and show that the metrics for which the Gauss-Codazzi system is Darboux-integrable coincides exactly with the set of metrics for which $\mathcal{I}_0$ is Darboux-integrable (see Theorem 3.5); the latter classification was obtained in [**3**]. In §4 we formulate the Darboux equation as an EDS $\mathcal{D}_0$, and show that its prolongation is Darboux-integrable for exactly the same set of metrics.

2. Exterior differential systems

In this section we review essential definitions used in later sections; more detailed discussion of these concepts may be found in [**9**].

Let M^m be a smooth manifold, let $\Omega^k(M)$ denote the $C^\infty(M)$-module of smooth differential k-forms on M, and let $\Omega^\bullet(M) = \oplus_{k=0}^m \Omega^k(M)$. This is an algebra with respect to the wedge product, and an *exterior differential system* on M is a homogeneous ideal in this algebra which is also closed under exterior differentiation. (For the sake of simplicity, we'll assume the ideal contains no 0-forms.) An EDS $\mathcal{I}$ may be specified by a giving a list of differential forms on M which are algebraic generators for the ideal, or more briefly giving a list of differential forms which generate the ideal along with their exterior derivatives. Ideals generated by 1-forms and their derivatives are known as *Pfaffian systems*. For such systems, we will the corresponding italic letter denote the vector sub-bundle of T^*M spanned by the 1-forms of the ideal; clearly, specifying the bundle I is equivalent to specifying the Pfaffian system $\mathcal{I}$, since the latter is algebraically generated by sections of I and their exterior derivatives. If r is the rank of the vector bundle then we say r is also the *rank* of the Pfaffian system. If the exterior derivatives of sections of I are already in the algebraic ideal generated by those sections, the Pfaffian system $\mathcal{I}$ is said to be Frobenius and (at least locally) can be generated by r closed 1-forms.

The analogue of a 'solution' for an EDS $\mathcal{I}$ is a submanifold $S \subset M$ such that every differential form in $\mathcal{I}$ pulls back, under the inclusion map, to vanish on S. These are called *integral submanifolds* of $\mathcal{I}$. Exterior differential systems are often accompanied by non-degeneracy conditions which take the form of an *independence condition* defined by a decomposable n-form $\Psi \in \Omega^n(M)$. In that case, an n-dimensional integral submanifold S is said to be *admissible* if Ψ pulls back to be nowhere vanishing on S. (The form Ψ need only be well-defined up to nonzero multiple, and up to the addition of n-forms in $\mathcal{I}$)

To define the prolongation of $\mathcal{I}$, we need to consider its set of integral elements. An n-dimensional subspace $\mathsf{E} \subset T_pM$ is said to be an *integral element* of EDS $\mathcal{I}$ if

$\mathsf{v} \lrcorner \varphi = 0$ for all vectors $\mathsf{v} \in \mathsf{E}$ and all $\varphi \in \mathcal{I}$. We define $\mathcal{E}_n(\mathcal{I}) \subset G_n(TM)$ as the set of all n-dimensional integral elements of $\mathcal{I}$, and when $\mathcal{I}$ has an independence condition we let $\mathcal{E}_n(\mathcal{I}, \Psi)$ denote the set of integral n-planes E satisfying the independence condition (i.e., $\Psi|_{\mathsf{E}} \neq 0$). In most cases, $\mathcal{E}_n(\mathcal{I}, \Psi)$ forms a smooth fiber bundle over M, so we will assume this.

The Grassmann bundle $G = G_n(TM)$ carries a natural Pfaffian system $\mathcal{C}$, known as the canonical contact system. The 1-forms of $\mathcal{C}$ are sections of a vector bundle $C \subset T^*G$ whose fiber at a point $(p, \mathsf{E}) \in G$ is $\pi^*\mathsf{E}^\perp$, where $\pi : G \to M$ is the bundle map and $\mathsf{E}^\perp$ denotes the annihilator of E inside $T^*_p M$. Then the *prolongation* of $\mathcal{I}$ is the pullback of $\mathcal{C}$ to $\mathcal{E}_n(\mathcal{I}, \Psi)$, with independence condition given by $\pi^*\Psi$. In practice, we calculate the prolongation by introducing fiber coordinates on $\mathcal{E}_n(\mathcal{I}, \Psi)$ and writing the 1-forms that annihilate an admissible integral n-plane in terms of these coordinates.

In subsequent sections, we will re-formulate the problem of isometrically immersed a surface into Euclidean space as the problem of constructing admissible integral surfaces for an exterior differential system $\mathcal{I}_0$. We will see that the prolongation $\mathcal{I}_1$ of this system incorporates the Gauss-Codazzi equations (1.2)-(1.3), and so we will refer to this prolongation as the *extended Gauss-Codazzi system.* We will recast the Darboux equation as an equivalent Pfaffian system. Then the relationship between the Darboux equation and the Gauss-Codazzi system will be made precise (see Proposition 4.1 below) through the notion of *integrable extension.*

Integrable extensions. Let $\mathcal{I}$ be an EDS on manifold M, let $\pi : N \to M$ be a submersion. Then the pullbacks to N of all differential forms of $\mathcal{I}$ form an EDS on N which we denote by $\pi^*\mathcal{I}$. But it is sometimes also possible to add to this ideal some new 1-forms which are not basic for π, and which form a new EDS without adding any generators of higher degree. Specifically, an EDS $\mathcal{E}$ on N is said to be an *integrable extension* of $\mathcal{I}$ if $\pi^*\mathcal{I} \subset \mathcal{E}$ and there exists a sub-bundle $C \subset T^*N$ whose sections are transverse to the fibers of π (i.e., the pairing between sections of C and vector fields tangent to the fiber is nondegenerate at each point) and such that $\mathcal{E}$ is algebraically generated by sections of C and the differential forms in $\pi^*\mathcal{I}$. It follows that if $S \subset M$ is an integral submanifold of $\mathcal{I}$, then $\mathcal{E}$ pulls back to $\pi^{-1}(S)$ to be a Frobenius system.

Darboux integrability for Monge systems. We define a *Monge exterior differential system of class* k as an EDS defined on a manifold of dimension $k+4$ and which, in the vicinity of any point, is algebraically generated by 1-forms $\psi_1, \ldots, \psi_k$ and two 2-forms Ω_1, Ω_2 (both sets assumed to be pointwise linearly independent). For example, any second order PDE for one function of two variables is equivalent to a Monge system of class 3, and if the PDE is a Monge-Ampère equation then it is equivalent to a Monge system of class 1 (see, e.g., Chapter 7 in [**9**]). Thus, Monge systems of class 1 also called Monge-Ampère exterior differential systems. Mirroring the typology of the corresponding PDE systems, Monge systems are divided into types as follows.

Again letting $I \subset T^*M$ denote the bundle spanned by the generator 1-forms of a class k Monge system $\mathcal{I}$, we let $W \subset \Lambda^2(T^*M/I)$ be the bundle spanned by the generator 2-forms. Then for $\Upsilon \in C^\infty(W)$ we define

$$Q(\Upsilon) = \Upsilon \wedge \Upsilon \wedge \psi_1 \wedge \ldots \wedge \psi_k.$$

This is only well-defined up to multiple at each point of M, as we could choose a different set of 1-form generators. Using the identification of the fiber of $\Lambda^{k+4}T^*M$ with $\mathbb{R}$ (provided, for example, by the choice of a volume form), Q becomes a scalar-valued quadratic form on the fiber of W which is well-defined up to a nonzero multiple. We say the Monge system is *elliptic, hyperbolic, or parabolic* at $p \in M$ according to whether this quadratic form is respectively nondegenerate and definite, nondegenerate and indefinite, or degenerate on the fiber W_p. We will limit our attention to Monge systems where the type is constant.

For hyperbolic Monge systems we let Υ_+ and Υ_- denote local sections of W which are null directions for Q. These 2-forms are decomposable, modulo sections of I, and we define the associated *characteristic Pfaffian systems* $\mathcal{V}_+$ (resp. $\mathcal{V}_-$) as generated by the factors of Υ_+ (resp. Υ_-) together with $\psi_1, \dots \psi_k$. A hyperbolic Monge system is defined to be *Darboux-integrable* if each characteristic system contains a pair of closed 1-forms which are linearly independent from each other and the sections of I. Such systems were classically known to be solvable by ODE methods (see, e.g., Chapter IX in [**6**]), but the understanding of these systems, and their solvability by means of group actions, has been greatly advanced by the recent work of Anderson, Fels and Vassiliou [**1**, **2**]. (In those papers, the characteristic systems are called *singular systems.*)

For an elliptic Monge system $\mathcal{I}$, there are complex conjugate local sections $\Upsilon, \overline{\Upsilon}$ of $W \otimes \mathbb{C}$ which are null directions for the $\mathbb{C}$-linear extension of Q. We analogously define the characteristic systems $\mathcal{V}, \overline{\mathcal{V}}$ to be generated by the sub-bundles of $T^*M \otimes \mathbb{C}$ spanned by $\psi_1, \dots, \psi_k$ and the factors of Υ (resp. $\overline{\Upsilon}$), and say that $\mathcal{I}$ is Darboux-integrable if $\mathcal{V}$ contains a pair linearly independent closed (complex-valued) 1-forms (which, by complex conjugation, implies the same for $\overline{\mathcal{V}}$).

For parabolic Monge systems, Darboux integrability is not applicable, since there is only one characteristic system and the existence of first integrals for that system does not enable us to reduce to solving ODE systems.

We also note that Darboux integrability is relatively easy to check. Recall that, for a Pfaffian system generated by sections of $I \subset T^*M$, the *derived system* of $\mathcal{I}$ is generated by the sections of I that are in the kernel of the map $\delta_I : \theta \mapsto d\theta$ mod I – in other words, by 1-forms whose exterior derivatives are in the algebraic ideal generated purely by the 1-forms of $\mathcal{I}$. (Notice that δ_I is a $C^\infty(M)$-linear mapping, so has a well-defined rank on each fiber of I; we assume this rank is constant.) Applying this construction iteratively results in a sequence of vector bundles $I \supset I^{(1)} \supset I^{(2)} \dots$, where each member of which is the derived system of the previous one, and may be calculated using only differentiation and linear algebra. Since this *derived flag* terminates in the highest-rank Frobenius system contained in $\mathcal{I}$, the Darboux integrability condition requires that the derived flag of each characteristic system terminate in a Frobenius system of rank at least two.

The prolongation of a Monge system of class k is a Monge system of class $k+2$ of the same type, and while the initial system may not be Darboux-integrable, it may turn out that its prolongation is. For example, we will see in §4 that the PDE (1.6) is equivalent to a Monge-Ampère system $\mathcal{D}_0$ that, for certain metrics, becomes integrable after one prolongation. Since this system $\mathcal{D}_0$ lives on the 5-dimensional space $J^1(M, \mathbb{R})$ of 1-jets of functions on M, whereas its prolongation lives on the space of 2-jets of such functions, we say that (1.6) becomes Darboux-integrable at the 2-jet level. There are examples given in Ch. VII of [**6**] of second-order PDE that become Darboux-integrable at the n-jet level for arbitrarily large n.

3. Isometric immersions via moving frames

For an oriented Riemannian surface M, we will let $\mathcal{F}_M$ denote the oriented orthonormal frame bundle of M. This is a principal $SO(2)$-bundle and carries canonical 1-forms η^1, η^2 which are semibasic for the projection to M. These are uniquely characterized by the property that if σ is a local section of $\mathcal{F}_M$, defined by an oriented orthonormal frame $\mathsf{v}_1, \mathsf{v}_2$, then $\tilde{\eta}^a = \sigma^* \eta^a$ is the dual coframe, where $1 \le a \le 2$. The 1-forms η^1, η^2, together with the connection 1-form η^1_2, comprise a global coframing on $\mathcal{F}_M$, and these 1-forms satisfy the structure equations

$$d\eta^1 = -\eta^1_2 \wedge \eta^2, \quad d\eta^2 = -\eta^2_1 \wedge \eta^1, \quad d\eta^1_2 = K\eta^1 \wedge \eta^2, \tag{3.1}$$

where $\eta^2_1 = -\eta^1_2$, and K is the Gauss curvature which is a well-defined function on M.

On the Euclidean side, we let $\mathcal{F}_{\mathbb{R}}$ denote the oriented orthonormal frame bundle of $\mathbb{R}^3$, which is a principal $SO(3)$-bundle. Here, the canonical 1-forms ω^i and connection forms $\omega^i_j = -\omega^j_i$ for $1 \le i, j \le 3$ are readily defined as coefficients in the derivatives of the vector-valued basepoint function x and functions $\mathsf{e}_1, \mathsf{e}_2, \mathsf{e}_3$ giving the members of the frame:

$$d\mathsf{x} = \mathsf{e}_i \omega^i, \tag{3.2}$$

$$d\mathsf{e}_i = \mathsf{e}_j \omega^j_i, \tag{3.3}$$

where we adopt the usual summation convention from now on. Differentiating these equations leads in turn to the structure equations

$$d\omega^i = -\omega^i_j \wedge \omega^j, \qquad d\omega^i_j = -\omega^i_k \wedge \omega^k_j.$$

An isometric immersion $\mathsf{f} : M \to \mathbb{R}^3$ induces a lift $\mathsf{f}' : \mathcal{F}_M \to \mathcal{F}_{\mathbb{R}}$ that maps any pair of oriented orthonormal vectors $\mathsf{v}_1, \mathsf{v}_2$ at $p \in M$ to the orthonormal frame at $\mathsf{f}(p)$ given by $\mathsf{e}_1 = \mathsf{f}_* \mathsf{v}_1$, $\mathsf{e}_2 = \mathsf{f}_* \mathsf{v}_2$ and $\mathsf{e}_3 = \mathsf{e}_1 \times \mathsf{e}_2$. The graphs of these lifts may be characterized as integral surfaces of an exterior differential system $\mathcal{I}_0$ which we will refer to as the *isometric immersion system*:

PROPOSITION 3.1 (Prop. 1 in [**3**]). *Let $\mathcal{I}_0$ be the Pfaffian system on $\Sigma_0 := \mathcal{F}_M \times \mathcal{F}_{\mathbb{R}}$ generated by the 1-forms*

$$\theta_0 := \omega^3, \quad \theta_1 := \omega^1 - \eta^1, \quad \theta_2 := \omega^2 - \eta^2, \quad \theta_3 := \omega^1_2 - \eta^1_2.$$

Then a surface $S \subset \Sigma_0$ is the graph of a lift f' of an isometric immersion defined on an open subset of M if and only if S is an integral surface of $\mathcal{I}_0$ satisfying the independence condition $\eta^1 \wedge \eta^2 \neq 0$.

REMARK 3.2. Our approach here to setting up the isometric immersion system is different from the approaches employed in [**7**] and in [**9**]. For example, in [**7**] the Pfaffian system generated by $\theta_0, \theta_1, \theta_2$ is called the *naive isometric embedding system*, and those authors mostly work with the full prolongation of this system, while the system we use here is the smallest partial prolongation of the naive system that is involutive.

Although $\mathcal{I}_0$ is defined on the 9-dimensional manifold Σ_0, it is the pullback of a Monge system of class 4 defined on an 8-dimensional quotient. To see this, note that there is a natural $SO(2)$-action on the fibers of $\mathcal{F}_M$ that takes frame $(\mathsf{v}_1, \mathsf{v}_2)$ to a rotated frame $(\cos\phi\,\mathsf{v}_1 - \sin\phi\,\mathsf{v}_2, \cos\phi\,\mathsf{v}_2 + \sin\phi\,\mathsf{v}_1)$. When we similarly define an $SO(2)$-action on the fibers of $\mathcal{F}_{\mathbb{R}}$ that rotates vectors $\mathsf{e}_1, \mathsf{e}_2$ through an angle ϕ, the diagonal action of $SO(2)$ on Σ_0 gives a Cauchy characteristic symmetry of $\mathcal{I}_0$. Thus, $\mathcal{I}_0$ is the pullback of a well-defined EDS on the quotient manifold $\Sigma_0/SO(2)$, and this EDS is a Monge system. [1] For the sake of convenience, however, we will calculate the structure of $\mathcal{I}_0$ on Σ_0. For example, its generator 2-forms are

$$\Upsilon_1 = \omega_1^3 \wedge \eta^1 + \omega_2^3 \wedge \eta^2, \quad \Upsilon_2 = \omega_2^3 \wedge \omega_1^3 - K\eta^1 \wedge \eta^2.$$

It is easy now to check that the Monge system of elliptic, hyperbolic or parabolic type at a point of Σ_0 according to whether $K > 0$, $K < 0$ and $K = 0$ respectively at the underlying point on M. Again assuming that the system is of constant type, we can characterize those metrics on M for which $\mathcal{I}_0$ is Darboux-integrable:

THEOREM 3.3 (Theorem 2 in [**3**]). *Let M be a Riemannian surface with either strictly positive or strictly negative curvature K; in the first case let $\varepsilon = 1$, in the second case let $\varepsilon = -1$, and let $k = \sqrt{\varepsilon K}$. Then the system $\mathcal{I}_0$ is Darboux integrable if and only if k satisfies*

$$k_{11} = \frac{5}{2}\frac{k_1^2}{k} - 2\varepsilon k^3, \quad k_{12} = \frac{5}{2}\frac{k_1 k_2}{k}, \quad k_{22} = \frac{5}{2}\frac{k_2^2}{k} - 2\varepsilon k^3, \tag{3.4}$$

where k_a and k_{ab} for $1 \le a, b \le 2$ denote the components of the first and second covariant derivatives of k with respect to an orthonormal coframe on M.

THEOREM 3.4 (Theorem 5 in [**3**]). *Let M is a Riemannian surface with metric g satisfying the conditions in Theorem 3.3. For $K > 0$, there are local coordinates x, y on M in which the metric is one of the following:*

$$\begin{aligned} g_1 &= (\cosh^4 x)dx^2 + (\sinh^2 x)dy^2, \\ g_2 &= (\sinh^4 x)dx^2 + (\cosh^2 x)dy^2, \\ g_3 &= x^2(dx^2 + dy^2) \end{aligned}$$

for $x > 0$. For $K < 0$, there are local coordinates in which the metric is

$$g_4 = (\cos^4 x)dx^2 + (\sin^2 x)dy^2, \qquad x \in (0, \pi/2).$$

As we will see below, this is exactly the set of metrics for which the Darboux equation (1.6) is integrable by the method of Darboux. In order to establish this, we need to calculate the prolongation $\mathcal{I}_1$ of $\mathcal{I}_0$, which we will in turn connect to the Darboux equation in the next section.

If $\mathsf{E} \subset T_q\Sigma_0$ is an admissible integral 2-plane of $\mathcal{I}_0$, then the vanishing of $\Upsilon_1|_{\mathsf{E}}$ shows that there are numbers $h_{ab} = h_{ba}$ such that

$$(\omega_a^3 - h_{ab}\eta^b)|_{\mathsf{E}} = 0, \qquad 1 \le a, b \le 2, \tag{3.5}$$

[1] Another way of obtaining a Monge system would be to 'break' the Cauchy characteristic symmetry by fixing a choice of orthonormal frame on M. However, then we could only work on the open subset of M where the frame is defined, thus excluding globally defined immersions from the purview of Proposition 3.1. As well, we would have to introduce the components of the connection form η_1^2 with respect to the local frame as extra variables.

and the vanishing of $\Upsilon_2|_{\mathsf{E}}$ shows that these numbers must satisfy

$$h_{11}h_{22} - h_{12}^2 = K. \tag{3.6}$$

If S is an admissible integral surface of $\mathcal{I}_0$ then the h_{ab} are functions along S that give the components of the second fundamental form II of $f(M)$ under the corresponding isometric immersion f, with respect the orthonormal frame field $\mathsf{e}_1, \mathsf{e}_2$. Then we see that (3.6) is nothing but the Gauss equation. Moreover, the h_{ab} uniquely determine the image of E under the quotient map to $\Sigma_0/SO(2)$. Thus, we may regard the h_{ab} as fiber coordinates[2] for the prolongation $\mathcal{I}_1$, and the generator 1-forms for $\mathcal{I}_1$ are the pullbacks of the 1-forms of $\mathcal{I}_0$ with the addition of the 1-forms on the left in (3.5) which annihilate E.

More formally, we let h_{ab} be coordinates on the space $S^2(\mathbb{R}^2)$ of symmetric 2×2 matrices, let $\mathcal{H} \subset S^2(\mathbb{R}^2) \times \mathcal{F}_M$ be the hypersurface defined by the Gauss equation (3.6), let $\Sigma_1 := \mathcal{H} \times \mathcal{F}_{\mathbb{R}}$, and let $\mathbf{p}$ denote the restriction to Σ_1 of the projection from $S^2(\mathbb{R}^2) \times \Sigma_0$ onto the second factor Σ_0. Then we define the prolongation as the Pfaffian system $\mathcal{I}_1$ on Σ_1 generated by the pullbacks $\mathbf{p}^*\theta_0, \ldots, \mathbf{p}^*\theta_3$ of the generators from above, and the new 1-forms

$$\nu_1 := \omega_1^3 - h_{1a}\eta^a, \quad \nu_2 := \omega_2^3 - h_{2a}\eta^a.$$

The 2-form generators of $\mathcal{I}_1$ are computed to be

$$Dh_{1a} \wedge \eta^a \text{ and } Dh_{2a} \wedge \eta^a, \tag{3.7}$$

where

$$Dh_{ab} := dh_{ab} - h_{ac}\eta_b^c - h_{cb}\eta_a^c, \qquad 1 \le a, b, c \le 2.$$

Because the vanishing of the 2-forms in (3.7) is easily seen to be equivalent to the condition that the covariant derivative ∇II is a totally symmetric $(3,0)$-tensor, we see that $\mathcal{I}_1$ incoporates the Gauss-Codazzi equations for the second fundamental form of the immersion.

In general, the prolongation of a Darboux-integrable system is also Darboux-integrable, but (as mentioned earlier) a system that is not Darboux-integrable may become Darboux-integrable after sufficiently many prolongations. In the case of $\mathcal{I}_0$ we find that no additional metrics exist for which the prolongation $\mathcal{I}_1$ is Darboux-integrable:

THEOREM 3.5. *The extended Gauss-Codazzi system $\mathcal{I}_1$ is Darboux-integrable only when the conditions* (3.4) *in Theorem* 3.3 *hold.*

PROOF. The proof involves some lengthy calculations of derived systems which unfortunately cannot be included here for reasons of space; however, we will summarize the steps in the proof.

Let $\mathcal{V}_+^{(k)}$ (respectively, $\mathcal{V}_-^{(k)}$) denote the members of the derived flag of the characteristic system $\mathcal{V}_+$ (resp., $\mathcal{V}_-$) of $\mathcal{I}_0$, and similarly let $\widehat{\mathcal{V}}_\pm^{(k)}$ denote members of the derived flags of the characteristic systems of the prolongation. In [**3**], it was shown that $\mathcal{V}_+^{(1)}, \mathcal{V}_+^{(2)}$ have ranks 5 and 3 respectively for any metric where K

[2] Of course these are not really coordinates, since the Gauss equation makes them functionally dependent. When necessary, we may introduce independent fiber coordinates λ, μ by setting

$$h_{11} = ke^{\lambda}\cosh\mu, \quad h_{12} = k\sinh\mu, \quad h_{22} = ke^{-\lambda}\cosh\mu$$

in the elliptic case, or a similar parametrization with cosh and sinh interchanged in the hyperbolic case.

is not constant, and $\mathcal{V}_+^{(3)}$ has rank 2 if and only if the conditions (3.4) hold, in which case $\mathcal{V}_+^{(3)}$ is a Frobenius system. (All statements for the 'plus' systems carry over without change to the 'minus' systems.) Similarly, we find that $\widehat{\mathcal{V}}_+^{(1)}, \widehat{\mathcal{V}}_+^{(2)}, \widehat{\mathcal{V}}_+^{(3)}$ generically have ranks 7, 6 and 4 respectively. It is standard that $\mathbf{p}^*\mathcal{V}_+ \subset \widehat{\mathcal{V}}_+$, and consequently $\mathbf{p}^*\mathcal{V}_+^{(k)} \subset \widehat{\mathcal{V}}_+^{(k)}$, but in this case we also find that $\mathbf{p}^*\mathcal{V}_+ \subset \widehat{\mathcal{V}}_+^{(1)}$, and hence $\mathbf{p}^*\mathcal{V}_+^{(k)} \subset \widehat{\mathcal{V}}_+^{(k+1)}$. In particular, $\widehat{\mathcal{V}}_+^{(3)}$ contains the rank 3 subsystem $\mathbf{p}^*\mathcal{V}_+^{(2)}$. For the sake of brevity, let $\mathcal{W} = \mathbf{p}^*\mathcal{V}_+^{(2)}$ and $\mathcal{Z} = \widehat{\mathcal{V}}_+^{(3)}$, and let W and Z denote the corresponding bundles.

Suppose that the conditions (3.4) do not hold, but nevertheless $\mathcal{Z}$ contains a Frobenius subsystem $\mathcal{F}$ of rank 2. Then δ_Z has nullity at least 2. For dimensional reasons the nullspace of δ_Z in each fiber must have at least a one-dimensional intersection with the fiber of W. Because $\mathcal{W}$ is basic for $\mathbf{p}$, then the restriction of δ_Z to the fiber of W has the same nullspace as δ_W. Hence the nullspace of δ_W has dimension at least one – in other words, $\mathcal{V}_+^{(3)}$ must have rank at least one. However, by similar reasoning, if δ_Z has nullity equal to 3 then $\mathcal{V}_+^{(3)}$ would have rank 2, but this is impossible since the conditions in (3.4) do not hold. Thus, $\mathcal{V}_+^{(3)}$ must have rank exactly one and $\widehat{\mathcal{V}}_+^{(4)}$ must have rank exactly 2. Hence $\widehat{\mathcal{V}}_+^{(4)} = \mathcal{F}$ and $\mathbf{p}^*\mathcal{V}_+^{(3)}$ is contained in $\mathcal{F}$. Suppose $\mathcal{V}_+^{(3)}$ is spanned by a 1-form α. By writing $\mathbf{p}^*\alpha$ as a linear combination of two exact differentials in $\mathcal{F}$, one sees that α must have Pfaff rank at most one, i.e., $\alpha \wedge d\alpha$ may be nonzero but $\alpha \wedge (d\alpha)^2 = 0$. However, we can check that whenever $\mathcal{V}_+^{(3)}$ has rank one, its generator satisfies $\alpha \wedge (d\alpha)^2 \neq 0$. $\square$

4. Integrability of the Darboux Equation

In this section we will set up the Darboux equation as a Monge-Ampère EDS $\mathcal{D}_0$, show how it is related to the extended Gauss-Codazzi system $\mathcal{I}_1$, and determine the metrics for which the prolongation $\mathcal{D}_1$ of $\mathcal{D}_0$ is Darboux-integrable.

Given a function u on M, the components of the first and second covariant derivatives of u with respect to an orthonormal coframe depend on the choice of coframe, and are thus well-defined as functions on $\mathcal{F}_M$ rather than M. In terms of the canonical and connection forms on $\mathcal{F}_M$ these components are defined

$$du = u_a \eta^a, \qquad du_a = u_b \eta_a^b + u_{ab}\eta^b, \tag{4.1}$$

where, in the left-hand side of the first equation, u is pulled back to $\mathcal{F}_M$ via the projection to M. In terms of these components, the Darboux equation (1.6) is written as

$$u_{11}u_{22} - u_{12}^2 = (1 - u_1^2 - u_2^2)K. \tag{4.2}$$

The combinations of components on both sides are invariant under pointwise rotations of the frame, so this gives a well-defined condition on M.

As stated above, we wish to define a Monge-Ampére system that is equivalent to the partial differential equation (4.2). We note that, in light of (4.1), the equation (4.2) is equivalent to the vanishing of the 2-form

$$\Upsilon := (du_1 - u_2\eta_1^2) \wedge (du_2 - u_1\eta_2^1) - (1 - u_1^2 - u_2^2)K\, \eta^1 \wedge \eta^2. \tag{4.3}$$

Based on this observation, we will define $\mathcal{D}_0$ by beginning with $\mathcal{F}_M$ and adding u, u_1, u_2 as extra variables which are constrained by the nondegeneracy condition $|du|^2 < 1$. In other words, $\mathcal{D}_0$ is defined on the 6-dimensional manifold $\mathcal{F}_M \times \mathcal{W}$, where $\mathcal{W}$ denotes the open subset of $\mathbb{R}^3$ where the coordinates u, u_1, u_2 satisfy $u_1^2 + u_2^2 < 1$. The algebraic generators of $\mathcal{D}_0$ are the 1-form

$$\psi_0 := du - u_1\eta^1 - u^2\eta^2, \tag{4.4}$$

and the 2-forms $d\psi_0$ and Υ. This system is elliptic at points where $K > 0$ and hyperbolic at points where $K < 0$. Moreover, as with systems $\mathcal{I}_0$ and $\mathcal{I}_1$ above, $\mathcal{D}_0$ has a Cauchy characteristic symmetry corresponding to pointwise rotations of the frame, and thus descends to be a well-defined Monge-Ampère system on the five-dimensional quotient of $\mathcal{F}_M \times \mathcal{W}$ under the diagonal action of $SO(2)$. (Alternatively, we may break the symmetry by specializing to a particular choice of frame, which we will do when we work with specific metrics below.)

We are able to define an EDS encoding the Darboux equation on a relatively low-dimensional space, using only u and its first derivative components as variables, because the equation is only mildly nonlinear in the second derivatives of u. However, to make the connection with the extended Gauss-Codazzi system we must pass to the prolongation $\mathcal{D}_1$, and thus it is necessary to introduce the second covariant derivative components as new variables, and define additional 1-forms whose vanishing is equivalent to the equations (4.1). Accordingly, on $\mathcal{F}_M \times \mathbb{R}^6$ (where we take u, u_a and $u_{ab} = u_{ba}$ as coordinates on the second factor) define the 1-forms

$$\psi_0 := du - u_a\eta^a, \quad \psi_a := du_a - u_b\eta_a^b - u_{ab}\eta^b, \qquad 1 \le a, b \le 2, \tag{4.5}$$

let $\Sigma_2 \subset \mathcal{F}_M \times \mathbb{R}^6$ be the 8-dimensional submanifold where (4.2) holds and $u_1^2 + u_2^2 < 1$ (note that this condition ensures that Σ_2 is smooth), and let $\mathcal{D}_1$ be the Pfaffian system on Σ_2 generated by ψ_0, ψ_1, ψ_2 and their exterior derivatives. Then $\mathcal{D}_1$ is a Monge system of class 3 and is the prolongation of $\mathcal{D}_0$.

We now define a mapping $\mathbf{r}$ that takes integral surfaces of $\mathcal{I}_1$ to integral surfaces of $\mathcal{D}_1$. (This mapping will merely formalize the derivation of the Darboux equation from the Gauss-Codazzi equations that we gave earlier in §1.) As before, we fix a unit vector $\hat{\mathsf{v}}$ in $\mathbb{R}^3$, and let $\tilde{\Sigma}_1$ be the open subset of Σ_1 where $\mathsf{e}_3 \cdot \hat{\mathsf{v}} \neq 0$. We will first define $\mathbf{r} : \tilde{\Sigma}_1 \to \mathcal{F}_M \times \mathbb{R}^6$ by requiring $\mathbf{r}$ to be the identity on the $\mathcal{F}_M$-factor, and letting

$$u = \mathsf{x} \cdot \hat{\mathsf{v}}, \qquad u_a = \mathsf{e}_a \cdot \hat{\mathsf{v}}, \qquad u_{ab} = h_{ab}(\mathsf{e}_3 \cdot \hat{\mathsf{v}}). \tag{4.6}$$

Then, because the h_{ab} satisfy the Gauss equation (3.6), and $(u_1)^2 + (u_2)^2 = 1 - (\mathsf{e}_3 \cdot \hat{\mathsf{v}})^2 < 1$, the image of $\mathbf{r}$ lies in Σ_2.

PROPOSITION 4.1. *The mapping $\mathbf{r} : \tilde{\Sigma}_1 \to \Sigma_2$ makes $\mathcal{I}_1$ an integrable extension of $\mathcal{D}_1$.*

PROOF. We must show first that the pullbacks of the generator 1-forms of $\mathcal{D}_1$ lie in $\mathcal{I}_1$, and then define a complementary sub-bundle C such that $\mathcal{I}_1$ is generated algebraically by sections of C and the pullbacks of forms in $\mathcal{D}_1$.

First, using (3.2), (3.3) and (4.6) we compute

$$\mathbf{r}^* du = (\mathsf{e}_a \cdot \hat{\mathsf{v}})\omega^a, \quad \mathbf{r}^* du_a = (\mathsf{e}_i \cdot \hat{\mathsf{v}})\omega_a^i.$$

From these it follows, respectively, that

$$\mathbf{r}^*\psi_0 = (\mathsf{e}_a \cdot \hat{\mathsf{v}})\theta_a,$$

and

$$\begin{aligned} \mathbf{r}^*\psi_1 &= -(\mathsf{e}_2 \cdot \hat{\mathsf{v}})\theta_3 + (\mathsf{e}_3 \cdot \hat{\mathsf{v}})\nu_1, \\ \mathbf{r}^*\psi_2 &= (\mathsf{e}_1 \cdot \hat{\mathsf{v}})\theta_3 + (\mathsf{e}_3 \cdot \hat{\mathsf{v}})\nu_2. \end{aligned} \tag{4.7}$$

Thus, the pullbacks of the generator 1-forms of $\mathcal{D}_1$ span a rank-3 sub-bundle within I_1. We define C as spanned by θ_0, θ_3 and $(\mathsf{e}_1 \cdot \hat{\mathsf{v}})\theta_2 - (\mathsf{e}_2 \cdot \hat{\mathsf{v}})\theta_1$. Thus, the 1-forms of $\mathcal{I}_1$ may be expressed as linear combinations of sections of C and $\mathbf{r}^*\psi_0, \mathbf{r}^*\psi_1, \mathbf{r}^*\psi_2$. Since $\mathcal{I}_1$ is a prolongation, its first derived system is spanned by the pullbacks from Σ_1 of the 1-forms $\theta_0, \ldots, \theta_3$ of $\mathcal{I}_0$, and thus its 2-form generators are $d\nu_1$ and $d\nu_2$. Because the rightmost coefficients in (4.7) are nonzero, $\mathbf{r}^*d\psi_1$ and $\mathbf{r}^*d\psi_2$ give an equivalent pair of 2-form generators. Thus, $\mathcal{I}_1$ is algebraically generated by $\mathbf{r}^*\mathcal{D}_1$ and sections of C. □

COROLLARY 4.2. If $\mathcal{D}_1$ is Darboux-integrable, then $\mathcal{I}_1$ is Darboux-integrable.

PROOF. Because $\mathcal{I}_1$ is an integrable extension of $\mathcal{D}_1$, this follows by Theorem 5.1 in [**1**]. □

This lets us prove our main result:

THEOREM 4.3. *The Darboux equation* (1.6) *associated to a metric g on surface M is Darboux-integrable at the 2-jet level if and only if g is locally isometric to one of the metrics listed in Theorem* 3.4.

PROOF. From Corollary 4.2 it follows that if the system $\mathcal{D}_1$ is Darboux-integrable then $\mathcal{I}_1$ is Darboux-integrable, and hence by Theorem 3.5 the metric must satisfy the conditions (3.4). It only remains to check that, for each of the metrics listed in Theorem 3.4, the system $\mathcal{D}_1$ is in fact Darboux-integrable.

We begin with metric g_4, since in this case $\mathcal{D}_0, \mathcal{D}_1$ are hyperbolic and the characteristic systems are real. Before showing that $\mathcal{D}_1$ is Darboux-integrable, we will calculate generators and characteristic systems for $\mathcal{D}_0$ on $M \times \mathcal{W}$, pulling the previously-defined system back by a specific choice of orthonormal coframe on M:

$$\tilde{\eta}^1 = \cos^2 x\, dx, \quad \eta^2 = \sin x\, dy,$$

Then the structure equations (3.1) imply that $\tilde{\eta}^1_2 = -\sec x\, dy$ and $K = -\sec^4 x$. In place of the coordinates u_1, u_2 on $\mathcal{W}$ we will use alternate coordinates θ, ϕ related to them by

$$u_1 = \sin\phi\cos\theta, \quad u_2 = \sin\phi\sin\theta, \qquad \phi \in (0, \pi/2). \tag{4.8}$$

Using (4.3) and (4.4) we now express the generators of $\mathcal{D}_0$ as

$$\psi_0 = du - \sin\phi\,(\cos\theta\,\tilde{\eta}^1 + \sin\theta\,\tilde{\eta}^2) \tag{4.9}$$

and

$$\Upsilon = \cos\phi\sin\phi\, d\phi \wedge (d\theta + \sec x\, dy) + \cos^2\phi\sin x\sec^2 x\, dx \wedge dy.$$

By solving for the null directions of the quadratic form Q defined in §2 above, we find that the decomposable 2-form generators of the ideal are $\Upsilon \mp (\cos\phi\sec^2 x)d\psi_0$, and the characteristic systems are

$$V_\pm = \{\psi_0, d\phi \pm (\sec^2 x)\sigma, d\theta + \sec x\, dy \pm (\cot\phi\sec^2 x)\tau\}$$

where, for the sake of convenience, we have set

$$\sigma := (\sin\theta)\tilde{\eta}^1 - (\cos\theta)\tilde{\eta}^2, \qquad \tau := (\cos\theta)\tilde{\eta}^1 + (\sin\theta)\tilde{\eta}^2. \tag{4.10}$$

On any admissible integral surface of $\mathcal{D}_0$, each system V_+, V_- pulls back to have rank one. In particular, the last two members of each system restrict to be linearly dependent. Thus, the prolongation $\mathcal{D}_1$ is generated by ψ_0 and

$$\psi_1' := \tan\phi(d\theta + \sec x\, dy) + (\sec^2 x)\tau - p(d\phi + (\sec^2 x)\sigma),$$
$$\psi_2' := \tan\phi(d\theta + \sec x\, dy) - (\sec^2 x)\tau - q(d\phi - (\sec^2 x)\sigma),$$

where we introduce p, q as new variables that give the coefficients of these linear dependencies, and are thus fiber coordinates for the prolongation. After subtracting off suitable multiples of ψ_1', ψ_2', the exterior derivatives $d\psi_1'$ and $d\psi_2'$ become decomposable 2-forms. Each characteristic system of $\mathcal{D}_1$ is generated by the 1-forms of $\mathcal{D}_1$ and the factors of one of these decomposable 2-forms; for example, the factors in $d\psi_1'$ lead to

$$\widehat{V}_+ = \{\psi_0, \psi_1', \psi_2', \tau - q\sigma, \cot x(\cos^2 x\, dp - \cot\phi(1+p^2)\tau) + (\sin\theta + p\cos\theta)(\tau - p\sigma)\},$$

while the generators of the other characteristic system are obtained by switching p and q, and changing angle ϕ to its complement $\pi/2 - \phi$.

Each characteristic system turns out to contain a rank 2 Frobenius system, so that $\mathcal{D}_1$ is Darboux-integrable. In the case of $\widehat{V}_+$, the Frobenius system is

$$\widehat{V}_+^{(\infty)} = \{\cos\theta\, d\theta + (\sin\theta\cot\phi + \cot x)d\phi + (\sin\theta\cot x + \cot\phi)dx,$$
$$dp + p\tan x(\sin\theta + p\cos\theta)d\phi + \left((p+\tan\theta)\tan x - (1+p^2)\sec\theta\cot\phi\right)(dx + \sin\theta\, d\phi)\}.$$

We note that first 1-form does not involve the prolongation variables p, q, and thus is the pullback of a 1-form defined on $M \times \mathcal{W}$; in fact, this 1-form is integrable, and spans the terminal derived system of V_+. The Frobenius system inside $\widehat{V}_-$ is obtained from $\widehat{V}_+^{(\infty)}$ by the same switching of variables mentioned above, and also contains the pullback of the rank one terminal derived system of V_-.

We now turn to the metrics g_1, g_2, g_3, for which the Gauss curvature is positive, and hence the systems $\mathcal{D}_0, \mathcal{D}_1$ are of elliptic type. The decomposable 2-forms in this case will be complex-valued, and the same goes for the 1-forms generating the characteristic systems, but for $\mathcal{D}_0$ and $\mathcal{D}_1$ each pair of characteristic systems will be related by complex conjugation, so that it is only necessary to do calculations for one of the pair.

Other than the complexification, the calculation for metric g_1 resembles that for g_4, with $\cos x$ and $\sin x$ being replaced by their hyperbolic counterparts. The orthonormal coframe on M is

$$\tilde{\eta}^1 = \cosh^2 x\, dx, \quad \tilde{\eta}^2 = \sinh x\, dy,$$

resulting in $\tilde{\eta}^1_2 = -\operatorname{sech} x\, dy$ and $K = \operatorname{sech}^4 x$. We make the same change of variable as in (4.8), so that the 1-form generator of $\mathcal{D}_0$ takes the same form as in (4.9). The 2-form generator of $\mathcal{D}_0$ is

$$\Upsilon = \cos\phi\sin\phi\, d\phi \wedge (d\theta + \operatorname{sech} x\, dy) + \cos^2\phi\sinh x\operatorname{sech}^2 x\, dx \wedge dy.$$

The decomposable 2-forms in the ideal are $\Upsilon \mp \mathrm{i}(\cos\phi\operatorname{sech}^2 x)d\psi_0$; taking the factors in the case of the upper sign gives the characteristic system

$$V_+ = \{\psi_0, d\phi + \mathrm{i}(\operatorname{sech}^2 x)\sigma, d\theta + \operatorname{sech} x\, dy + \mathrm{i}(\cot\phi\operatorname{sech}^2 x)\tau\}$$

where we again take σ, τ as in (4.10).

The prolongation $\mathcal{D}_1$ is generated by ψ_0 and the real and imaginary parts of

$$\psi_1' := \tan\phi(d\theta + \operatorname{sech} x\, dy) + \mathrm{i}(\operatorname{sech}^2 x)\tau - (p + \mathrm{i}q)(d\phi + \mathrm{i}(\operatorname{sech}^2 x)\sigma)\}$$

where fiber coordinates p, q are real-valued. Computing the exterior derivative $d\psi_1'$ modulo the 1-form generators of $\mathcal{D}_1$ yields a decomposable 2-form whose factors allow us to form the characteristic system

$$\widehat{V}_+ = \{\psi_0, \operatorname{Re}\psi_1', \operatorname{Im}\psi_1', \tau - \overline{z}\sigma, \coth x(\cosh^2 x\, dz - \mathrm{i}\cot\phi(1+z^2)\tau) + (\sin\theta + z\cos\theta)(z\sigma - \tau)\},$$

where we have set $z = p + iq$. (Again, the other characteristic system is the complex conjugate of this one.) Each characteristic system contains a rank 2 Frobenius system, for example

$$\widehat{V}_+^{(\infty)} = \{\cos\theta\, d\theta + (\sin\theta\cot\phi - \mathrm{i}\coth x)d\phi + (\sin\theta\coth x + \mathrm{i}\cot\phi)dx,$$
$$dz + \mathrm{i}z\tanh x(\sin\theta + z\cos\theta)d\phi - \big((z + \tan\theta)\tanh x + \mathrm{i}(1 + z^2)\sec\theta\cot\phi\big)\,(dx - \mathrm{i}\sin\theta d\phi)\}.$$

Again, the first 1-form in this system is integrable, and is the pullback of a 1-form spanning the terminal derived system of V_+.

The calculation for metric g_2 is similar, but with $\cosh x$ and $\sinh x$ interchanged.

The calculation for metric g_3 is similar to g_1 and g_2 but simpler in form. The orthonormal coframe on M is

$$\tilde{\eta}^1 = x\, dx, \quad \eta^2 = x\, dy,$$

resulting in $\eta^1_2 = -(1/x)dy$ and $K = 1/x^4$. The variable change (4.8) and form of ψ_0 are the same. The 2-form generator of $\mathcal{D}_0$ is

$$\Upsilon = \cos\phi\sin\phi\, d\phi \wedge (d\theta + x^{-1}dy) + x^{-2}\cos^2\phi\, dx \wedge dy.$$

We define σ, τ as in (4.10), and in terms of these one characteristic system is

$$V_+ = \{\psi_0, x^2 d\phi + \mathrm{i}\sigma, x^2 d\theta + x\, dy + \mathrm{i}(\cot\phi)\tau\}.$$

The prolongation $\mathcal{D}_1$ is again generated by ψ_0 and the real and imaginary parts of

$$\psi_1' := \tan\phi(x^2 d\theta + x\, dy) + \mathrm{i}\tau - (p + \mathrm{i}q)(x^2 d\phi + \mathrm{i}\sigma).$$

The corresponding characteristic system of the prolongation is

$$\widehat{V}_+ = \{\psi_0, \operatorname{Re}\psi_1', \operatorname{Im}\psi_1', \tau - \overline{z}\sigma, x^2 dz - \mathrm{i}\cot\phi(1 + z^2)\tau + (\sin\theta + z\cos\theta)(z\sigma - \tau)\},$$

and this contains a rank 2 Frobenius system

$$\widehat{V}_+^{(\infty)} = \{\cos\theta d\theta + (\sin\theta\cot\phi - \mathrm{i})d\phi + (\sin\theta + \mathrm{i}\cot\phi)x^{-1}dx,$$
$$dz + \mathrm{i}z(\sin\theta + z\cos\theta)d\phi - (z + \tan\theta + \mathrm{i}(1 + z^2)\sec\theta\cot\phi)(x^{-1}dx - \mathrm{i}\sin\theta d\phi).$$

□

Acknowledgments

The author is grateful to the organizers of a special session on geometry of submanifolds, held in honor of the 70th birthday of Professor Bang-Yen Chen, for inviting this submission. The author also acknowledges the support of the National Science Foundation, through an Independent Research and Development plan undertaken by the author while an NSF employee. Finally, the author is grateful to the referee for helpful suggestions for improving the first draft of this submission.

References

[1] I. M. Anderson and M. E. Fels, *Bäcklund transformations for Darboux integrable differential systems*, Selecta Math. (N.S.) **21** (2015), no. 2, 379–448, DOI 10.1007/s00029-014-0159-5. MR3338681

[2] I. M. Anderson, M. E. Fels, and P. J. Vassiliou, *Superposition formulas for exterior differential systems*, Adv. Math. **221** (2009), no. 6, 1910–1963, DOI 10.1016/j.aim.2009.03.010. MR2522831

[3] J. N. Clelland, T. A. Ivey, N. Tehseen, and P. J. Vassiliou, *Isometric embedding and Darboux integrability*, Geom. Dedicata **203** (2019), 353–388, DOI 10.1007/s10711-019-00441-5. MR4027599

[4] G. Darboux, *Leçons sur la théorie générale des surfaces et les applications géométriques du calcul infinitésimal. Quatrième partie.* Déformation infiniment petite et representation sphérique; Réimpression de la première édition de 1896; Chelsea Publishing Co., Bronx, N. Y., 1972. MR0396214

[5] L. P. Eisenhart, *A treatise on the differential geometry of curves and surfaces*, Ginn, 1909; Dover Publications, Inc., New York, 1960. MR0115134

[6] E. Goursat, *Leçons sur l'intégration des équations aux dérivées partielles du second ordre*, Gauthier-Villars, 1890.

[7] P. A. Griffiths and G. R. Jensen, *Differential systems and isometric embeddings*, Annals of Mathematics Studies, vol. 114, Princeton University Press, Princeton, NJ, 1987. The William H. Roever Lectures in Geometry. MR890959

[8] Q. Han, J-X. Hong, *Isometric Embedding of Riemannian Manifolds*, American Mathematical Society, 2006.

[9] T. A. Ivey and J. M. Landsberg, *Cartan for beginners: Differential geometry via moving frames and exterior differential systems*, Graduate Studies in Mathematics, vol. 175, American Mathematical Society, Providence, RI, 2016. Second edition [of MR2003610]. MR3586335

[10] M. Spivak, *A comprehensive introduction to differential geometry. Vol. I*, 2nd ed., Publish or Perish, Inc., Wilmington, Del., 1979. MR532830

[11] D. J. Struik, *Lectures on classical differential geometry*, 2nd ed., Dover Publications, Inc., New York, 1988. MR939369

Dept. of Mathematics, College of Charleston, 66 George St., Charleston SC 29424
Email address: iveyt@cofc.edu

Contemporary Mathematics
Volume **756**, 2020
https://doi.org/10.1090/conm/756/15204

On the geometry of Einstein spaces: a note on their curvature symmetries

Adela Mihai

This paper is dedicated to Professor Bang-Yen Chen on the occasion of his 75-th birthday.

ABSTRACT. We recall few justifications for the study and therefore for the definition of an Einstein manifold. Recent results from the author and U. Simon on curvature symmetries on Einstein spaces of arbitrary dimension $n \geq 4$ are given.

1. Motivation

In the Introduction of the well-known monograph of A. Besse [**B**], the following question asked in 1995 by René Thom at Strasbourg Mathematical Library was recalled:

Are there any best (or nicest, or distinguished) Riemannian structures on M? (where M is a compact C^∞-differentiable manifold, $\dim M = n$.)

For $n = 2$, the best Riemannian structures on a given compact surface M are those of constant (Gauss) curvature. For $n \geq 3$, a Riemannian manifold (M, g) of dimension n has 3 main notions of curvature:

1. The *Riemann curvature tensor* $\mathcal{R}$ (equivalent to the sectional curvature function κ, defined on the set of all tangent planes) - a bi-quadratic form giving complete information, at the curvature level, on the metric g.
2. The *Ricci curvature* $\mathcal{R}ic$, the trace of $\mathcal{R}$ with respect to g.
3. The *scalar curvature* τ, a scalar function on M, the trace of $\mathcal{R}ic$ with respect to g.

Remark that if $n = 2$ all the above three curvatures are equivalent (and this is the reason to consider $n \geq 3$). On the other hand, for $n = 3$, the Ricci curvature contains as much information as the Riemann curvature tensor.

It follows that the interest starts with $n = 4$.

A natural generalization of the concept of constant sectional curvature would be the notion of *constant Ricci curvature.*

The *Ricci curvature* of an n-dimensional Riemannian manifold (M, g) is a quadratic differential form, denoted by $\mathcal{R}ic$ (or $\mathcal{R}ic_g$ if the metric needs to be specified),

$$\mathcal{R}ic(X) := \text{trace}(Z \mapsto \mathcal{R}(X, Z)X).$$

2010 *Mathematics Subject Classification.* Primary 53C25; Secondary 53B20.
Key words and phrases. Einstein manifold; Ricci curvature; curvature symmetries.
The author was supported in part by Grant # PN-III-P1-1.1-MC-2018-2760 of the Romanian Ministry of Research and Innovation.

Equivalently, for $X \in UM$ (the unit tangent bundle of M), it represents the sum

$$\mathcal{R}ic(X) = \sum_{i=2}^{n} \kappa(X, X_i)$$

of the sectional curvatures $\kappa(X, X_i)$, for any completion of X to an orthonormal basis $\{X, X_2, ..., X_n\}$.

DEFINITION 1.1. A Riemannian manifold (M, g) of dimension $n \geq 3$ is called an *Einstein space* if $\mathcal{R}ic = \lambda \cdot id$.

Trivially, λ is the normalized scalar curvature. If $n \geq 3$ one easily proves that λ has to be constant for any Einstein manifold. Any 2-dimensional Riemannian n-manifold satisfies the relation $\mathcal{R}ic = \lambda \cdot id$, but for $n = 2$ the function λ is not necessarily a constant! It is well known that any 3-dimensional Einstein space is of constant sectional curvature.

It follows that the interest in Einstein spaces starts (again) with dimension $n = 4$.

Next, we will recall some justifications for the fact that the Einstein metric is a good candidate for a privileged metric (see [**B**]).

First justification.

Constancy of sectional curvature is too strong (many manifolds of dimension greater than 4 cannot admit such a metric).

Constancy of scalar curvature is too weak (on any compact manifold of any dimension there exist such Riemannian metrics and they form an infinite dimensional family for $n \geq 3$).

Constancy of the Ricci curvature is more appropriate; both the metric and the Ricci curvature depend on the same number of parameters $\frac{n(n+1)}{2}$ and they form a finite dimensional family, for which many examples are known.

Second justification.

The Einstein condition for a compact Riemannian manifold (M, g_0) is equivalent to the following: on the space $\mathcal{M}_1(M)$ of metrics of volume one, the functional

$$\mathbf{S} : g \mapsto \int_M \tau_g dvol_g$$

(the total scalar curvature) admits g_0 as critical point, i.e., for any variation $g_0 + th$, where h belongs to the space of bilinear symmetric differential forms on M, the derivative

$$\frac{d}{dt}\mathbf{S}(g_0 + th) \mid_{t=0} = 0.$$

Thus, the Einstein metrics can be considered as privileged!

Third justification.

The constant Ricci curvature is equivalent to the constant curvature in dimension 2. The best Riemannian structures on a given compact surface are those of constant curvature. For any compact surface, there exist metrics with constant curvature and they form a finite dimensional set.

A Riemannian metric on a 3-dimensional manifold is Einstein if and only if it has constant sectional curvature. In contrast with the 2-dimensional case, every 3-manifold does not admit a metric with constant sectional curvature.

Justification from Physics.

In Relativity, one considers Lorentzian manifolds instead of Riemannian manifolds. Einstein (1913) proposed the famous field equation

$$\mathcal{R}ic - \frac{1}{2}\tau g = T,$$

where T is the energy-momentum tensor.

No mass corresponds to Ricci-flat manifolds: $\mathcal{R}ic = 0$, from where the Euler-Lagrange equation was derived. In fact, $\mathcal{R}ic - \frac{1}{2}\tau g$ is the gradient of the total scalar curvature functional. If one considers only metrics of volume equal to 1, one gets the *Einstein metrics.*

At the end this section we recall the decomposition (linear representation) of the curvature tensor $\mathcal{R}$, $\mathcal{R} = U + Z + W$, into pieces corresponding to the irreducible components of this action, where W is the Weil conformal curvature tensor, U is equivalent to the scalar curvature τ, Z is the traceless part of the Ricci curvature. $Z = 0$ is equivalent to the *Einstein condition.*

For the approach to the subject, we recommend also [**Ber**] and [**Bou**].

2. The "feel" of Ricci curvature

To get the "feel" of Ricci curvature, we will follow the explanations given by A. Besse in [**B**].

Let $M \subset \mathbf{R}^{n+1}$ be a hypersurface embedded in a Euclidean space; its second fundamental form is a quadratic differential form on M. Its eigenvalues (with respect to the metric g) are the principal curvatures of M and its eigendirections yield the lines of curvature of M. If (M, g) is an abstract Riemannian manifold, there is no such quadratic form, no privileged directions.

Then G.C. Ricci introduced the trace of the curvature tensor.

The Ricci eigendirections in abstract Riemannian geometry do not appear to have geometrical properties, with the exception of conformal Riemannian geometry [**E**] and some other special manifolds (see [**B**], Chapter 16).

Trying to discover the geometric meaning of the Ricci curvature, one can think on *volume.*

The Ricci curvature is a trace and the trace appears as a derivative of a determinant. The Ricci curvature plays an important role in the proof of the following formula.

Let $B(p, t)$ be a geodesic ball (the set of points in (M, g) at the distance from the point p smaller than t). Then its volume it is expressed by

$$vol(B(p,t)) = \omega(n)t^n \left[1 - \frac{1}{6n+12}\tau(p)t^2 + o(t^2)\right],$$

where $\omega(n)$ denotes the volume of the unit ball of $\mathbf{R}^n$ and $\tau(p)$ is the scalar curvature at p.

On the other hand, to work with a measure on (M, g) one transfers things to the tangent space T_pM by means of the exponential map. From this, one can find that the Ricci curvature measures the defect of the solid angle (i.e. a measure of the amount of the field of view from some particular point that a given object covers; that is, it is a measure of how large the object appears to an observer looking from that point) in (M, g) from being equal to that of Euclidean space.

In the case of Kaehler manifolds, the Ricci curvature, transformed by the complex structure into a 2-form of type $(1,1)$ is nothing but the $\partial\bar{\partial}$ second derivative of the volume function.

Another important discovery was made by R. Bishop [**Bis**]: the Ricci curvature yields a global inequality! A short story on Bishop's theorem can be found in [**B**].

3. Basic definitions and notational conventions

Let (M,g) be a connected, oriented Riemannian manifold of dimension $n \geq 2$. For $p \in M$ we denote its tangent space by T_pM, and a local frame around p by $\{e_1,...,e_n\}$. We assume the frame to be orthonormal. If $L \subset T_pM$ is a subspace, we denote its orthogonal complement by $L^\perp$.

By $\mathcal{R}$ we denote the $(0,4)$-Riemannian curvature tensor, by Ric its $(0,2)$-Ricci tensor and by $\mathcal{R}ic$ its associated Ricci operator with eigenvalues $\rho_1,...,\rho_n$. Its trace gives rise to the *normalized scalar curvature* κ, namely

$$n \cdot \kappa := \tau := \operatorname{trace} \mathcal{R}ic = \sum_{j=1}^{n} \rho_j,$$

where τ is the *scalar curvature.*

For the 2-plane $\operatorname{span}(e_i,e_j) \subset T_pM$ the sectional curvature is denoted by

$$\kappa_{ij} := \kappa(e_i,e_j),$$

where $i \neq j$; formally we set $\kappa_{ii} = 0$.

We have the relation

$$\rho_r = \sum_s \kappa_{rs}.$$

In the preceding relations the ranges of indices are $i,j,r,s = 1,...,n$.

Consider an r-dimensional subspace $L \subset T_pM$ for $r \geq 2$. Let $\{e_1,...,e_r\}$ be an orthonormal basis of L. Define

$$\tau_r(L) := \sum_{1 \leq \alpha < \beta \leq r} \kappa(e_\alpha, e_\beta).$$

By analogy to the above notation we write

$$\kappa_{\alpha\beta} := \kappa(e_\alpha, e_\beta).$$

$2\tau_r(L)$ is called the *scalar r-curvature* of the subspace L. For $r = n$ we get the scalar curvature of (M,g) at p:

$$2\tau_n(L) = \tau,$$

while for $r = 2$ the expression $\tau_2(L)$ gives the sectional curvature of the plane $L = \operatorname{span}(e_\alpha, e_\beta) \subset T_p(M)$.

Below we give 3 concrete examples of Einstein manifolds.

EXAMPLE 3.1. Any Riemannian space form of arbitrary dimension $n \geq 2$ is an Einstein manifold.

EXAMPLE 3.2. The Schwarzschild space time [**ONeill**] is an example of Einstein manifold ($\mathcal{R}ic = 0$) which has not constant sectional curvature (an Einstein manifold has sense in the pseudo-Riemannian geometry, too; see the Justification from Physics from Section 1).

EXAMPLE 3.3. Let $N_1^k(a)$, $N_2^p(b)$ $(k, p \geq 2)$ be Riemannian space forms and $M = N_1^k(a) \times N_2^p(b)$.

Consider $\{e_1, ..., e_k, e_{k+1}, ..., e_{k+p}\}$ an orthonormal frame on M, such that $e_1, ..., e_k$ tangent to $N_1^k(a)$ and $e_{k+1}, ..., e_{k+p}$ tangent to $N_2^p(b)$.

Then $\forall i = 1, ..., k$, $\mathcal{R}ic(e_i) = (k-1)a$ and $\forall j = 1, ..., p$, $\mathcal{R}ic(e_{k+j}) = (p-1)b$.

It follows that M is an Einstein space if and only if $(k-1)a = (p-1)b$.

In particular:

$a = 0 \Leftrightarrow b = 0$ trivial (Euclidean space, of null sectional curvature);

$k = p \Leftrightarrow a = b$ (Einstein space of even dimension and non-constant sectional curvature);

$a \neq 0 \Rightarrow b = \frac{k-1}{p-1}a$; remark $ab > 0$ (Einstein space of arbitrary dimension and non-constant sectional curvature).

There are many other interesting spaces admitting Einstein metrics (some even more then one!) [**B**].

4. Symmetry of sectional curvatures on Einstein spaces

Singer and Thorpe [**ST**] discovered a symmetry of sectional curvatures which characterizes 4-dimensional Einstein spaces. Later this result was generalized to Einstein spaces of even dimensions $n = 2r \geq 4$, by Chen, Dillen, Verstraelen and Vrancken [**CDVV**].

Our main result [**MS**] establishes curvature symmetries for Einstein spaces of arbitrary dimension $n \geq 4$.

More precisely, Singer and Thorpe considered 4-dimensional Einstein spaces and started the study of two interesting topics:

i) The irreducible decomposition of the Riemannian $(0,4)$-curvature tensor.

This study initialized generalizations:

- first to algebraic curvature tensors (see [**B**]);
- later to other types of curvature tensors (see [**BGNS**]).

ii) Symmetry properties of certain curvature functions.

Here we are interested in the second topic (ii).

We use the following notational convention for 4-dimensional Riemannian manifolds: for $\text{span}(e_i, e_j)$ the orthogonal complement $\text{span}(e_i, e_j)^\perp$ is well defined; for simplicity we denote its sectional curvature by $\kappa_{ij}^\perp$.

Singer and Thorpe proved the following

THEOREM 4.1. [**ST**] *Let (M, g) be a Riemannian manifold of dimension $n = 4$. Then the following assertions are equivalent:*

(i) *(M, g) is an Einstein space;*

(ii) *at any point $p \in M$ and for every 2-plane $\text{span}(e_i, e_j) \subset T_pM$, where $i, j \in \{1, 2, 3, 4\}$ and $i \neq j$, we have the following equality of sectional curvatures:*

$$\kappa_{ij} = \kappa_{ij}^\perp .$$

B.-Y. Chen, F. Dillen, L. Verstraelen and L. Vrancken [**CDVV**] gave a generalization to even-dimensional Riemannian manifolds.

THEOREM 4.2. [**CDVV**] *Let $\dim M = n = 2r$ for $r \geq 2$. Then the following statements are equivalent:*

(i) *(M, g) is an Einstein space;*

(ii) *for any* $p \in M$ *and any subspace* $L \subset T_pM$ *with* $dim(L) = r = dim(L^\perp)$ *and* $L \oplus L^\perp = T_pM$ *we have* $\tau_r(L) = \tau_r(L^\perp)$.

Our result generalizes the results cited above and is valid in any arbitrary dimensions $n \geq 4$.

We will use the following notations:

(i) Let (V, g) be a Euclidean vector space of dimension $n \geq 4$ with inner product g. Let $L \subset V$ be a subspace of dimension r, where $2 \leq r = dim\, L \leq n-2$, and consider the orthogonal decomposition

$$L \oplus L^\perp = V$$

with $2 \leq s := dim\, L^\perp \leq n-2$, thus $r + s = n$.

(ii) Additionally we introduce the following notation: set $N := \{1, ..., n\}$ and let $\sigma : N \to N$ with $\{1, ..., n\} \mapsto \{i_1, ..., i_n\}$ be a permutation. We consider a disjoint decomposition $N = N_r^\sigma \cup N_s^\sigma$, with $N_r^\sigma := \{i_1, ..., i_r\}$ and $N_s^\sigma := \{i_{r+1}, ..., i_{r+s}\}$.

(iii) On a Riemannian manifold (M, g) with $p \in M$ consider an r-dimensional subspace $L \subset T_pM$ with orthonormal basis $\{e_{i_1}, ..., e_{i_r}\}$; we extend this basis to an orthonormal basis $\{e_{i_1}, ..., e_{i_r}, e_{i_{r+1}}, ..., e_{i_n}\}$ of T_pM, thus $\{e_{i_{r+1}}, ..., e_{i_n}\}$ is an orthonormal basis of the subspace $L^\perp \subset T_pM$.

From the foregoing sections, the curvature invariants

$$2\tau(L) = \sum_{p,q \in N_r^\sigma} \kappa_{pq}; \qquad 2\tau(L^\perp) = \sum_{p,q \in N_s^\sigma} \kappa_{pq}$$

are well defined. Then, to relate the scalar curvatures of subspaces we add up the Ricci curvatures:

$$\sum_{p \in N_r^\sigma} \rho_p = \sum_{p \in N_r^\sigma} \sum_{i \in N} \kappa_{pi},$$

$$\sum_{p \in N_s^\sigma} \rho_p = \sum_{p \in N_s^\sigma} \sum_{i \in N} \kappa_{pi}.$$

We have:

$$\sum_{p \in N_r^\sigma} \rho_p - \sum_{p \in N_s^\sigma} \rho_p = \sum_{p \in N_r^\sigma} \sum_{i \in N_r^\sigma} \kappa_{pi} + \sum_{p \in N_r^\sigma} \sum_{i \in N_s^\sigma} \kappa_{pi} -$$

$$- \sum_{p \in N_s^\sigma} \sum_{i \in N_s^\sigma} \kappa_{pi} - \sum_{p \in N_s^\sigma} \sum_{i \in N_r^\sigma} \kappa_{pi} = 2\big[\tau_r(L) - \tau_s(L^\perp\big];$$

to prove the last equality we interchanged some sub-indices and used the symmetry $\kappa_{pi} = \kappa_{ip}$.

Based on the above notations and calculations, our main result is the following

THEOREM 4.3. [**MS**] *Let* (M, g) *be a Riemannian manifold of dimension* $n \geq 4$. *Using the notations of the foregoing subsections, we have equivalence of the following two conditions:*

(i) (M, g) *is an Einstein space.*

(ii) *For any* $r, s \in N$ *satisfying the relations* $r + s = n$ *and* $2 \leq r, s \leq n-2$ *and any* $p \in M$, *there exists a real* $c := c_p$, *independent of* r, *such that for any subspace* $L \subset T_pM$ *of dimension* r, *we have*

$$(r - s)c_p = 2\big[\tau_r(L) - \tau_s(L^\perp)\big].$$

The complete Proof can be found in [**MS**].

REMARK 4.4. If $n = 2r = 2s$ we get the even-dimensional result of Chen-Dillen-Verstraelen-Vrancken. If $r = s = 2$ we get the result of Singer and Thorpe.

REMARK 4.5. (i) Recall the fact that, for any Riemannian manifold, the sectional curvature function at $p \in M$ determines the Riemannian curvature tensor at p; see for example [**KN**], p. 198, Proposition 1.2.

(ii) From the preceding Theorem, for an Einstein space, at any $p \in M$, the scalar $(n-2)$-curvatures $\tau_{n-2}(T_pM)$ and the scalar curvature $\tau = \tau_n(T_pM)$ together determine the scalar 2-curvatures $\tau_2(T_pM)$, that means the sectional curvatures at p.

Then (i) and (ii) together imply that at any $p \in M$, where (M, g) is an Einstein space of dimension $n \geq 5$, the scalar $(n-2)$-curvatures $\tau_{n-2}(T_pM)$ together with the scalar curvature $\tau = \tau_n(T_pM)$ determine the Riemannian curvature tensor.

We would also like to point out the following

REMARK 4.6. Let (M, g) be an Einstein space of dimension $n = 4$. Following the result of Singer and Thorpe, at any $p \in M$ we have 6 sectional curvatures, that means 3 pairs as $\kappa_{ij} = \kappa_{ij}^{\perp}$; as three representatives of the 3 pairs we can fix an arbitrary index $i \in \{1, 2, 3, 4\}$ and consider the 3 sectional curvatures κ_{ij}, where $i \neq j \in \{1, 2, 3, 4\}$. Thus, for arbitrary $i \in \{1, 2, 3, 4\}$, these three representatives determine the Riemannian curvature tensor at $p \in M$.

For $n = 4$ this fact suggests the following procedure:

Choose $i = 1$ for a curve starting at p with e_1 as prescribed tangent vector at p. By a parallel displacement a frame $\{e_1, e_2, e_3, e_4\}$ moves from p along the curve. The sectional curvatures κ_{1j} for $j = 2, 3, 4$ together determine the curvature tensor along the curve.

5. Final Remarks

The study of Einstein manifolds represents an important and very large topic in Differential Geometry, studied by many geometers. Of course, there are many important related topics: Kaehler-Einstein spaces, Riemannian submersions, homogeneous Riemannian spaces, Riemannian functionals and their critical points, Yang-Mills theory, holonomy groups, quaternionic manifolds, algebraic geometry, etc. The monograph of A. Besse [**B**] represents an excellent book, even including this related material. Also, it contains a huge collection of books and papers listed at References.

This note, beside a short motivation of the study of Einstein manifolds, intends to present only a generalization of results of Singer-Thorpe and Chen-Dillen-Verstraelen-Vrancken on curvature symmetries for Einstein manifolds of arbitrary dimension, in the hope that it would be useful to researchers in this direction.

In particular, the contribution of Professor Chen in the field of Einstein manifolds is well-known and inspired many specialists in this field. I am very glad that I had the chance to meet Professor Chen and collaborate with him and I am honored that I was invited to present my work at the Special Session of the AMS Meeting, organized to celebrate his 75-birthday. This short article is based on the talk I gave.

Therefore, I would like to congratulate again Professor Bang-Yen Chen on this occasion and to address him many thanks for his major influence on my scientific career.

References

[B] A. L. Besse, *Einstein Manifolds*, Classics in Mathematics, Springer-Verlag, Berlin, 2008. Reprint of the 1987 edition. MR2371700

[Bis] R. L. Bishop, *A relation between volume, mean curvature and diameter*, Notices AMS **10** (1963), 364.

[Ber] M. Berger, *Rapport sur les variétés d'Einstein* (French, with English summary), Analysis on manifolds (Conf., Univ. Metz, Metz, 1979), Astérisque, vol. **80**, Soc. Math. France, Paris, 1980, pp. 3, 5–19. MR620166

[BGNS] N. Blažić, P. Gilkey, S. Nikčević, and U. Simon, *Algebraic theory of affine curvature tensors*, Arch. Math. (Brno) **42** (2006), no. suppl., 147–168. MR2322404

[Bou] J.-P. Bourguignon, *A review of Einstein manifolds*, Proceedings of the 1981 Shanghai symposium on differential geometry and differential equations (Shanghai/Hefei, 1981), Sci. Press Beijing, Beijing, 1984, pp. 11–45. MR825255

[CDVV] B.-Y. Chen, F. Dillen, L. Verstraelen, and L. Vrancken, *Characterizations of Riemannian space forms, Einstein spaces and conformally flat spaces*, Proc. Amer. Math. Soc. **128** (2000), no. 2, 589–598, DOI 10.1090/S0002-9939-99-05332-0. MR1664333

[E] L. P. Eisenhart, *Riemannian Geometry*, Princeton University Press, Princeton, N. J., 1949. 2d printing. MR0035081

[GNS] P. Gilkey, S. Nikčević, and U. Simon, *Geometric theory of equiaffine curvature tensors*, Results Math. **56** (2009), no. 1-4, 275–317, DOI 10.1007/s00025-009-0438-1. MR2575863

[KN] S. Kobayashi and K. Nomizu, *Foundations of Differential Geometry. Vol I*, Interscience Publishers, a division of John Wiley & Sons, New York-London, 1963. MR0152974

[MS] A. Mihai and U. Simon, *Curvature symmetries characterizing Einstein spaces*, Colloq. Math. **152** (2018), no. 1, 23–28, DOI 10.4064/cm7234-5-2017. MR3778893

[ONeill] B. O'Neill, *Semi-Riemannian Geometry. With Applications to Relativity*, Pure and Applied Mathematics, vol. 103, Academic Press, Inc. [Harcourt Brace Jovanovich, Publishers], New York, 1983. MR719023

[ST] I. M. Singer and J. A. Thorpe, *The curvature of 4-dimensional Einstein spaces*, Global Analysis (Papers in Honor of K. Kodaira), Univ. Tokyo Press, Tokyo, 1969, pp. 355–365. MR0256303

Department of Mathematics and Computer Science, Technical University of Civil Engineering Bucharest, Romania, Bvd. Lacul Tei 122-124, 020396 Bucharest, Romania

Email address: adela.mihai@utcb.ro

Contemporary Mathematics
Volume **756**, 2020
https://doi.org/10.1090/conm/756/15206

Statistical manifolds and their submanifolds. Results on Chen-like invariants

Ion Mihai

Dedicated to Professor Bang-Yen Chen on the occasion of his 75th birthday.

ABSTRACT. Statistical manifolds were introduced by S. Amari [Springer, 1985]. They generalize the Hessian manifolds. The geometry of statistical manifolds and their submanifolds is an actual topic of research in pure and applied mathematics. M.E. Aydin, A. Mihai and the present author [Filomat, 2015] obtained geometric inequalities for the scalar curvature and Ricci curvature associated to the dual connections for submanifolds in statistical manifolds of constant curvature. The same authors [Bull. Math. Sci., 2017] proved a generalized Wintgen inequality for such submanifolds. Recently, in co-operation with A. Mihai [Mathematics, 2018], we initiated the study of the geometry of submanifolds in Hessian manifolds of constant Hessian curvature and established a Euler inequality and a Chen-Ricci inequality for such submanifolds. We continue the study of Chen-like invariants on submanifolds in Hessian manifolds of constant Hessian curvature.

1. Introduction

Statistical manifolds were introduced in 1985 by Amari [**1**]. They have been studied in terms of information geometry. Since the geometry of such manifolds includes the notion of dual connections, also called conjugate connections in affine geometry, it is closely related to affine differential geometry. Further, a Hessian manifold is a statistical manifold of constant curvature 0, then the geometry of statistical manifolds generalizes the Hessian geometry.

On the other hand, it is well-known that curvature invariants play the most fundamental role in Riemannian geometry. Curvature invariants provide the intrinsic characteristics of Riemannian manifolds which affect the behavior in general of the Riemannian manifold. They are the main Riemannian invariants and the most natural ones. Curvature invariants also play key roles in physics. For instance, the magnitude of a force required to move an object at constant speed, according to Newton's laws, is a constant multiple of the curvature of the trajectory. The motion of a body in a gravitational field is determined, according to Einstein's general

2010 *Mathematics Subject Classification.* Primary 53C40; Secondary 53C05.
Key words and phrases. Statistical manifolds, Hessian manifolds, statistical submanifolds, curvature invariants, Euler inequality, Chen-Ricci inequality, Wintgen inequality, Chen invariants.
Partially supported by the grant PN-III-P1-1.1-MC-2018-2768 of the Romanian Ministry of Research and Innovation.

theory of relativity, by the curvatures of spacetime. All sort of shapes, from soap bubbles to red cells are determined by various curvatures.

We investigate certain curvature invariants, e.g., scalar curvature, Ricci curvature, normal scalar curvature and Chen first invariant, on statistical submanifolds in statistical manifolds of constant curvature, in particular in Hessian manifolds of constant Hessian curvature.

2. Statistical manifolds and their submanifolds

A *statistical manifold* is an m-dimensional Riemannian manifold $(\tilde{M}^m, g)$ endowed with a pair of torsion-free affine connections $\tilde{\nabla}$ and $\tilde{\nabla}^*$ satisfying

$$Zg(X,Y) = g(\tilde{\nabla}_Z X, Y) + g(X, \tilde{\nabla}^*_Z Y), \tag{2.1}$$

for any $X, Y, Z \in \Gamma(T\tilde{M}^m)$. The connections $\tilde{\nabla}$ and $\tilde{\nabla}^*$ are called *dual connections* (see [**1, 18, 20**]), and it is easily shown that $(\tilde{\nabla}^*)^* = \tilde{\nabla}$. The pair $(\tilde{\nabla}, g)$ is called a *statistical structure*. If $(\tilde{\nabla}, g)$ is a statistical structure on $\tilde{M}^m$, then $(\tilde{\nabla}^*, g)$ is a statistical structure too [**1**].

On the other hand, any torsion-free affine connection $\tilde{\nabla}$ always has a dual connection given by

$$\tilde{\nabla} + \tilde{\nabla}^* = 2\tilde{\nabla}^0, \tag{2.2}$$

where $\tilde{\nabla}^0$ is the Levi-Civita connection on $\tilde{M}^m$ and the curvature tensor associated with $\tilde{\nabla}^0$ is called the Riemannian curvature tensor.

Denote by $\tilde{R}$ and $\tilde{R}^*$ the curvature tensor fields with respect to $\tilde{\nabla}$ and $\tilde{\nabla}^*$, respectively. A statistical structure $(\tilde{\nabla}, g)$ is said to be *of constant curvature* $\varepsilon \in \mathbb{R}$ if

$$\tilde{R}(X,Y)Z = \varepsilon\{g(Y,Z)X - g(X,Z)Y\},\ \forall X, Y, Z \in \Gamma(T\tilde{M}^m). \tag{2.3}$$

A statistical structure $(\tilde{\nabla}, g)$ of constant curvature 0 is known as a *Hessian structure*.

The curvature tensor fields $\tilde{R}$ and $\tilde{R}^*$ of the dual connections satisfy

$$g(\tilde{R}^*(X,Y)Z, W) = -g(Z, \tilde{R}(X,Y)W). \tag{2.4}$$

From (2.4) it follows immediately that if $(\tilde{\nabla}, g)$ is a statistical structure of constant curvature ε, then $(\tilde{\nabla}^*, g)$ is also a statistical structure of constant curvature ε. In particular, if $(\tilde{\nabla}, g)$ is Hessian, $(\tilde{\nabla}^*, g)$ is also Hessian [**19**].

On a Hessian manifold $(\tilde{M}^m, \tilde{\nabla})$, let $\gamma = \tilde{\nabla} - \tilde{\nabla}^0$. The tensor field $\tilde{Q}$ of type (1,3) defined by

$$\tilde{Q}(X,Y) = [\gamma_X, \gamma_Y], \quad X, Y \in \Gamma(T\tilde{M}^m),$$

is called the *Hessian curvature tensor* for $\tilde{\nabla}$ (see [**19**], [**17**]). It satisfies

$$\tilde{R}(X,Y) + \tilde{R}^*(X,Y) = 2\tilde{R}^0(X,Y) + 2\tilde{Q}(X,Y).$$

By using the Hessian curvature tensor $\tilde{Q}$, Hessian sectional curvature can be defined on a Hessian manifold. In fact, let $p \in \tilde{M}^m$ and π a plane section in $T_p\tilde{M}^m$. Take an orthonormal basis $\{X, Y\}$ of π. Then the Hessian sectional curvature is defined by

$$\tilde{K}(\pi) = g(\tilde{Q}(X,Y)Y, X),$$

which is independent of the choice of an orthonormal basis.

A Hessian manifold has constant Hessian sectional curvature c if and only if (see [**19**])

$$\tilde{Q}(X,Y,Z,W)=\frac{c}{2}\{g(X,Y)g(Z,W)+g(X,W)g(Y,Z)\},$$

for all vector fields X,Y,Z,W on $\tilde{M}^m$.

It is known in [**19**] that a Hessian manifold of constant Hessian sectional curvature c is a Riemannian space form of constant sectional curvature $-c/4$.

If $(\tilde{M}^m,g)$ is a statistical manifold and if M^n is an n-dimensional submanifold of $\tilde{M}^m$, then (M^n,g) is also a statistical manifold with the induced connection ∇ by $\tilde{\nabla}$ and induced metric g. In the case that $(\tilde{M}^m,g)$ is a semi-Riemannian manifold, the induced metric g has to be non-degenerate. For details, see [**21**].

In the geometry of Riemannian submanifolds, the fundamental equations are the Gauss and Weingarten formulae and the equations of Gauss, Codazzi and Ricci (see [**5**]).

Let denote the set of the sections of the normal bundle to M^n by $\Gamma(T^{\perp}M^n)$. In our case, according to [**21**], for any $X,Y\in\Gamma(TM^n)$, the corresponding Gauss formulae are

$$\tilde{\nabla}_XY=\nabla_XY+h(X,Y), \tag{2.5}$$

$$\tilde{\nabla}^*_XY=\nabla^*_XY+h^*(X,Y), \tag{2.6}$$

where h, $h^*:\Gamma(TM^n)\times\Gamma(TM^n)\to\Gamma(T^{\perp}M^n)$ are symmetric and bilinear, called the *imbedding curvature tensor* of M^n in $\tilde{M}^m$ for $\tilde{\nabla}$ and the *imbedding curvature tensor* of M^n in $\tilde{M}^m$ for $\tilde{\nabla}^*$, respectively. In [**21**], it is also proved that (∇,g) and (∇^*,g) are dual statistical structures on M^n.

Since h and h^* are bilinear, there exist linear transformations A_ξ and A^*_ξ on TM^n defined by

$$g(A_\xi X,Y)=g(h(X,Y),\xi), \tag{2.7}$$

$$g(A^*_\xi X,Y)=g(h^*(X,Y),\xi), \tag{2.8}$$

for any $\xi\in\Gamma(T^{\perp}M^n)$ and $X,Y\in\Gamma(TM^n)$. Further, the corresponding Weingarten formulas are (see [**21**])

$$\tilde{\nabla}_X\xi=-A^*_\xi X+\nabla^{\perp}_X\xi, \tag{2.9}$$

$$\tilde{\nabla}^*_X\xi=-A_\xi X+\nabla^{*\perp}_X\xi, \tag{2.10}$$

for any $\xi\in\Gamma(T^{\perp}M^n)$ and $X\in\Gamma(TM^n)$. The connections $\nabla^{\perp}$ and $\nabla^{*\perp}$ defined by (2.9) and (2.10) are Riemannian dual connections with respect to induced metric on $\Gamma(T^{\perp}M^n)$.

Let $\{e_1,...,e_n\}$ and $\{e_{n+1},...,e_m\}$ be orthonormal tangent and normal frames on M^n, respectively. Then the mean curvature vector fields are defined by

$$H=\frac{1}{n}\sum_{i=1}^{n}h(e_i,e_i)=\frac{1}{n}\sum_{\alpha=n+1}^{m}\left(\sum_{i=1}^{n}h^{\alpha}_{ii}\right)e_\alpha,\ h^{\alpha}_{ij}=g(h(e_i,e_j),e_\alpha), \tag{2.11}$$

and

$$H^*=\frac{1}{n}\sum_{i=1}^{n}h^*(e_i,e_i)=\frac{1}{n}\sum_{\alpha=n+1}^{m}\left(\sum_{i=1}^{n}h^{*\alpha}_{ii}\right)e_\alpha, h^{*\alpha}_{ij}=g(h^*(e_i,e_j),e_\alpha), \tag{2.12}$$

for $1\le i,j\le n$ and $n+1\le\alpha\le m$.

The corresponding Gauss, Codazzi and Ricci equations are given by the following result.

PROPOSITION 2.1. [**21**] *Let $\tilde{\nabla}$ and $\tilde{\nabla}^*$ be dual connections on a statistical manifold $\tilde{M}^m$ and ∇ the induced connection by $\tilde{\nabla}$ on a statistical submanifold M^n. Let $\tilde{R}$ and R be the Riemannian curvature tensors for $\tilde{\nabla}$ and ∇, respectively. Then*

$$\begin{aligned} g(\tilde{R}(X,Y)Z,W) &= g(R(X,Y)Z,W) + g(h(X,Z),h^*(Y,W)) - \\ &\quad - g(h^*(X,W),h(Y,Z)), \\ (\tilde{R}(X,Y)Z)^\perp &= \nabla^\perp_X h(Y,Z) - h(\nabla_X Y,Z) - h(Y,\nabla_X Z) - \\ &\quad - \{\nabla^\perp_Y h(Y,Z) - h(\nabla_Y X,Z) - h(X,\nabla_Y Z)\}, \\ g(R^\perp(X,Y)\xi,\eta) &= g(\tilde{R}(X,Y)\xi,\eta) + g([A^*_\xi, A_\eta]X,Y), \end{aligned}$$

where $R^\perp$ is the Riemannian curvature tensor of $\nabla^\perp$ on $T^\perp M^n$, $\xi,\eta \in \Gamma(T^\perp M^n)$, and $[A^*_\xi, A_\eta] = A^*_\xi A_\eta - A_\eta A^*_\xi$.

The equations of Gauss, Codazzi and Ricci with respect to the connection $\tilde{\nabla}^*$ on M^n are given by

PROPOSITION 2.2. [**21**] *Let $\tilde{\nabla}$ and $\tilde{\nabla}^*$ be dual connections on a statistical manifold $\tilde{M}^m$ and ∇^* the induced connection by $\tilde{\nabla}^*$ on a statistical submanifold M^n. Let $\tilde{R}^*$ and R^* be the Riemannian curvature tensors for $\tilde{\nabla}^*$ and ∇^*, respectively. Then*

$$\begin{aligned} g(\tilde{R}^*(X,Y)Z,W) &= g(R^*(X,Y)Z,W) + g(h^*(X,Z),h(Y,W)) - \\ &\quad - g(h(X,W),h^*(Y,Z)), \\ (\tilde{R}^*(X,Y)Z)^\perp &= \nabla^{*\perp}_X h^*(Y,Z) - h^*(\nabla^*_X Y,Z) - h^*(Y,\nabla^*_X Z) - \\ &\quad - \{\nabla^{*\perp}_Y h^*(Y,Z) - h^*(\nabla^*_Y X,Z) - h^*(X,\nabla^*_Y Z)\}, \\ g(R^{*\perp}(X,Y)\xi,\eta) &= g(\tilde{R}^*(X,Y)\xi,\eta) + g([A_\xi, A^*_\eta]X,Y), \end{aligned}$$

where $R^{*\perp}$ is the Riemannian curvature tensor of $\nabla^{\perp*}$ on $T^\perp M^n$, $\xi,\eta \in \Gamma(T^\perp M^n)$, and $[A_\xi, A^*_\eta] = A_\xi A^*_\eta - A^*_\eta A_\xi$.

3. Curvature invariants on Riemannian submanifolds

Classically, among the curvature invariants on a Riemannian manifold, the most studied were sectional, scalar and Ricci curvatures.

For surfaces M^2 of the Euclidean space $\mathbb{E}^3$, the Euler inequality

$$G \leq \|H\|^2$$

is fulfilled, where G is the (intrinsic) Gauss curvature of M^2 and $\|H\|^2$ is the (extrinsic) squared mean curvature of M^2.

Furthermore, $G = \|H\|^2$ everywhere on M^2 if and only if M^2 is totally umbilical, or still, by a theorem of Meusnier, if and only if M^2 is (a part of) a plane $\mathbb{E}^2$ or, it is (a part of) a round sphere S^2 in $\mathbb{E}^3$.

B.-Y. Chen [**7**] established a generalized Euler inequality for submanifolds in real space forms.

THEOREM 3.1. *Let $x : M^n \to R^m(c)$ be an isometric immersion of a Riemannian n-manifold M^n with normalized scalar curvature ρ into an m-dimensional real-space-form $R^m(c)$ of constant sectional curvature c. Then we have*

$$\|H\|^2 \geq \rho - c,$$

equality holding at a point $p \in M^n$ if and only if p is a totally umbilical point.

Also a sharp relationship between the Ricci curvature and the squared mean curvature for any Riemannian submanifold of a real space form was proved in [**8**], which is known as the Chen-Ricci inequality.

THEOREM 3.2. *Let $x : M^n \to R^m(c)$ be an isometric immersion of a Riemannian n-manifold M^n into a Riemannian space form $R^m(c)$. Then, for each unit tangent vector $X \in T_pM^n$, we have*

$$\|H\|^2(p) \geq \frac{4}{n^2}[Ric(X) - (n-1)c],$$

where $Ric(X)$ the Ricci curvature of M^n at X.

The equality case was investigated.

In 1979, P. Wintgen [**22**] proved that the Gauss curvature G, the squared mean curvature $\|H\|^2$ and the normal curvature $G^\perp$ of any surface M^2 in $\mathbb{E}^4$ always satisfy the inequality

$$G \leq \|H\|^2 - |G^\perp|$$

and that actually the equality holds if and only if the ellipse of curvature of M^2 in $\mathbb{E}^4$ is a circle.

P.J. De Smet, F. Dillen, L. Verstraelen and L. Vrancken [**12**], conjectured, in 1999, a generalized Wintgen inequality for submanifolds of arbitrary dimension and codimension in Riemannian space forms. This conjecture was proved by Z. Lu [**14**] and by J. Ge and Z. Tang [**13**], independently.

Let M^n be an n-dimensional Riemannian submanifold of an $(n+m)$-dimensional Riemannian manifold and $\{e_1, ..., e_n\}$ and $\{\xi_1, ..., \xi_m\}$ orthonormal frames tangent and normal to M^n respectively.

One denotes by K and $R^\perp$ the sectional curvature function and the normal curvature tensor on M^n, respectively. Then the normalized scalar curvature (an intrinsic invariant) is given by

$$\rho = \frac{2\tau}{n(n-1)} = \frac{2}{n(n-1)} \sum_{1 \leq i < j \leq n} K(e_i \wedge e_j),$$

where τ is the scalar curvature, and the normalized scalar normal curvature (an extrinsic invariant) by

$$\rho^\perp = \frac{2\tau^\perp}{n(n-1)} = \frac{2}{n(n-1)} \sqrt{\sum_{1 \leq i < j \leq n} \sum_{1 \leq \alpha < \beta \leq m} (R^\perp(e_i, e_j, \xi_\alpha, \xi_\beta))^2}.$$

THEOREM 3.3. *Let $f : M^n \to \tilde{M}^{n+m}(c)$ be an isometric immersion, where $\tilde{M}^{n+m}(c)$ is a real space form of constant sectional curvature c. Then*

$$\rho \leq \|H\|^2 - \rho^\perp + c.$$

The equality case was characterized in terms of the shape operators.

4. Curvature invariants on statistical submanifolds

In [**2**], M. E. Aydin, A. Mihai and the present author obtained geometrical inequalities for the scalar curvature and the Ricci curvature associated to the dual connections for submanifolds in statistical manifolds of constant curvature.

Let $\tilde{M}^{n+k}(\varepsilon)$ be an $(n+k)$-dimensional statistical manifold of constant curvature $\varepsilon \in \mathbb{R}$ and M^n an n-dimensional statistical submanifold of $\tilde{M}^{n+k}(\varepsilon)$.

Consider $\{e_1, ..., e_n\}$ and $\{e_{n+1}, ..., e_m\}$ orthonormal tangent and normal frames on M^n, respectively.

We point-out that the dual connections ∇ and ∇^* on M^n are not metric. However

$$\tau = \sum_{1 \le i < j \le n} g(R(e_i, e_j)e_j, e_i)$$

is well defined and we call it the *scalar curvature* of M^n.

Similarly

$$\tau^* = \sum_{1 \le i < j \le n} g(R^*(e_i, e_j)e_j, e_i).$$

REMARK 4.1. On any statistical submanifold M^n of $\tilde{M}^{n+k}(\varepsilon)$ one has $\tau = \tau^*$.

An estimate of the scalar curvature was established.

PROPOSITION 4.1. *Let $\tilde{M}^{n+k}(\varepsilon)$ be an $(n+k)$-dimensional statistical manifold of constant curvature $\varepsilon \in \mathbb{R}$ and M^n an n-dimensional statistical submanifold of $\tilde{M}^{n+k}(\varepsilon)$. We have*

$$2\tau \ge n(n-1)\varepsilon + n^2 g(H, H^*) - \|h\| \, \|h^*\| .$$

Let $p \in M^n$, $X \in T_pM^n$ unit. We denote by Ric and Ric^0 the Ricci curvature with respect to the dual connection ∇ and the Levi-Civita connection ∇^0 on M^n, respectively.

THEOREM 4.2. *Let M^n be an n-dimensional statistical submanifold of an $(n+k)$-dimensional statistical manifold $\tilde{M}^{n+k}(\varepsilon)$. For each $X \in T_pM^n$ unit, we have*

$$Ric(X) \ge 2Ric^0(X) - \frac{n^2}{8} g(H, H) - \frac{n^2}{8} g(H^*, H^*) + (n-1)\varepsilon - 2(n-1) \max \tilde{K}^0(X \wedge \cdot),$$

where $\tilde{K}^0$ is the Riemannian sectional curvature of $\tilde{M}^{n+k}(\varepsilon)$.

In particular, if M^n is a minimal submanifold, one derives the following.

COROLLARY 4.1. *Let M^n be a minimal n-dimensional statistical submanifold of an $(n+k)$-dimensional statistical manifold $\tilde{M}(c)$. For each $X \in T_pM^n$ unit, we have*

$$Ric(X) \ge 2Ric^0(X) + \frac{n^2}{4} g(H, H^*) + (n-1)\varepsilon - 2(n-1) \max \tilde{K}^0(X \wedge \cdot).$$

REMARK 4.3. Similar inequalities can be stated for the Ricci curvature Ric^*.

Let (M^n, g) be a statistical manifold of dimension n endowed with dual connections ∇ and ∇^*. Unfortunately, the $(0,4)$-tensor field $g(R(X,Y)Z,W)$ is not skew-symmetric with respect to Z and W (the dual connections are non-metric). Then we cannot define a sectional curvature on M^n by the standard definition.

We define a skew-symmetric $(0,4)$-tensor field on M^n by

$$T(X,Y,Z,W) = \frac{1}{2}[g(R(X,Y)Z,W) + g(R^*(X,Y)Z,W)],$$

for all $X,Y,Z,W \in \Gamma(TM^n)$.

Then we are able to define a sectional curvature on M^n by the formula

$$K(X \wedge Y) = \frac{T(X,Y,X,Y)}{g(X,X)g(Y,Y) - g^2(X,Y)},$$

for any linearly independent tangent vectors X, Y at $p \in \tilde{M}$ (see also [**17**], [**16**]).

The Wintgen inequality for statistical manifolds in statistical manifolds of constant curvature was proved in [**4**].

For a statistical surface M^2, we can define a Gauss curvature by

$$G = K(e_1 \wedge e_2),$$

for any orthonormal frame $\{e_1, e_2\}$ on M^2.

A normal curvature of a statistical surface M^2 in an orientable 4-dimensional statistical manifold $\tilde{M}^4$ can be defined as follows.

Let $\{e_1, e_2, e_3, e_4\}$ be a positive oriented orthonormal frame on $\tilde{M}^4$ such that e_1, e_2 are tangent to M^2.

$$G^{\perp} = \frac{1}{2}\left[g\left(R^{\perp}(e_1,e_2)e_3,e_4\right) + g\left(R^{*\perp}(e_1,e_2)e_3,e_4\right)\right].$$

THEOREM 4.4. *Let M^2 be a statistical surface in a 4-dimensional statistical manifold $\tilde{M}^4(\varepsilon)$ of constant curvature ε. Then*

$$G + \left|G^{\perp}\right| + 2G^0 \leq \frac{1}{2}\left(\|H\|^2 + \|H^*\|^2\right) - \varepsilon + 2\tilde{K}^0(e_1 \wedge e_2).$$

In particular, for $\varepsilon = 0$ we derive the following.

COROLLARY 4.2. *Let M^2 be a statistical surface of a Hessian 4-dimensional statistical manifold $\tilde{M}^4$. Then:*

$$G + \left|G^{\perp}\right| + 2G^0 \leq \frac{1}{2}\left(\|H\|^2 + \|H^*\|^2\right).$$

In [**3**] the authors established a generalized Wintgen inequality for statistical submanifolds in statistical manifolds of constant curvature by using a different sectional curvature than in [**17**].

THEOREM 4.5. *Let M^n be a submanifold in a statistical manifold $\tilde{M}^{n+m}(\varepsilon)$ of constant curvature ε. Then*

$$\rho^{\perp} + 2\rho \leq 5\|H\|^2 + 5\|H^*\|^2 + 8g(H,H^*) - 2\varepsilon + 20\left(\tilde{\rho}^0 - \rho^0\right).$$

Submanifolds in Hessian manifolds were considered first in [**15**]. The authors dealt with statistical submanifolds in Hessian manifolds of constant Hessian curvature c. It is known in [**19**] that such a manifold is a statistical manifold of null constant curvature and also a Riemannian space form of constant sectional curvature $-c/4$ (with respect to the sectional curvature defined by the Levi-Civita connection). A Euler inequality and a Chen-Ricci inequality were proved in [**15**].

THEOREM 4.6. *Let M^n be a statistical submanifold of a Hessian manifold $\tilde{M}^m(c)$ of constant Hessian curvature c. Then the scalar curvature satisfies:*

$$2\tau \geq -\frac{n^2}{4}\|H - H^*\|^2 + n(n-1)\frac{c}{4}.$$

Moreover, the equality holds at any pont $p \in M^n$ if and only if $h = h^*$. In this case, the scalar curvature is constant, $2\tau = n(n-1)\frac{c}{4}$.

We want to point-out that τ is non-positive on standard examples of Hessian manifolds.

THEOREM 4.7. *Let M^n be a statistical submanifold of a Hessian manifold $\tilde{M}^m(c)$ of constant Hessian curvature c. Then the Ricci curvature of a unit vector $X \in T_pM^n$ satisfies:*

$$Ric(X) \geq (n-1)\frac{c}{2} - \frac{n^2}{8}\|H\|^2 - \frac{n^2}{8}\|H^*\|^2 + Ric^0(X).$$

Moreover, the equality case holds if and only if

$$\begin{cases} 2h(X,X) = nH(p),\ h(X,Y) = 0,\ \forall Y \in T_pM^n\, orthogonal\, to\, X, \\ 2h^*(X,X) = nH^*(p),\ h^*(X,Y) = 0,\ \forall Y \in T_pM^n\, orthogonal\, to\, X. \end{cases}$$

5. A Chen first inequality for statistical submanifolds

Until recently, Chen inequalities for statistical submanifolds in statistical manifolds were not known.

We point-out that the dual connections on a statistical manifold are non-metric. Then one cannot define a sectional curvature by the standard definition of the sectional curvature in the Riemannian case.

The main difficulty is to find suitable sectional curvatures on statistical manifolds. Moreover, for the chosen sectional curvature to establish algebraic inequalities which will be used in the proof of Chen inequalities.

A Chen first inequality for statistical submanifolds in Hessian manifolds of constant Hessian curvature was proved by B.Y. Chen, A. Mihai and the present author.

THEOREM 5.1. *Let M^n be a statistical submanifold in a Hessian manifold $\tilde{M}^m(c)$ of constant Hessian curvature c. Then, for any $p \in M^n$ and any plane section π at p, we have:*

$$\tau_0 - K_0(\pi) \leq \tau - K(\pi) + \frac{n^2(n-2)}{4(n-1)}(\|H\|^2 + \|H^*\|^2) - (n-2)(n+1)\frac{c}{4},$$

where τ_0 and K_0 are the scalar curvature and the sectional curvature of M^n with respect to the Riemann curvature tensor and τ and K with respect to the Hessian curvature tensor Q.

Moreover, the equality holds if and only if for any $\alpha \in \{n+1, ..., m\}$,

$$h^\alpha_{11} + h^\alpha_{22} = h^\alpha_{33} = ... = h^\alpha_{nn},$$
$$h^{*\alpha}_{11} + h^{*\alpha}_{22} = h^{*\alpha}_{33} = ... = h^{*\alpha}_{nn},$$
$$h^\alpha_{ij} = h^{*\alpha}_{ij} = 0, \quad \forall 1 \leq i \neq j \leq n, \{i,j\} \neq \{1,2\}.$$

The proof is based on the following algebraic lemma and the Gauss equation (see [**11**]).

LEMMA 5.2. *Let $n \geq 3$ be an integer and let $a_1, ..., a_n$ be n real numbers. Then one has:*

$$\sum_{1 \leq i < j \leq n} a_i a_j - a_1 a_2 \leq \frac{n-2}{2(n-1)} \left(\sum_{i=1}^{n} a_i \right)^2 .$$

Moreover, the equality holds if and only if $a_1 + a_2 = a_3 = ... = a_n$.

An immediate consequence of Theorem 3.1 is the following.

COROLLARY 5.1. *Let M^n be a statistical submanifold in a Hessian manifold $\tilde{M}^m(c)$ of constant Hessian curvature c. If there exist a point $p \in M^n$ and a plane section π at p such that*

$$(\tau - K(\pi)) - (\tau_0 - K_0(\pi)) < (n-2)(n+1)\frac{c}{4},$$

then M is non-minimal in $\tilde{M}^m(c)$, i.e., either $H \neq 0$ or $H^* \neq 0$.

This research can be continued by considering other Chen invariants on statistical submanifolds in Hessian manifolds of constant Hessian curvature.

References

[1] S.-i. Amari, *Differential-geometrical methods in statistics*, Lecture Notes in Statistics, vol. 28, Springer-Verlag, New York, 1985. MR788689

[2] M. E. Aydin, A. Mihai, and I. Mihai, *Some inequalities on submanifolds in statistical manifolds of constant curvature*, Filomat **29** (2015), no. 3, 465–476, DOI 10.2298/FIL1503465A. MR3359440

[3] M. E. Aydin, A. Mihai, and I. Mihai, *Generalized Wintgen inequality for statistical submanifolds in statistical manifolds of constant curvature*, Bull. Math. Sci. **7** (2017), no. 1, 155–166, DOI 10.1007/s13373-016-0086-1. MR3625853

[4] M. E. Aydin and I. Mihai, *Wintgen inequality for statistical surfaces*, Math. Inequal. Appl. **22** (2019), no. 1, 123–132, DOI 10.7153/mia-2019-22-09. MR3905975

[5] B.-Y. Chen, *Geometry of submanifolds*, Marcel Dekker, Inc., New York, 1973. Pure and Applied Mathematics, No. 22. MR0353212

[6] B.-Y. Chen, *Some pinching and classification theorems for minimal submanifolds*, Arch. Math. (Basel) **60** (1993), no. 6, 568–578, DOI 10.1007/BF01236084. MR1216703

[7] B.-Y. Chen, *Mean curvature and shape operator of isometric immersions in real-space-forms*, Glasgow Math. J. **38** (1996), no. 1, 87–97, DOI 10.1017/S001708950003130X. MR1373963

[8] B.-Y. Chen, *Relations between Ricci curvature and shape operator for submanifolds with arbitrary codimensions*, Glasg. Math. J. **41** (1999), no. 1, 33–41, DOI 10.1017/S0017089599970271. MR1689730

[9] B.-Y. Chen, *Some new obstructions to minimal and Lagrangian isometric immersions*, Japan. J. Math. (N.S.) **26** (2000), no. 1, 105–127, DOI 10.4099/math1924.26.105. MR1771434

[10] B.-Y. Chen, *Pseudo-Riemannian geometry, δ-invariants and applications*, World Scientific Publishing Co. Pte. Ltd., Hackensack, NJ, 2011. With a foreword by Leopold Verstraelen. MR2799371

[11] B.-Y. Chen, A. Mihai, and I. Mihai, *A Chen first inequality for statistical submanifolds in Hessian manifolds of constant Hessian curvature*, Results Math. **74** (2019), no. 4, Art. 165, 11, DOI 10.1007/s00025-019-1091-y. MR3998760

[12] P. J. De Smet, F. Dillen, L. Verstraelen, and L. Vrancken, *A pointwise inequality in submanifold theory*, Arch. Math. (Brno) **35** (1999), no. 2, 115–128. MR1711669

[13] J. Ge and Z. Tang, *A proof of the DDVV conjecture and its equality case*, Pacific J. Math. **237** (2008), no. 1, 87–95, DOI 10.2140/pjm.2008.237.87. MR2415209

[14] Z. Lu, *Normal scalar curvature conjecture and its applications*, J. Funct. Anal. **261** (2011), no. 5, 1284–1308, DOI 10.1016/j.jfa.2011.05.002. MR2807100

[15] A. Mihai and I. Mihai, *Curvature invariants for statistical submanifolds of Hessian manifolds of constant Hessian curvature*, Mathematics **6** (2018), Art. 44.

[16] B. Opozda, *Bochner's technique for statistical structures*, Ann. Global Anal. Geom. **48** (2015), no. 4, 357–395, DOI 10.1007/s10455-015-9475-z. MR3422914

[17] B. Opozda, *A sectional curvature for statistical structures*, Linear Algebra Appl. **497** (2016), 134–161, DOI 10.1016/j.laa.2016.02.021. MR3466639

[18] K. Nomizu and T. Sasaki, *Affine differential geometry*, Cambridge Tracts in Mathematics, vol. 111, Cambridge University Press, Cambridge, 1994. Geometry of affine immersions. MR1311248

[19] H. Shima, *The geometry of Hessian structures*, World Scientific Publishing Co. Pte. Ltd., Hackensack, NJ, 2007. MR2293045

[20] U. Simon, *Affine differential geometry*, Handbook of differential geometry, Vol. I, North-Holland, Amsterdam, 2000, pp. 905–961, DOI 10.1016/S1874-5741(00)80012-6. MR1736860

[21] P. W. Vos, *Fundamental equations for statistical submanifolds with applications to the Bartlett correction*, Ann. Inst. Statist. Math. **41** (1989), no. 3, 429–450, DOI 10.1007/BF00050660. MR1032591

[22] P. Wintgen, *Sur l'inégalité de Chen-Willmore* (French, with English summary), C. R. Acad. Sci. Paris Sér. A-B **288** (1979), no. 21, A993–A995. MR540375

Faculty of Mathematics and Computer Science, University of Bucharest, Str. Academiei 14, 010014 Bucharest, Romania

Email address: imihai@fmi.unibuc.ro

Contemporary Mathematics
Volume **756**, 2020
https://doi.org/10.1090/conm/756/15207

Warped product hypersurfaces in pseudo-Riemannian real space forms

Marilena Moruz and Luc Vrancken

Dedicated to Professor Bang-Yen Chen at the occasion of his 75th anniversary

Abstract. We study hypersurfaces in pseudo-Riemannian real space forms of non-zero sectional curvature, which write as a warped product of a 1-dimensional base with an $(n-1)$-manifold of constant sectional curvature. We show that either they have constant sectional curvature or they are contained in a rotational hypersurface. Therefore, we first define rotational hypersurfaces in indefinite real space forms of non-zero sectional curvature.

1. Introduction

Warped product manifolds are a generalization of product manifolds. The metric on such a manifold is a warped product metric, which roughly speaking, rescales the metric on the second factor of the product manifold. The fact that the metric warps on the product manifold creates an interesting geometry which mathematicians have studied since the second half of the twentieth century, when they were introduced ([**3**, **8**]).

Warped product manifolds play a very important role in physics as well ([**2**, **10**]), not only in mathematics. For instance, the space around a massive star or black hole is given as a warped product in the Schwarzschild space-time relativistic model ([**3**], p. 364-367). Also in string theory, in the context of the classical instability of extra spatial dimensions, warped product manifolds provide a more accurate description of the spacetime than just product manifolds ([**11**]). Moreover, a reasonably realistic cosmological model is the Robertson-Walker spacetime model. It is given as a warped product manifold $(I \times_f M^3(\kappa), g)$, $g = -dt^2 + f^2(t) g_{M^3}$, where I is a real interval and M^3 is a 3-dimensional Riemannian manifold with constant sectional curvature κ and Riemannian metric g_{M^3}. The warping function f describes the expanding or contracting of the Universe.

An interesting point of view for the study of warped product manifolds has been their immersability in different ambient spaces, such as Riemannian manifolds or Kähler manifolds. In this context, we study in the present paper hypersurfaces M^n in indefinite real space forms $\tilde{M}^{n+1}_{\tilde{s}}(\kappa)$, for $\kappa \neq 0$, which write as a warped product hypersurface of the form $M^n = I \times_f M^{n-1}(c)$, where $I \subset \mathbb{R}$ is an open

2010 *Mathematics Subject Classification.* Primary 53B25, 53B30, 53C42.
Key words and phrases. Warped product hypersurfaces, rotational hypersurfaces, pseudo-Riemannian real space forms.

interval and $M^{n-1}(c)$ has constant sectional curvature c. The work of M. Dajczer and R. Tojeiro ([**7**]) gives a characterization for such hypersurfaces in the positive definite case. We address this problem in the general situation when the metric in the ambient space is indefinite, of signature $s \geq 0$.

The work of M. do Carmo and M. Dajczer in [**6**] is of great interest for the problem which we study. They have extended in 1983 the rotation surfaces of the three-dimensional Euclidean space $\mathbb{E}^3$ to rotation hypersurfaces in spaces of constant sectional curvature of arbitrary dimension. In this sense, roughly speaking, a rotation hypersurface M^n in a real space form $\tilde{M}^{n+1}(\kappa)$ is generated by moving an $(n-1)$-dimensional submanifold $\Sigma \in \tilde{M}^{n+1}(\kappa)$ along some curve. The authors of [**6**] find sufficient conditions for hypersurfaces in real space forms to be rotational hypersurfaces and give examples of such rotational hypersurfaces. One of these conditions is expressed in terms of the principal curvatures of the immersed hypersurface and states that if the principal curvatures $k_1, k_2, \ldots, k_n$ satisfy $k_1 = \ldots = k_{n-1} = \lambda$, $k_n = \mu = \mu(\lambda)$ and $\lambda \neq \mu$, then the immersed hypersurface is contained in a rotation hypersurface, provided $n \geq 3$ (see Theorem 4.2. in [**6**]).

In the present paper, we define rotational hypersurfaces in pseudo-Riemannian real space forms of non-zero sectional curvature and we show that warped product hypersurfaces of the form mentioned above either must have constant sectional curvature or are rotational hypersurfaces. We prove the following theorem.

THEOREM 1. *Let $M^n = I \times_f M^{n-1}(c)$ be a warped product of a real interval $I \subset \mathbb{R}$ and an $(n-1)$-dimensional real space form $M^{n-1}(c)$ with constant sectional curvature c. Suppose that*

$$F : M^n \longrightarrow \tilde{M}^{n+1}_{\tilde{s}}(\kappa)$$

is an isometric immersion into an $n+1$-dimensional indefinite real space form $\tilde{M}^{n+1}_{\tilde{s}}(\kappa)$ of constant sectional curvature κ. Assume, moreover, that the metric on $\tilde{M}^{n+1}_{\tilde{s}}(\kappa)$ is indefinite, of signature $\tilde{s} \geq 0$ and its restriction to M^n is non-degenerate.

Then M^n has constant sectional curvature or it is a rotational hypersurface.

In this paper we consider $\kappa = \pm 1$. Notice that the case $\kappa = 0$ is treated in [**9**]. We organize the present work as follows. In section 2 we give preliminary notions on warped product manifolds and hypersurfaces in pseudo-Riemannian real space forms. In particular, we define what rotational hypersurfaces are in this setting. In the next section, we prove Theorem 1.

2. Preliminaries

Hypersurfaces in pseudo-Riemannian real space forms. Let $\tilde{M}^{n+1}_{\tilde{s}}(\kappa)$ be a pseudo-Riemannian real space form of sectional curvature κ, endowed with an indefinite metric of signature $\tilde{s}$. Let M^n be a hypersurface isometrically immersed in $\tilde{M}^{n+1}_{\tilde{s}}(\kappa)$. Denote by X, Y tangent vector fields and by ξ the normal vector field on M^n. Then the Gauss and Weingarten formulas write out as

$$\tilde{\nabla}_X Y = \nabla_X Y + h(X,Y),$$

$$\tilde{\nabla}_X \xi = -SX,$$

where $\tilde{\nabla}$ is the Levi-Civita connection on $\tilde{M}^{n+1}_{\tilde{s}}(\kappa)$, ∇ is the Levi-Civita connection of the induced metric on M, h is the second fundamental form of M and $S := A_\xi$

is the shape operator of ξ. It holds that $\langle h(X,Y),\xi\rangle = \langle SX,Y\rangle$. The equations of Gauss and Codazzi are

$$R(X,Y)Z = \kappa(\langle Y,Z\rangle X - \langle X,Z\rangle Y) + [A_{h(Y,Z)}X, A_{h(X,Z)}Y], \tag{1}$$

$$(\nabla h)(X,Y,Z) = (\nabla h)(Y,X,Z). \tag{2}$$

We recall that the covariant derivative of h is given by

$$(\nabla h)(X,Y,Z) = \nabla^{\perp}_X h(Y,Z) - h(\nabla_X Y,Z) - h(Y,\nabla_X Z).$$

The following Ricci identity holds

$$(\nabla^2 h)(X,Y,Z,W) - (\nabla^2 h)(Y,X,Z,W) = h(R(X,Y)Z,W) - h(Z,R(X,Y)W), \tag{3}$$

where the second covariant derivative of h is defined as

$$\begin{aligned}(\nabla^2 h)(X,Y,Z,W) =& \nabla^{\perp}_X((\nabla h)(Y,Z,W)) - (\nabla h)(\nabla_X Y,Z,W)\\ &- (\nabla h)(Y,\nabla_X Z,W) - (\nabla h)(Y,Z,\nabla_X W).\end{aligned} \tag{4}$$

Rotational hypersurfaces in pseudo-Riemannian real space forms. Let $\tilde{M}^{n+1}_{\tilde{s}}(\kappa)$ endowed with an indefinite metric of signature $\tilde{s}$ denote the $(n+1)$-dimensional sphere $\mathbb{S}^{n+1}_{\tilde{s}}(1)$ or the $(n+1)$-dimensional hyperbolic space $\mathbb{H}^{n+1}_{\tilde{s}}(-1)$, centered in the origin in $\mathbb{R}^{n+2}_s$. We have $\tilde{s} = s$ for $\kappa = 1$, and $\tilde{s} = s-1$ for $\kappa = -1$. In the following part we define what rotational hypersurfaces in $\tilde{M}^{n+1}_{\tilde{s}}(\kappa)$ are. Notice that, given the way they are constructed, the condition of non-degeneracy of the metric restricted to a 3-dimensional space implies the discussion of a few cases.

The rotation hypersurface will be determined by rotating a space curve α around a 2-dimensional axis Π^2. We fix Π^2 to be the plane which passes through the origin of $\mathbb{R}^{n+2}_s$ and contains a geodesic γ of $\tilde{M}^{n+1}_{\tilde{s}}(\kappa)$. Let us now fix the space curve α. We choose a 3-dimensional space Π^3 such that $\Pi^2 \subset \Pi^3$. Let α be a curve lying at the intersection of $\tilde{M}^{n+1}_{\tilde{s}}(\kappa)$ with Π^3. Assume the metric restricted to Π^3 is non-degenerate. We generate our rotational hypersurface by looking at the image of α under all isometries which preserve Π^2 (and hence the geodesic) point-wisely.

Let $\{e_1, e_2, \ldots, e_{n+2}\}$ be an orthogonal basis of $\mathbb{R}^{n+2}_s$, for which the lengths of the vectors are ± 1. We will assume, for more convenience in our reasoning, possibly after reordering the vectors, that the s " $-$ " signs of the metric correspond to the last s vectors $\{e_{n+3-s}, \ldots, e_{n+2}\}$ in the basis. So, a way to express the metric is $(+,+,\ldots,-,-)$.

Let $\mathcal{O}$ denote an isometry of $\mathbb{R}^{n+2}_s$. We know that the plane Π^2 is preserved by $\mathcal{O}$. We have that the metric on Π^3 is non-degenerate. Then, the metric restricted to Π^2 can be as in the following cases:

$$a.)\begin{pmatrix}1&0\\0&0\end{pmatrix}\quad b.)\begin{pmatrix}-1&0\\0&0\end{pmatrix}\quad c.)\begin{pmatrix}1&0\\0&1\end{pmatrix}\quad d.)\begin{pmatrix}1&0\\0&-1\end{pmatrix}\quad e.)\begin{pmatrix}-1&0\\0&-1\end{pmatrix}.$$

a.) Assume the metric restricted to Π^2 is of the form

$$\begin{pmatrix}1&0\\0&0\end{pmatrix}. \tag{5}$$

Let e_1 be one of the generating vectors of Π^3, contained as well in Π^2. The orthogonal space to e_1 may be assumed to be spanned by e_2 and e_{n+2}. Then, we choose a basis on this orthogonal space to be given by $n_1 = (0,1,0,\ldots,0,1)$ and $n_2 = (0,1,0,\ldots,0,-1)$, which are two null vectors. This way, the restriction of the

metric to Π^2 corresponds to the degenerate metric assumed in (5). At the same time, the fact that Π^2 is preserved by isometries means that e_1 is preserved and a second direction preserved may be chosen to be precisely n_1.
The generating curve $\alpha \subset \Pi^3$ of the hypersurface is described by

$$\alpha(t) = f(t)\begin{pmatrix}1\\0\\0\\\vdots\\0\end{pmatrix} + g(t)\begin{pmatrix}0\\1\\0\\\vdots\\1\end{pmatrix} + h(t)\begin{pmatrix}0\\1\\0\\\vdots\\-1\end{pmatrix}, \tag{6}$$

with f, g, h real functions. We have $\mathcal{O}e_1 = e_1$, $\mathcal{O}n_1 = n_1$ and $\mathcal{O}n_2 = v^t$, where $v \in \mathbb{R}^{n+2}_s$ and v^t is the transposed of $v = (v_1, \dots, v_{n+2})$. Since the metric is preserved by $\mathcal{O}$, we have

$$\langle e_1, v\rangle = 0, \quad \langle n_1, v\rangle = 2, \quad \langle v, v\rangle = 0$$

and therefore $v_1 = 0$, $v_2 = 1 - \frac{1}{4}\sum_{i=3}^{n+1} \varepsilon_i v_1^2$, $v_{n+2} = -1 - \frac{1}{4}\sum_{i=3}^{n+1} \varepsilon_i v_i^2$, where $\varepsilon_i = \pm 1$ and are determined according to the metric on $\mathbb{R}^{n+2}_s$. Therefore, the rotation hypersurface is given by

$$(t, v_3, \dots, v_{n+1}) \mapsto f(t)\begin{pmatrix}1\\0\\0\\\vdots\\0\end{pmatrix} + g(t)\begin{pmatrix}0\\1\\0\\\vdots\\1\end{pmatrix} + h(t)\begin{pmatrix}0\\1 - \frac{1}{4}\sum_{i=3}^{n+1}\varepsilon_i v_i^2\\v_3\\\vdots\\v_{n+1}\\-1 - \frac{1}{4}\sum_{i=3}^{n+1}\varepsilon_i v_i^2\end{pmatrix},$$

or, equivalently, as

$$(t, v_3, \dots, v_{n+1}) \mapsto \Big(f(t), g(t)+h(t)[1-\frac{1}{4}\sum_{i=3}^{n+1}\varepsilon_i v_i^2], h(t)v_3, \dots, h(t)v_{n+1}, g(t)+h(t)[-1-\frac{1}{4}\sum_{i=3}^{n+1}\varepsilon_i v_i^2]\Big). \tag{7}$$

b.) Assume the metric restricted to Π^2 is of the form

$$\begin{pmatrix}-1 & 0\\0 & 0\end{pmatrix}.$$

Let e_{n+2} give one generating direction of Π^3. Assume this is one of the two directions in Π^2 which stays fixed under isometries of $\mathbb{R}^{n+2}_s$. Let the orthogonal space to this direction be spanned by e_1 and e_{n+1}. Let $n_1 = (1, 0, \dots, 0, 1, 0)$ and $n_2 = (1, 0, \dots, 0, -1, 0)$ be two null vectors in this orthogonal space such that a direction given by a combination of n_1 and n_2 will stay fixed under isometries of $\mathbb{R}^{n+2}_s$. Without loss of generality we may assume that this direction is precisely n_1. Hence, we have $\mathcal{O}e_{n+2} = e_{n+2}$, $\mathcal{O}n_1 = n_1$ and $\mathcal{O}n_2 = v^t$, where $v = (v_1, \dots, v_{n+2}) \in \mathbb{R}^{n+2}_s$. Additionally, $\langle n_2, n_2\rangle = 0$, $\langle n_2, e_{n+2}\rangle = 0$, $\langle n_1, n_2\rangle = 2$ and since the metric is preserved by $\mathcal{O}$, we get

$$v_{n+2} = 0, \quad v_1 - v_{n+1} = 2, \quad v_1^2 + \sum_{i=2}^{n} \varepsilon_i v_i^2 - v_{n+1}^2 = 0.$$

Therefore, $v_1 = 1 - \frac{1}{4}\sum_{i=2}^{n} \varepsilon_i v_i^2$, $v_{n+1} = -1 - \frac{1}{4}\sum_{i=2}^{n} \varepsilon_i v_i^2$ and the rotation hypersurface in this case is given by

$$(t, v_2, \ldots, v_n) \mapsto f(t)\begin{pmatrix} 0 \\ 0 \\ \vdots \\ 0 \\ 1 \end{pmatrix} + g(t)\begin{pmatrix} 1 \\ 0 \\ \vdots \\ 1 \\ 0 \end{pmatrix} + h(t)\begin{pmatrix} 1 - \frac{1}{4}\sum_{i=2}^{n} \varepsilon_i v_i^2 \\ v_2 \\ \vdots \\ v_n \\ -1 - \frac{1}{4}\sum_{i=2}^{n} \varepsilon_i v_i^2 \\ 0 \end{pmatrix},$$

or equivalently :

(8)

$$(t, v_2, \ldots, v_n) \mapsto \Big(g(t) + h(t)[1 - \frac{1}{4}\sum_{i=2}^{n} \varepsilon_i v_i^2], h(t)v_2, \ldots, h(t)v_n, g(t) + h(t)[-1 - \frac{1}{4}\sum_{i=2}^{n} \varepsilon_i v_i^2], f(t)\Big).$$

c.) Assume the metric restricted to Π^2 is the non-degenerate metric of the form

$$\begin{pmatrix} 1 & 0 \\ 0 & 1 \end{pmatrix}.$$

Let Π^2 be spanned by e_1 and e_2. A space curve in Π^3 may then be described by an additional direction given by e_3 or e_{n+2}:

$$\alpha(t) = (f(t), g(t), h(t), 0, \ldots, 0) \quad \text{or} \quad \alpha(t) = (f(t), g(t), 0, \ldots, 0, h(t)).$$

An isometry preserving Π^2 is of the form

$$\text{(9)} \qquad \begin{pmatrix} 1 & 0 & & & \\ 0 & 1 & & 0 & \\ & & v_1 & \ldots & u_1 \\ & 0 & \vdots & & \vdots \\ & & v_n & \ldots & u_n \end{pmatrix}$$

with $v_1^2 \pm \ldots - v_n^2 = 1$ and $u_1^2 \pm \ldots - u_n^2 = -1$. The rotation hypersurface in this case is

$$\text{(10)} \quad (t, v_1, \ldots, v_{n-1}) \mapsto \Big(f(t), g(t), h(t)v_1, \ldots, h(t)v_{n-1}, \pm\, h(t)\sqrt{\sum_{i=1}^{n-1} \varepsilon_i v_i^2 - 1}\Big)$$

or

$$\text{(11)} \quad (t, u_1, \ldots, u_{n-1}) \mapsto \Big(f(t), g(t), h(t)u_1, \ldots, h(t)u_{n-1}, \pm\, h(t)\sqrt{\sum_{i=1}^{n-1} \varepsilon_i u_i^2 + 1}\Big),$$

with ε_i determined according to the metric on $\mathbb{R}_s^{n+2}$.

d.) Assume the metric restricted to Π^2 is the non-degenerate metric of the form

$$\begin{pmatrix} 1 & 0 \\ 0 & -1 \end{pmatrix}.$$

Let Π^2 be spanned by e_1 and e_{n+2}. A space curve in Π^3 may then be described by an additional direction given by e_2 or e_{n+1} :

$$\alpha(t) = (f(t), h(t), 0, \ldots, g(t)) \quad \text{or} \quad \alpha(t) = (f(t), 0, \ldots, h(t), g(t)).$$

An isometry preserving Π^2 is of the form $\begin{pmatrix} 1 & 0 & \dots & 0 & 0 \\ 0 & v_1 & \dots & u_1 & 0 \\ \vdots & \vdots & & \vdots & \vdots \\ 0 & v_n & \dots & u_n & 0 \\ 0 & 0 & \dots & 0 & 1 \end{pmatrix}$, where $\sum_{i=1}^{n} \varepsilon_i v_i^2 = 1$ and $\sum_{i=1}^{n} \varepsilon_i u_i^2 = -1$. Then, a rotation hypersurface is given by

$$(t, v_1, \dots, v_{n-1}) \mapsto \Big(f(t), h(t)v_1, \dots, h(t)v_{n-1}, \pm h(t)\sqrt{\sum_{i=1}^{n-1} \varepsilon_i v_i^2 - 1}, g(t)\Big) \tag{12}$$

or

$$(t, u_1, \dots, u_{n-1}) \mapsto \Big(f(t), h(t)u_1, \dots, h(t)u_{n-1}, \pm h(t)\sqrt{\sum_{i=1}^{n-1} \varepsilon_i u_i^2 + 1}, g(t)\Big). \tag{13}$$

e.) Assume the metric restricted to Π^2 is the non-degenerate metric of the form

$$\begin{pmatrix} -1 & 0 \\ 0 & -1 \end{pmatrix}.$$

Let e_{n+1} and e_{n+2} span Π^2. A space curve in Π^3 may then be described by an additional direction given by e_1 or e_n :

$$\alpha(t) = (h(t), 0, \dots, 0, f(t), g(t)) \quad \text{or} \quad \alpha(t) = (0, \dots, 0, h(t), f(t), g(t)).$$

An isometry preserving Π^2 is of the form

$$\begin{pmatrix} v_1 & \dots & u_1 & & \\ \vdots & & \vdots & \multicolumn{2}{c}{0} \\ v_n & \dots & u_n & & \\ & & & 1 & 0 \\ & 0 & & 0 & 1 \end{pmatrix} \tag{14}$$

where $v_1^2 + \sum_{i=2}^{n} \varepsilon_i v_i^2 = 1$ and $u_1^2 + \sum_{j=2}^{n} \varepsilon_j u_j^2 = -1$, with $\varepsilon_i = \pm 1, \varepsilon_j = \pm 1$. Then, a rotation hypersurface is given by

$$(t, v_2, \dots, v_n) \mapsto \Big(\pm h(t)\sqrt{1 - \sum_{i=2}^{n} \varepsilon_i v_i^2}, h(t)v_2, \dots, h(t)v_n, f(t), g(t)\Big) \tag{15}$$

or

$$(t, u_2, \dots, u_n) \mapsto \Big(\pm h(t)\sqrt{-(1 + \sum_{i=2}^{n} \varepsilon_i u_i^2)}, h(t)u_2, \dots, h(t)u_n, f(t), g(t)\Big). \tag{16}$$

Warped product manifolds ([5]). Let B and F be two pseudo-Riemannian manifolds of positive dimensions, endowed with pseudo-Riemannian metrics g_B and g_F, respectively. Let f be a positive smooth function on B. On the product manifold $B \times F$, on which we have the natural projections

$$\pi : B \times F \longrightarrow B \text{ and } \eta : B \times F \longrightarrow F,$$

we consider the pseudo-Riemannian structure such that

$$\langle X, Y\rangle = \langle \pi_\star(X), \pi_\star(Y)\rangle + f^2(\pi(x))g(\eta_\star(X), \eta_\star(Y)), \tag{17}$$

for any tangent vectors $X, Y \in TM$. This way $B \times F$ becomes a warped product manifold $M = B \times_f F$, for which the metric g satisfies $g = g_B + f^2 g_F$.

The function f is called the warping function of the warped product, B is called the base and F the fiber. If f is constant, then the warped product $B \times_f F$ is called trivial. In this case $B \times_f F$ is the Riemannian product $B \times F_f$, where F_f is the Riemannian manifold equipped with the metric $f^2 g_F$.

The leaves $B \times \{q\} = \eta^{-1}(q)$ and the fibers $\{p\} \times F = \pi^{-1}(p)$ are Riemannian submanifolds of M. Vectors tangent to leaves are called horizontal and those tangent to fibers are called vertical. If $u \in T_pB$, $p \in B$ and $q \in F$, then the lift $\bar{u}$ of u to (p, q) is the unique horizontal vector in $T_{(p,q)}M$ such that $\pi_\star(\bar{u}) = u$. For a vector field $X \in \mathfrak{X}(B)$, the lift of X to M is the vector field $\bar{X}$ whose value at (p, q) is the lift of X_p to (p, q). The set of all such horizontal lifts is denoted by $\mathcal{L}(B)$. Similarly, we denote by $\mathcal{L}(F)$ the set of all vertical lifts.

One may express the curvature tensor of the warped product manifold in terms of the warping function and the curvature tensors of the two components B and F. For a warped product manifold $M = B \times_f F$, we define the lift $\tilde{T}$ of a covariant tensor T on B to M as the pullback $\pi^*(T)$ via the projection $\pi : M \longrightarrow B$. We will denote by ${}^BR$ and ${}^FR$ the curvature tensors of B and F, respectively. We have the following proposition from [**5**].

PROPOSITION 1. *Let $M = B \times_f F$ be a warped product of two pseudo-Riemannian manifolds. If $X, Y, Z \in \mathcal{L}(B)$ and $U, V, W \in \mathcal{L}(F)$, then we have*

(1) *$R(X, Y)Z \in \mathcal{L}(B)$ is the lift of ${}^BR(X, Y)Z$ on B;*
(2) *$R(X, V)Y = \frac{H^f(X,Y)}{f}V$;*
(3) *$R(X, Y)V = R(V, W)X = 0$;*
(4) *$R(X, V)W = -\frac{\langle V,W\rangle}{f}\nabla_X(\nabla f)$;*
(5) *$R(V, W)U = {}^FR(V, W)U + \frac{\langle \nabla f, \nabla f\rangle}{f^2}\{\langle V, U\rangle W - \langle W, U\rangle V\}$,*

where R is the curvature tensor of M and H^f is the Hessian of f.

3. Proof of Theorem 1 for $\kappa \neq 0$

In this section we give the proof of Theorem 1 and we organize it in the following way. After fixing the notations, we use the curvature relations together with a new technique of computations in order to obtain relation (20). Next, we consider two cases as whether M^n has constant curvature (Case 1) or not (Case 2). In order to eventually determine the immersion F, the case when M^n does not have constant curvature requires a more lengthy proof. In a first step, we determine the form of the shape operator of M^n. Then, we treat two different sub-cases, where we construct three new vector fields using F, ξ and E_1, which we can determine explicitly thanks to their properties. This allows us to recover back the immersion F.

Let M^n be a warped product hypersurface isometrically immersed into a real space form $\tilde{M}^{n+1}_{\tilde{s}}(\kappa)$, of the form

$$M^n = I \times_f M^{n-1}(c) \xrightarrow{F} \tilde{M}^{n+1}_{\tilde{s}}(\kappa),$$

where $I \subset \mathbb{R}$, f is a function defined on I, while the second factor $M^{n-1}(c)$ is a real space form with constant sectional curvature c, endowed with the metric g and of dimension $n-1$. By $\langle\cdot,\cdot\rangle$ we denote the metric on the ambient space. Let $E_1 = \frac{\partial}{\partial t}$ be the unit vector tangent to I:

$$\langle E_1, E_1\rangle = 1.$$

Notice that we can always assume E_1 to have positive length, since, if necessary, we can change the signature of the metric on $\tilde{M}^{n+1}_{\tilde{s}}(\kappa)$ from $\tilde{s}$ to $n+1-\tilde{s}$ and work on the corresponding diffeomorphically anti-isometric space. Denote by ξ the unit normal to the hypersurface and let $\tilde{\varepsilon}$ such that:

$$\langle\xi,\xi\rangle = \tilde{\varepsilon} = \pm 1.$$

We denote by F as well the position vector on the image of the hypersurface. We have $\langle F, F\rangle = \kappa$. Let U, V, W be tangent vectors to M^{n-1}. Then from equation (17) it follows that:

$$\langle E_1, U\rangle = 0, \quad \langle U, V\rangle = f^2 g(U,V)$$

and from Proposition 1 we know that the curvature tensor on M is given as follows:

$$\begin{aligned}
&R(E_1,V)E_1 = \frac{f''}{f}V,\\
&R(W,V)E_1 = 0,\\
&R(E_1,V)W = -\langle V,W\rangle\frac{f''}{f}E_1,\\
&R(V,W)U = {}^{M^{n-1}}R(V,W)U + \frac{f'^2}{f^2}\{\langle V,U\rangle W - \langle W,U\rangle V\},
\end{aligned}\tag{18}$$

where ${}^{M^{n-1}}R$ is the curvature tensor on M^{n-1} and is given by

$$ {}^{M^{n-1}}R(U,V)W = c(g(V,W)U - g(U,W)V).\tag{19}$$

In order to investigate our hypersurfaces, we will use the same method as in a few other papers, initially introduced in [**1**], and which the authors called the Tsinghua principle. The following computations describe this method, which, eventually brings the advantage of providing a new relation for the data on the hypersurface, see (20). Let X, Y, Z, W be tangent vector fields on M^n. First, we take the derivative with respect to W in the Codazzi equation (2). This gives

$$(\nabla^2 h)(W,X,Y,Z) - (\nabla^2 h)(W,Y,X,Z) = 0.$$

We then cyclically permute W, X, Y in this relation and use additionally the Ricci identity as in (3), together with the first Bianchi identity to obtain

$$0 = h(Y, R(W,X)Z) + h(W, R(X,Y)Z) + h(X, R(Y,W)Z).\tag{20}$$

Let $W, X, Z \in T_qM^{n-1}$, for $q \in M^{n-1}$ and let $Y = E_1$. Then equation (20) implies that

$$0 = \left(c + f''f - f'^2\right)\Big(\langle X,Z\rangle h(W,E_1) - \langle W,Z\rangle h(X,E_1)\Big),\tag{21}$$

which yields two cases, according to whether $c + f''f - f'^2$ vanishes or not.

Case 1. Assume

$$c + f''f - f'^2 = 0.\tag{22}$$

We shall see that M^n has constant sectional curvature.

For more convenience, in this case we prefer to denote by E_1 the unit vector spanning the tangent space to I and by X_i, $i = 2,\dots,n$ the orthonormal vectors spanning the tangent space to M^{n-1}, with $\langle X_i, X_i\rangle = c_i = \pm 1$. Let $\Pi_{1i} = span\{E_1, X_i\}$ and $\Pi_{ij} = span\{X_i, X_j\}$ be 2-dimensional planes in the tangent space of T_xM^n, for $x \in M^n$. We get immediately that the sectional curvatures of M^n for all such planes are given by $K(\Pi_{1i}) = -\frac{f''}{f}$ and $K(\Pi_{ij}) = \frac{c}{f^2} - \frac{f'^2}{f^2}$.

By (22) we have $-\frac{f''}{f} = \frac{c}{f^2} - \frac{f'^2}{f^2}$ and we can prove straightforwardly that $\frac{f''}{f}$ is constant. Hence, M^n has constant sectional curvature.

Case 2. Assume

$$c + f''f - f'^2 \neq 0.$$

From (21) we have that

$$\langle X, Z\rangle h(W, E_1) - \langle W, Z\rangle h(X, E_1) = 0.$$

Next, we choose Z such that $\langle X, Z\rangle = 0$ and $\langle W, Z\rangle \neq 0$. It follows that $h(X, E_1) = 0$, for all $X \in T_qM^{n-1}$, i.e. $SE_1 \perp X$. Then there exists μ function of t such that

$$SE_1 = \mu E_1 \tag{23}$$

and we have $h(E_1, E_1) = \tilde{\varepsilon}\mu\xi$. We use the curvature relations in (18) and the Gauss equation in (1) to express $R(E_1, X)E_1$. We obtain

$$SX = -\frac{1}{\tilde{\varepsilon}\mu}\left(\frac{f''}{f} + \kappa\right)X. \tag{24}$$

We do similarly for $R(X, Y)X$, with X, Y tangent vectors to M^{n-1}, and obtain

$$(\langle X, X\rangle Y - \langle X, Y\rangle X)\left(\frac{f'^2}{f^2} - \frac{c}{f^2} + \kappa\right) = \langle SY, X\rangle SX - \langle SX, X\rangle SY. \tag{25}$$

In order to understand equations (24) and (25), we need to determine the form of the shape operator of M^{n-1}. We prove the following proposition.

PROPOSITION 2. *There exists a basis $\{X_i\}_{i=1,\dots,n-1}$ in T_qM^{n-1} for which the shape operator restricted to M^{n-1} is given by $SX_i = \lambda X_i$, for real functions $\lambda = \lambda(t)$.*

PROOF. We choose first a convenient notation for an orthonormal basis on T_qM^{n-1}, with vectors of length ± 1: $\{X_1, \dots, X_{n-1}\}$. We will use the result in [**12**], which states that for the shape operator S, which is symmetric with respect to the metric $\langle\cdot,\cdot\rangle$ (and so with respect to $g(\cdot,\cdot)$ too) there exists a basis on the tangent space of M^n such that we have, for some natural positive numbers $m_i \leq m \leq n$, the following:

$$S = \begin{pmatrix} S_1 & & \\ & \ddots & \\ & & S_m \end{pmatrix}, \quad g = \begin{pmatrix} g_1 & & \\ & \ddots & \\ & & g_m \end{pmatrix}, \tag{26}$$

where

$$S_i = \begin{pmatrix} \lambda_i & 1 & & \\ & \lambda_i & \ddots & \\ & & \ddots & 1 \\ & & & \lambda_i \end{pmatrix}_{m_i\times m_i}, \quad g_i = \begin{pmatrix} & & \epsilon_i \\ & \iddots & \\ \epsilon_i & & \end{pmatrix}_{m_i\times m_i}, \ \mid \epsilon_i \mid = 1. \tag{27}$$

Notice that the eigenvalues λ_i may be complex, in which case $\epsilon_i \in \mathbb{C}$.

We will prove by contradiction that the shape operator S cannot contain blocks S_i of dimension $m_i > 1$. Assume first that there exists such a block of dimension 3. Without loss of generality we may assume that $m_1 = 3$. Then, according to (27), we may write

$$SX_1 = \lambda_1 X_1,$$
$$SX_2 = X_1 + \lambda_1 X_2,$$
$$SX_3 = X_2 + \lambda_1 X_3.$$

In this case, using the symmetry of the shape operator, we find that $\langle X_2, X_2\rangle = \langle X_1, X_3\rangle$, whereas we know from the expression of the metric that $\langle X_1, X_3\rangle \neq 0$. Moreover, the equation following from (25) for $R(X_2, X_3)X_2$ gives that $\langle X_2, X_2\rangle = 0$, which is a contradiction. Hence there cannot be blocks of dimension 3.

When there is at least one block of dimension m_i higher than 3, we prove similarly a contradiction by picking X_α, X_β, $1 \leq \alpha, \beta < n$ in (25) to be given either by $\alpha = \frac{m_i}{2}, \beta = \alpha + 2$ if m_i is even, or by $\alpha = [\frac{m_i}{2}] + 1, \beta = \alpha + 1$ if m_i is odd. Therefore $m_i \in \{1, 2\}$.

We shall now prove there cannot be any blocks of dimension 2 neither. Assume there exists at least one block of dimension 2 in the matrix of the shape operator. We may assume, without loss of generality, that $m_1 = 2$. Therefore

$$SX_1 = \lambda_1 X_1,$$
$$SX_2 = X_1 + \lambda_1 X_2.$$

We can write equation (25) for $R(X_1, X_2)X_1$ and it follows that $\lambda_1^2 = \frac{c^2}{f^2} - \frac{f'^2}{f^2} - \kappa$.

Similarly, for $R(X_2, X_1)X_2$ we obtain that $\lambda_1 = 0$, which is a contradiction to the fact that $\frac{c^2}{f^2} - \frac{f'^2}{f^2} - \kappa \neq 0$.

Therefore, there are only blocks of dimension 1 in the matrix of the shape operator and so

$$SX_i = \lambda_i X_i,\ i = 1, \ldots, n-1.$$

Furthermore, we use (18) and (19) to evaluate $R(E_1, X_i)E_1$ and $R(X_j, X_i)X_j$. It follows, respectively, that

$$\begin{aligned} -\tilde{\varepsilon}\mu\lambda_i &= \frac{f''}{f} + \kappa, \\ \tilde{\varepsilon}\lambda_i\lambda_j &= -\frac{f'^2}{f^2} + \frac{c}{f^2} - \kappa. \end{aligned} \tag{28}$$

It is straightforward to check from the second relation in (28) that for indexes $1 \leq i, j \leq n$ we must have $\lambda_i = \lambda_j$, for all $i, j = 1 \ldots, n-1$ and we also notice that the eigenvalues of X_i are real. We may of course denote $\lambda := \lambda_i$. □

Equations (28) become

$$-\tilde{\varepsilon}\mu\lambda = \frac{f''}{f} + \kappa, \tag{29}$$

$$\tilde{\varepsilon}\lambda^2 = -\frac{f'^2}{f^2} + \frac{c}{f^2} - \kappa. \tag{30}$$

We identify a tangent vector $Z \in T_x M^n$ with its image through the differential dF and so $D_Z F = dF(Z) \equiv Z$. Let X be a tangent vector in $T_q M^{n-1}$. It follows that

$$D_X\xi = -\lambda X, \quad D_X E_1 = \frac{f'}{f} X.$$

Let us now define two vectors ψ_1 and ψ_2 as

$$\begin{aligned} \psi_1 &= \lambda F + \xi, \\ \psi_2 &= \frac{f'}{f} F - E_1. \end{aligned} \tag{31}$$

We get straightforwardly that both ψ_1 and ψ_2 only depend on t, since $D_X\psi_1 = D_X\psi_2 = 0$. Moreover, we can check that

$$D_{E_1}\psi_1 = -(\lambda-\mu)\psi_2 \quad \text{and} \quad D_{E_1}\psi_2 = -\frac{f'}{f}\psi_2 - \tilde{\varepsilon}\mu\psi_1. \tag{32}$$

Furthermore, we have that $\langle\psi_1,\psi_1\rangle = \tilde{\varepsilon}\kappa\left(\frac{c}{f^2} - \frac{f'^2}{f^2}\right)$, $\langle\psi_2,\psi_2\rangle = 1+\kappa\frac{f'^2}{f^2}$, $\langle\psi_1,\psi_2\rangle = \lambda\frac{f'}{f}$. The two vectors ψ_1 and ψ_2 span a non-degenerate plane Π, provided $c \neq 0$.

Case 2.1. Assume $c \neq 0$.

Let $u = \pm\frac{\sqrt{\varepsilon c}}{f}$, with $\varepsilon = 1$ if $c > 0$ and $\varepsilon = -1$ if $c < 0$. We may choose a vector $\varphi \perp \psi_1, \psi_2$ of length ± 1 given by

$$\varphi = \frac{1}{u}\left(\kappa F - \lambda\tilde{\varepsilon}\xi + \frac{f'}{f}E_1\right). \tag{33}$$

It follows that

$$\begin{aligned} D_{E_1}\varphi &= \pm\frac{\tilde{\varepsilon}}{\sqrt{\varepsilon c}}[(\mu-\lambda)f' - \lambda' f]\xi \\ &= 0 \end{aligned}$$

and

$$\begin{aligned} D_X\varphi &= \frac{1}{u}\left(\kappa + \tilde{\varepsilon}\lambda^2 + \frac{f'^2}{f^2}\right)X \\ &= \frac{1}{u}\frac{c}{f^2}X. \end{aligned}$$

Hence $D_X\varphi \neq 0$ and so φ is an $(n-1)$-dimensional immersion $\varphi : M^{n-1} \longrightarrow \tilde{M}_{\tilde{s}}^{n+1}$. In the following step, we determine explicitly the immersion φ and eventually the immersion F.

First, notice that from equations (31) and (33), we obtain

$$F = \frac{1}{c}(\tilde{\varepsilon}f^2\lambda\psi_1 + ff'\psi_2 \pm \frac{f}{\sqrt{\varepsilon c}}\varphi). \tag{34}$$

Equations (32) lead to solving a second order differential equation with non constant coefficients $\psi_1'' + \left[(\lambda-\mu)\left(\frac{1}{\lambda-\mu}\right)' + \frac{f'}{f}\right]\psi_1' - \tilde{\varepsilon}\mu(\lambda-\mu)\psi_1 = 0$. It follows that $\psi_1 = c_1 y_1(t) + c_2 y_2(t)$ and $\psi_2 = c_1 y_3(t) + c_2 y_4(t)$, where $y_1(t)$ and $y_2(t)$ are two solutions of the homogeneous differential equation above, $y_3 = -\frac{y_1'(t)}{\lambda-\mu}$ and $y_4 = -\frac{y_2'(t)}{\lambda-\mu}$ and c_1, c_2 are constants. In fact, we can choose c_1 and c_2 in the following way. We have

$\psi_1' = (\mu - \lambda)\psi_2$ and $(f\psi_2)' = \tilde{\varepsilon} f \mu \psi_1$. Consider the vector $C = \tilde{a}\psi_1 + \tilde{b} f \psi_2$ and let a and b be functions of t satisfying the following system of differential equations:

$$a' = \tilde{\varepsilon} f \mu \, b,$$

$$b' = \frac{\mu - \lambda}{f} \, a.$$

We have that for initial condition at $t = t_0$ and solutions $a = 1$, $b = 0$ and $a = 0$, $b = 1$ respectively, C is constant. We choose then $c_1 = \psi_1|_{t=t_0}$ and $c_2 = (f\psi_2)|_{t=t_0}$.

• If the metric restricted to Π is indefinite, we may fix the two constant vectors c_1 and c_2 as

$$c_1 = (1, 0, \dots, 0) \text{ and } c_2 = (0, \dots, 0, 1).$$

It follows that

$$\psi_1 = (y_1(t), 0, \dots, 0, y_2(t)), \quad \psi_2 = (y_3(t), 0, \dots, 0, y_4(t))$$

and

$$\varphi : (u_1, \dots, u_{n-1}) \mapsto (0, k_1, \dots, k_n, 0), \tag{35}$$

with $k_i = k_i(u_1, \dots, u_{n-1})$, $i = 1, \dots, n$, $k_1^2 + \dots + k_n^2 = \pm 1$.

The immersion F is given by

$$(t, u_1, \dots, u_{n-1}) \mapsto (\tilde{\varepsilon} f^2 \lambda y_1(t) + f f' y_3(t), \tilde{\varepsilon} f k_1, \dots, \tilde{\varepsilon} f k_n, \tilde{\varepsilon} f^2 \lambda y_2(t) + f f' y_4(t)), \tag{36}$$

as stated in case $d.$) in Preliminaries.

• If the metric restricted to Π is positive definite, we may fix the two constant vectors c_1 and c_2 as

$$c_1 = (1, 0, 0, \dots, 0) \text{ and } c_2 = (0, 1, 0, \dots, 0).$$

It follows that

$$\psi_1 = (y_1(t), y_2(t), 0, \dots, 0), \quad \psi_2 = (y_3(t), y_4(t), 0, \dots, 0)$$

and

$$\varphi : (u_1, \dots, u_{n-1}) \mapsto (0, 0, k_1, \dots, k_n),$$

with $k_i = k_i(u_1, \dots, u_{n-1})$ and $k_1^2 + \dots + k_n^2 = \pm 1$. Then the immersion F is given by

$$(t, u_1, \dots, u_{n-1}) \mapsto \frac{1}{c}(\tilde{\varepsilon} f^2 \lambda y_1(t) + f f' y_3(t), \tilde{\varepsilon} f^2 \lambda y_2(t) + f f' y_4(t), \frac{\pm f}{\sqrt{\varepsilon c}} k_1, \dots, \frac{\pm f}{\sqrt{\varepsilon c}} k_n), \tag{37}$$

which corresponds to case $c.$) in Preliminaries.

• If the metric restricted to Π is negative definite, we may fix the two constant vectors c_1 and c_2 as

$$c_1 = (0, \dots, 0, 1) \text{ and } c_2 = (0, \dots, 1, 0).$$

We then obtain that

$$\psi_1 = (0, \dots, 0, y_2(t), y_1(t)), \quad \psi_2 = (0, \dots, 0, y_4(t), y_3(t))$$

and

$$\varphi : (u_1, \dots, u_{n-1}) \mapsto (k_1, \dots, k_n, 0, 0),$$

with $k_i = k_i(u_1, \ldots, u_{n-1})$ and $k_1^2 + \ldots + k_n^2 = \pm 1$. Finally, the immersion F is given in this case by

(38)

$$F : (t, u_1, \ldots, u_{n-1}) \mapsto \frac{1}{c}(\frac{\pm f}{\sqrt{\varepsilon c}}k_1, \ldots, \frac{\pm f}{\sqrt{\varepsilon c}}k_n, \tilde{\varepsilon} f^2 \lambda y_2(t) + ff' y_4(t), \tilde{\varepsilon} f^2 \lambda y_1(t) + ff' y_3(t)),$$

which corresponds to case $e.$) in Preliminaries.

Case 2.2. Assume $c = 0$. In this case, the metric restricted to the plane $span\{\psi_1, \psi_2\}$ is degenerate. Let $a = \frac{1}{2}\frac{(f^2+\kappa f'^2)'}{f^2+\kappa f'^2}$, $b = -\frac{\kappa f f'}{f^2+\kappa f'^2}$ and $h = h(t)$ such that $ln|Ah + B| = -At + C_0$, where $A = \frac{-f'(f^3+f'f''+ff'^2\kappa+f^2f''\kappa)}{(f^2+\kappa f'^2)^2}$, $B = 1 - \frac{ff'}{f'^2}$, $C_0 \in \mathbb{R}$. Consider $\tilde{\psi}_1 = \psi_1 + b\lambda\psi_2$. We obtain immediately that $a\tilde{\psi}_1$ is a c,onstant null vector. Moreover $\tilde{\psi}_2 = \frac{f}{f'}\psi_1 + h\tilde{\psi}_1$ is a constant vector of length ± 1 orthogonal to $a\tilde{\psi}_1$. Therefore, we can fix the following, according to the length of $\tilde{\psi}_2$.

$$\begin{aligned} a\tilde{\psi}_1 &= (0, 1, \ldots, 1), \\ \tilde{\psi}_2 &= (1, 0 \ldots, 0). \end{aligned} \quad \text{or} \quad \begin{aligned} a\tilde{\psi}_1 &= (1, 0, \ldots, 1, 0), \\ \tilde{\psi}_2 &= (0, \ldots, 0, 1). \end{aligned} \tag{39}$$

Next we choose $\varphi = (\varphi_1, \ldots, \varphi_{n+2})$ null vector such that $\varphi \perp \tilde{\psi}_2$ and $\langle \varphi, a\tilde{\psi}_1 \rangle = 2$. We must have that $\varphi_1 = 0$, $\varphi_2 = 1 - \frac{1}{4}\sum_{i=3}^{n+1} \varepsilon_i \varphi_i^2$, $\varphi_{n+2} = -1 - \frac{1}{4}\sum_{i=3}^{n+1} \varepsilon_i \varphi_i^2$ or $\varphi_1 = 1 - \frac{1}{4}\sum_{i=2}^{n} \varepsilon_i \varphi_i^2$, $\varphi_{n+1} = -1 - \frac{1}{4}\sum_{i=2}^{n} \varepsilon_i \varphi_i^2$, $\varphi_{n+2} = 0$. Since $a\tilde{\psi}_1$, $\tilde{\psi}_2, \varphi$ are vectors in the plane spanned by ψ_1, ψ_2, we may express them with respect to F, E_1, ξ. Hence, we obtain that F has one of the following forms:

$$F : (t, \varphi_3, \ldots, \varphi_{n+1}) \mapsto f_1(t)\begin{pmatrix}1\\0\\0\\\vdots\\0\end{pmatrix} + f_2(t)\begin{pmatrix}0\\1\\0\\\vdots\\1\end{pmatrix} + f_3(t)\begin{pmatrix}0\\1-\frac{1}{4}\sum_{i=3}^{n+1}\varepsilon_i\varphi_i^2\\0\\\varphi_3\\\vdots\\\varphi_{n+1}\\-1-\frac{1}{4}\sum_{i=3}^{n+1}\varepsilon_i\varphi_i^2\end{pmatrix} \tag{40}$$

or

$$F : (t, \varphi_2, \ldots, \varphi_n) \mapsto f_1(t)\begin{pmatrix}0\\0\\\vdots\\0\\1\end{pmatrix} + f_2(t)\begin{pmatrix}1\\0\\\vdots\\1\\0\end{pmatrix} + f_3(t)\begin{pmatrix}1-\frac{1}{4}\sum_{i=2}^{n}\varepsilon_i\varphi_i^2\\\varphi_2\\\vdots\\\varphi_n\\-1-\frac{1}{4}\sum_{i=2}^{n}\varepsilon_i\varphi_i^2\\0\end{pmatrix}, \tag{41}$$

which correspond to cases $a.$) and $b.$) in Preliminaries. We have therefore proven Theorem 1.

Acknowledgments

The first author is a Postdoctoral Fellow of *The Research Foundation – Flanders (FWO)*.

References

[1] M. Antić, H. Li, L. Vrancken, X. Wang, Affine hypersurfaces with constant sectional curvature, preprint.

[2] J. K. Beem, P. E. Ehrlich, and Th. G. Powell, *Warped product manifolds in relativity*, Selected studies: physics-astrophysics, mathematics, history of science, North-Holland, Amsterdam-New York, 1982, pp. 41–56. MR662851

[3] R. L. Bishop and B. O'Neill, *Manifolds of negative curvature*, Trans. Amer. Math. Soc. **145** (1969), 1–49, DOI 10.2307/1995057. MR251664

[4] B.-y. Chen, *Interaction of Legendre curves and Lagrangian submanifolds*, Israel J. Math. **99** (1997), 69–108, DOI 10.1007/BF02760677. MR1469088

[5] B.-Y. Chen, Warped Product Manifolds and Submanifolds, World Scientific, 2017.

[6] M. do Carmo and M. Dajczer, *Rotation hypersurfaces in spaces of constant curvature*, Trans. Amer. Math. Soc. **277** (1983), no. 2, 685–709, DOI 10.2307/1999231. MR694383

[7] M. Dajczer and R. Tojeiro, *Isometric immersions in codimension two of warped products into space forms*, Illinois J. Math. **48** (2004), no. 3, 711–746. MR2114248

[8] G. I. Kručkovič, *On semireducible Riemannian spaces* (Russian), Dokl. Akad. Nauk SSSR (N.S.) **115** (1957), 862–865. MR0096265

[9] M. Moruz, Warped product hypersurfaces in pseudo-Euclidean space, submitted preprint.

[10] B. O'Neill, *Semi-Riemannian geometry: With applications to relativity*, Pure and Applied Mathematics, vol. 103, Academic Press, Inc. [Harcourt Brace Jovanovich, Publishers], New York, 1983. MR719023

[11] R. Penrose, *On the instability of extra space dimensions*, The future of the theoretical physics and cosmology (Cambridge, 2002), Cambridge Univ. Press, Cambridge, 2003, pp. 185–201. MR2033264

[12] A. Z. Petrov, *Einstein spaces*, Translated from the Russian by R. F. Kelleher. Translation edited by J. Woodrow, Pergamon Press, Oxford-Edinburgh-New York, 1969. MR0244912

Katholieke Universiteit Leuven, Departement Wiskunde, BE-3001 Leuven, Belgium
Email address: marilena.moruz@kuleuven.be

Université Polytechnique Hauts-de-France, LMI, F-59313 Valenciennes, France; Katholieke Universiteit Leuven, Departement Wiskunde, BE-3001 Leuven, Belgium
Email address: luc.vrancken@uphf.fr

Contemporary Mathematics
Volume **756**, 2020
https://doi.org/10.1090/conm/756/15208

The development of rectifying submanifolds

Yun Myung Oh

This paper is dedicated to professor Bang-Yen Chen's 75th Birthday

ABSTRACT. In this paper we will review the recent development of rectifying submanifolds, first starting with rectifying curves in $\mathbb{E}^3$ in 2003 by B.-Y. Chen. He provided several characterizations and explicit formula for a space rectifying curve [**1**]. Then, in 2016, he extended the idea of rectifying curves to rectifying submanifold in the higher dimensional Euclidean spaces $\mathbb{E}^n$ [**2**] and after a year later he defined rectifying submanifold in pseudo-Euclidean spaces $\mathbb{E}^m_i$ [**9**]. The notion of rectifying submanifold in an arbitrary Riemannian manifolds was introduced associated with torqued vector fields [**3**] along with some applications and related results [**4**], [**16**]. At the end, a weaker version of rectifying submanifold in a Riemannian manifold was introduced with some properties.

1. Introduction

We consider a unit-speed regular curve $x : I \to \mathbb{E}^3$, where $I = (a, b)$ is a real open interval. We assume that it is arc length reparametrized, and we can define a unit tangent vector $T(s) = x'(s)$ and principal normal vector $N(s) = \frac{T'(s)}{\kappa(s)}$ where $\kappa(s) = ||T'(s)||$ is the curvature of the curve. The vector $B(s) = T(s) \times N(s)$ is called the binormal vector field, and the torsion τ is defined by the equation $B' = -\tau N$.

The Frenet-Serret apparatus is given by $\{\kappa, \tau, T, N, B\}$, and we have Frenet-Serret equations as below.

$$\begin{bmatrix} \vec{T}' \\ \vec{N}' \\ \vec{B}' \end{bmatrix} = \begin{bmatrix} 0 & \kappa & 0 \\ -\kappa & 0 & \tau \\ 0 & -\tau & 0 \end{bmatrix} \begin{bmatrix} \vec{T} \\ \vec{N} \\ \vec{B} \end{bmatrix}.$$

Along the curve, there are some associated planes spanned by $\{T, N, B\}$, an osculating plane spanned by $\{T, N\}$, a normal plane spanned by $\{N, B\}$ and a rectifying plane spanned by $\{T, B\}$. It is known that at every point on the curve, if the position vector lies in its osculating plane, then the curve lies in a plane. Also, if the position vector lies in normal plane at every point, then it is known to be a sphere curve. In 2003, B.-Y. Chen provided the solution for the case if the

2010 *Mathematics Subject Classification.* Primary 53A04.
Key words and phrases. Rectifying submanifolds, normal vector field.

position vector field lies in its rectifying plane at every point along the curve. He defined such curves as rectifying curves. Several characterization results have been obtained since then by many researchers [**8**], [**16**], [**17**].

The idea of a rectifying space curve was extended to rectifying submanifolds in n-dimensional Euclidean space and in pseudo-Euclidean space [**2**], [**9**]. In 2003 B.-Y. Chen published two papers on constant-ratio space-like submanifolds in pseudo-Euclidean space and convolution of Riemannian manifolds. It was shown that proper rectifying submanifold of pseudo-Euclidean space is an N-submanifold, and several other characterization results have been produced.

In [**3**], he introduced the notion of rectifying submanifolds in an arbitrary Riemannian manifold with torqued vector fields [**3**]. He also has classified the rectifying submanifolds and has provided applications to Ricci solitons [**4**].

In this paper, we will investigate some important facts about rectifying submanifolds in different ambient spaces and connections with different types of associated vector fields. At the end, a weak version of rectifying submanifold will be presented with properties.

2. Rectifying curves in $\mathbb{E}^3$ and $\mathbb{E}^3_1$

By definition, for a rectifying curve $\mathrm{x} : I \to \mathbb{E}^3$, the position vector x satisfies

$$\mathrm{x}(s) = \lambda(s)T(s) + \mu(s)B(s),$$

for some functions λ and μ.

The following results provide some interesting characterizations of rectifying curves in $\mathbb{E}^3$.

THEOREM 2.1. [**1**] *Let $x : I \to \mathbb{E}^3$ be a rectifying curve with $\kappa > 0$, and let s be its arclength function. Then:*

(a) *The distance function $\rho = |x|$ satisfies $\rho^2 = s^2 + c_1 s + c_2$, for some constants c_1 and c_2.*
(b) *The tangential component of the position vector of the curve is given by $\langle x, t\rangle = s + b$, for some constant b.*
(c) *The normal component x^N of the position vector of the curve has constant length and the distance function ρ is non-constant.*
(d) *The torsion τ is nonzero, and the binormal component of the position vector is constant, i.e. $\langle x, b\rangle$ is constant.*

Conversely, if $x : I \to \mathbb{E}^3$ is a curve with $\kappa > 0$ and if one of $(a), (b), (c)$ or (d) holds, then x is a rectifying curve.

It is well known that a space curve in $\mathbb{E}^3$ is a generalized helix if and only if the ratio of torsion to curvature $\frac{\tau}{\kappa}$ is a nonzero constant. In terms of this ratio, there is another characterization for rectifying curves.

THEOREM 2.2. [**1**] *Let $x : I \to \mathbb{E}^3$ be a smooth curve with $\kappa > 0$. Then x is congruent to a rectifying curve if and only if the ratio of torsion to curvature $\frac{\tau}{\kappa}$ is a non-constant linear function in arc length s, i.e., $\frac{\tau}{\kappa} = c_1 s + c_2$ for some constants $c_1 \neq 0$ and c_2.*

Furthermore, the curve can be written explicitly as given below.

THEOREM 2.3. [1] *Let $x : I \to \mathbb{E}^3$ be a smooth curve with $\kappa > 0$. Then x is congruent to a rectifying curve if and only if, up to parameterization, it is given by*

$$x(t) = (a \sec t) y(t),$$

where a is a positive number and $y = y(t)$ is a unit speed curve in S^2.

B.-Y. Chen and others have provided more results on rectifying curves in $\mathbb{E}^3$(or $\mathbb{E}^4$) [5],[8],[11],[18]. Also, rectifying curves in Minkowski space $\mathbb{E}_1^3$ have been investigated by many others [12],[13],[14],[15].

3. Euclidean (pseudo-Euclidean) rectifying submanifolds

Let $(N^m, \tilde{g})$ denote an m-dimensional Riemannian manifold and $\phi : M^n \to N^m$ an isometric immersion from an n-dimensional Riemannian manifold (M^n, g) into $(N^m, \tilde{g})$. For tangent vector fields X, Y on M and a normal vector field ξ to M^n, the formula of Gauss and Weingarten are given by

$$\tilde{\nabla}_X Y = \nabla_X Y + h(X,Y), \tilde{\nabla}_X \xi = -A_\eta X + D_X \xi,$$

where ∇ and $\tilde{\nabla}$ are Levi-Civita connections on M and N and $\nabla_X Y$ and $h(X,Y)$ are tangential and normal components of $\tilde{\nabla}_X Y$. Also, $-A_X Y$ and $D_X \xi$ are the tangential and normal components of $\tilde{\nabla}_X \xi$. Here, we define that h is the second fundamental form, A is the shape operator and D is the normal connection of M in the ambient space N.

The first normal space at a point $p \in M$ is defined as

$$\mathrm{Im} h_p = \mathrm{Span}\{h(X,Y) : X, Y \in T_p M\}.$$

For a smooth curve $X : I \to \mathbb{E}^3$, $I = (a,b)$ an open interval, it can be shown that the first normal space at s_0 of a curve is the line spanned by the principal normal vector $N(s_0)$. It is clear that the rectifying plane at s_0 becomes the plane orthogonal to the first normal space. By extending this for a submanifold M of Euclidean space $\mathbb{E}^m$ and a point p in M, the subspace of $T_p\mathbb{E}^m$ that is an orthogonal complement to the first normal space is defined as a rectifying space of M at p. Now we have the following definition for a rectifying submanifold of Euclidean space $\mathbb{E}^m$.

DEFINITION 3.1. [2] A Riemannian submanifold M of n-dimensional Euclidean space $\mathbb{E}^m$ is called a rectifying submanifold if the position vector field x of M always lies in its rectifying space, i.e. M is a rectifying submanifold if and only if

$$\langle \mathrm{x}(p), \mathrm{Im} h_p \rangle = 0$$

holds for every point p in M.

One considers a position vector field $x : M \to \mathbb{E}^m$, there exists a natural decomposition

$$\mathrm{x} = \mathrm{x}^T + \mathrm{x}^N,$$

where x^T and x^N denote the tangential and normal component of x, respectively. A rectifying submanifold M is called proper if its position vector field x satisfies $\mathrm{x} \neq \mathrm{x}^T$ and $\mathrm{x} \neq \mathrm{x}^N$.

DEFINITION 3.2. A non-trivial vector Z on a Riemannian manifold M is called a concurrent vector field if it satisfies

$$\nabla_X Z = X$$

for any tangent vector X to M, where ∇ is the Levi-Civita connection on M.

The below are some of characterizations of Euclidean rectifying submanifolds.

THEOREM 3.3. [2] *If the position vector field x of a submanifold M in $\mathbb{E}^m$ satisfies $x^N \neq 0$, then M is a proper rectifying submanifold if and only if x^T is a concurrent vector field on M.*

Also, we have the following as well.

THEOREM 3.4. [2] *Let M be a proper rectifying submanifold of $\mathbb{E}^m$. Then*
(a) $|x^T| = s + b$ *for some constant* b.
(b) $|x|^2 = s^2 + c_1 s + c_2$ *for some constants* c_1 *and* c_2.
(c) x^N *is of constant length.*
(d) $A_{x^N} = 0$.

The statements $(a), (b)$, and (c) are extended results of Theorem 2.3.

REMARK 3.5. One may find another characterization of rectifying submanifolds in terms of the normal component of the position vector as shown in the theorem 2.1.

The definition of Euclidean rectifying submanifolds can be extended to a pseudo-Euclidean rectifying submanifold in the similar way. Let $\mathbb{E}^m_i$ denote the pseudo-Euclidean m-space equipped with the canonical pseudo-Euclidean metric g_0 of index i given by

$$g_0 = -\sum_{r=1}^{i} du_r^2 + \sum_{t=i+1}^{m} du_t^2,$$

where $(u_1, ..., u_m)$ is a rectangular coordinate system of $\mathbb{E}^m_i$. Several results in Theorem 3.3 are still true for rectifying submanifold in $\mathbb{E}^m_i$. Furthermore, every proper rectifying space-like submanifold M of $\mathbb{E}^m_i$ is a warped product manifold.

THEOREM 3.6. [9] *Let M be a proper rectifying space-like submanifold of $\mathbb{E}^m_i$. Then M is a warped product manifold $I \times_s F$ with warping metric*

$$g = ds^2 + s^2 g_F,$$

such that $x^T = s\frac{\partial}{\partial s}$ and g_F is the metric tensor of a Riemannian manifold F.

4. Rectifying submanifolds in Riemannian manifolds

In [3] B.-Y. Chen defined the rectifying submanifold of an arbitrary Riemannian manifold, extending the notion of Euclidean (or pseudo-Euclidean) rectifying submanifolds. Furthermore, he has provided the characterizations of rectifying submanifolds with respect to a torqued vector field.

DEFINITION 4.1. (a) Let V is a non-vanishing vector field on a (pseudo) Riemannian manifold $\tilde{M}$ and let M be a submanifold of $\tilde{M}$ such that the normal component V^N of V is nowhere zero on M. Then M is called a rectifying submanifold (with respect to V) if and only if

$$\langle V(p), \mathrm{Im} h_p \rangle = 0$$

holds at every $p \in M$.

(b) A nowhere zero vector field v on a (pseudo) Riemannian manifold M is called torse-forming if it satisfies

$$\nabla_X v = \phi X + \gamma(X) v,$$

for some function ϕ and 1-form γ and any vector $X \in TM$. The 1-form γ is called the generating form, and the function ϕ is called the conformal scalar of v. Here, if $\gamma = 0$, then it is called concircular. Also, if $\gamma = 0$ and $\phi = 1$, then it is called concurrent.

(c) A nowhere zero vector field v on a (pseudo) Riemannian manifold M is called a torqued vector field if it satisfies

$$\nabla_X v = \phi X + \gamma(X)v, \ \gamma(v) = 0$$

for some function ϕ and 1-form γ and any vector $X \in TM$. The 1-form γ is called the torqued form, and the function ϕ is called the torqued function of v.

Chen proved that if M is a submanifold of a Riemannain manfold $\tilde{M}$ with a torqued vector field V, and if the tangential component of V is nowhere zero on M, then the sufficient and necessary condition for M to a rectifying submanifold with respect to V is the tangential component of V is torse-forming vector field on M whose conformal scalar is the restiriction of the torqued function and whose generating form is the restriction of the torqued form of V on M. As an application, he has provided another characterization of a rectifying submanifold of a Riemannian manifold with concircular vector field.

THEOREM 4.2. [**3**] *Let M be a submanifold of a Riemannian manifold $\tilde{M}$ endowed with a concircular vector field $V \neq 0$ with $V^T \neq 0$ on M. Then M is a rectifying submanfold in $\tilde{M}$ with respect to V if and only if the tangential component V^T of V is a concircular vector field with the concircular function given by the restriction of the concircular function of V on M.*

Recently, he has been working on the submanifold of a Riemannian manifold with different kinds of vector fields (e.g., incompressible [**6**], concurrent [**2**], and conformal [**7**] canonical vector fields). B.-Y Chen provided two basic properties of rectifying submanifolds of $\tilde{M}$ with a concircular vector field V using the tangential and normal component of V.

THEOREM 4.3. [**3**] *Let $\tilde{M}$ be a Riemannian m-manifold endowed with a concircular vector field V. If M is a rectifying submanifold of $\tilde{M}$ with respect to V, then we have*

(a) *Z^N is of constant length $\neq 0$.*
(b) *the concircular function ϕ of V^T is given by $\phi = V^T(ln\rho)$, where $\rho = |V^T|$.*

Now, we have a natural question about another characterization of a rectifying submanifold in terms of the normal component of a circular vector field V. Also, as seen in the Theorem 4.3, Z^N has a constant length for a rectifying submanifold of a Riemannian manifold $\tilde{M}$. In fact, Theorem 2.1 shows that the normal component of the position vector of rectifying curve has a constant length and this is one of the characterization of rectifying curve in $\mathbb{E}^3$. Next section, we introduce a weak version of rectifying submanifold by reflecting this fact.

5. Special rectifying submanifolds

DEFINITION 5.1. As a subspace of the first normal space, we define another subspace as

$$\mathrm{Im}h_{X(p)} = Span\{h(X,Y) : Y \in TM\}.$$

It is clear that $\mathrm{Im}h_{X(p)}$ is the subspace of the first normal space $\mathrm{Im}h_p$.

DEFINITION 5.2. (a) A submanifold M is called a rectifying submanifold (with respect to V) if and only if there exists a non-vanishing vector field V on $\tilde{M}$ such that the normal component V^N of V is nowhere zero on M and

$$\langle V(p), \mathrm{Im} h_p \rangle = 0,$$

for every point $p \in M$.
(b) A submanifold M is called a V^T-rectifying submanifold if and only if there exists a non-vanishing vector field V on $\tilde{M}$ such that the normal component V^N of V is nowhere zero on M and

$$\langle V(p), \mathrm{Im} h_{V^T(p)} \rangle = 0,$$

where V^T is the tangential component of V and for every point $p \in M$.

Here, we note that for a V^T-rectifying submanifold M, V^T is an eigenvector of A_{V_N}, with zero as its eigenvalue. The next theorem shows the characterization of V^T-rectifying submanifold in terms of the normal component of a concircular vector field.

THEOREM 5.3. *Let $\tilde{M}$ be a Riemannian m-manifold endowed with a concircular vector field V. The submanifold M is a V^T-rectifying submanifold of $\tilde{M}$ if and only if V^N is of constant length.*

PROOF. Since V is a concircular vector field, we have the following.

$$D_Y V^N = -h(Y, V^T), \nabla_Y V^T = A_{V^N} Y + \phi Y,$$

for Y tangent to M.
Suppose that V^N is of constant length. Then

$$\langle D_Y V^N, V^N \rangle = 0 \Rightarrow \langle D_Y V^N, V \rangle = 0 \Rightarrow \langle h(Y, V^T), V \rangle = 0,$$

for any tangent vector Y on M. Thus, M is V^T-rectifying submanifold of $\tilde{M}$. The converse can be shown in the similar way as above. □

THEOREM 5.4. *Let $\tilde{M}$ be a Riemannian m-manifold endowed with a torse-forming vector field V. If M is a V^T-rectifying submanifold of $\tilde{M}$, then V^N is of constant length if and only if the generating form of V is zero. i.e. V is concircular.*

PROOF. Since V is torse-forming, we have

$$\tilde{\nabla}_Y V = \phi Y + \alpha(Y)(V^T + V^N), \tag{5.1}$$

for a function ϕ and a 1-form α on M.
On the other hand, by comparing the tangential and normal component of $\tilde{\nabla}_Y V$ in (5.1), we have

$$D_Y V^N = \alpha(Y) V^N - h(Y, V^T),$$

for Y tangent to M. Suppose V^N has constant length. Then for $Y \in TM$,

$$0 = Y \langle V^N, V^N \rangle = \langle D_Y V^N, V^N \rangle = \langle \alpha(Y) V^N - h(Y, V^T), V^N \rangle.$$

Since M is a V^T-rectifying submanifold, the above equation becomes

$$\alpha(Y) ||V^N||^2 = 0,$$

for any tangent Y on M. Since we assume V^N is nowhere zero on M, we have $\alpha = 0$, which implies that V is concircular.
The converse can be shown easily in the similar way as above. □

6. Acknowledgments

It is my pleasure and honor to write this paper on Professor Chen's recent work on rectifying submanifolds with associated different kinds of vector fields, i.e. concircular, concurrent, torse-forming, and torqued vector fields. It will be interesting to investigate the further development of V^T-rectifying submanifolds associated with the properties of a vector field V. I thank Bogdan Suceavă for many contributions and support to make this volume possible. Furthermore, I thank the two anonymous referees for their helpful comments.

References

[1] B.-Y. Chen, *When does the position vector of a space curve always lie in its rectifying plane?*, Amer. Math. Monthly **110** (2003), no. 2, 147–152, DOI 10.2307/3647775. MR1952443

[2] B.-Y. Chen, *Differential geometry of rectifying submanifolds*, Int. Electron. J. Geom. **9** (2016), no. 2, 1–8. MR3571425

[3] B.-Y. Chen, *Rectifying submanifolds of Riemannian manifolds and torqued vector fields*, Kragujevac J. Math. **41** (2017), no. 1, 93–103, DOI 10.5937/KgJMath1701093C. MR3668255

[4] B.-Y. Chen, *Classification of torqued vector fields and its applications to Ricci solitons*, Kragujevac J. Math. **41** (2017), no. 2, 239–250, DOI 10.5937/kgjmath1702239c. MR3741761

[5] B.-Y. Chen, *Rectifying curves and geodesics on a cone in the Euclidean 3-space*, Tamkang J. Math. **48** (2017), no. 2, 209–214, DOI 10.5556/j.tkjm.48.2017.2382. MR3659280

[6] B.-Y. Chen, *Euclidean submanifolds with incompressible canonical vector field*, Serdica Math. J. **43** (2017), no. 3-4, 321–334. MR3838809

[7] B.-Y. Chen and S. Deshmukh, *Euclidean submanifolds with conformal canonical vector field*, Bull. Korean Math. Soc. **55** (2018), no. 6, 1823–1834. MR3890906

[8] B.-Y. Chen and F. Dillen, *Rectifying curves as centrodes and extremal curves*, Bull. Inst. Math. Acad. Sinica **33** (2005), no. 2, 77–90. MR2151806

[9] B.-Y. Chen and Y. M. Oh, *Classification of rectifying space-like submanifolds in pseudo-Euclidean spaces*, Int. Electron. J. Geom. **10** (2017), no. 1, 86–95. MR3636649

[10] B.-Y. Chen and S. Wei, Riemannian submanifolds with concircular canonical field, Bull. Korean Math. **56**(6): 1525–1537, November 2019

[11] K. İlarslan and E. Nešović, *Some characterizations of rectifying curves in the Euclidean space* $\mathbb{E}^4$, Turkish J. Math. **32** (2008), no. 1, 21–30. MR2399839

[12] K. Ilarslan, E. Nešović, and M. Petrović-Torgašev, *Some characterizations of rectifying curves in the Minkowski 3-space*, Novi Sad J. Math. **33** (2003), no. 2, 23–32. MR2046159

[13] K. Ilarslan and E. Nešović, *On rectifying curves as centrodes and extremal curves in the Minkowski 3-space*, Novi Sad J. Math. **37** (2007), no. 1, 53–64. MR2402051

[14] K. İlarslan and E. Nešović, *Some relations between normal and rectifying curves in Minkowski space-time*, Int. Electron. J. Geom. **7** (2014), no. 1, 26–35. MR3198738

[15] L. Krzywon and Y. M. Oh, Rectifying curves in Minkowski space $\mathbb{E}^3_1$, Pi Mu Epsilon Journal **15**(2): 99–105, Spring 2020

[16] P. Lucas and J. A. Ortega-Yagües, *Rectifying curves in the three-dimensional sphere*, J. Math. Anal. Appl. **421** (2015), no. 2, 1855–1868, DOI 10.1016/j.jmaa.2014.07.079. MR3258354

[17] Y. M. Oh and Y. L. Seo, A curve satisfying $\frac{\tau}{\kappa} = s$ with constant $\kappa > 0$, American Journal of Undergraduate Research 12(2) (2015), 57-62

[18] G. Macsim, A. Mihai, and A. Olteanu, *On rectifying-type curves in a Myller configuration*, Bull. Korean Math. Soc. **56** (2019), no. 2, 383–390. MR3936473

Department of Mathematics, Andrews University, Berrien Springs, Michigan 49103
Email address: ohy@andrews.edu

Contemporary Mathematics
Volume **756**, 2020
https://doi.org/10.1090/conm/756/15209

Some recent work on biharmonic conformal maps

Ye-Lin Ou

ABSTRACT. This note reviews some of the recent work on biharmonic conformal maps. It will be focused on biharmonic conformal immersions and biharmonic conformal maps between manifolds of the same dimension and their links to isoparametric functions and Yamabe type equations, though biharmonic morphisms (maps that preserve solutions of bi-Laplace equations), generalized harmonic morphisms (maps that pull back germs of harmonic functions to germs of biharmonic functions), and biharmonic conformal and Riemannian submersions will also be touched.

1. Biharmonic conformal maps

A *weakly conformal map* is a conformal immersion $\phi : (M^m, g) \to (N^n, h)$ away from its branch points where $\mathrm{d}\phi_p = 0$. So, for a weakly conformal map, we have $\phi^* h = \lambda^2 g$ for a function $\lambda \geq 0$ called the conformal factor of ϕ, and $\lambda(p) = 0$ if and only if $p \in M$ is a branch point of ϕ. A weakly conformal map with conformal factor $\lambda > 0$ (respectively, $\lambda = C > 0$, or $\lambda = 1$) is called a *conformal immersion* (respectively, a *homothetic*, or an *isometric immersion*).

A *horizontally weakly conformal map* is a map $\phi : (M, g) \to (N, h)$ such that for any $p \in M$, either $\mathrm{d}\phi_p = 0$, or $\mathrm{d}\phi_p|_{H_p} : H_p \to T_{\phi(p)}N$ is conformal and onto, where $H_p = (\ker \mathrm{d}\phi_p)^\perp$ is the horizontal subspace of ϕ. It is well known (see, e.g., [**9**]) that for a horizontally weakly conformal map, there exists a function $\lambda \geq 0$, called the *dilation* of ϕ, such that $h(\mathrm{d}\phi(X), \mathrm{d}\phi(Y)) = \lambda^2(p) g(X, Y)$ for any horizontal vectors $X, Y \in H_p$. Note that $\lambda(p) = 0$ if and only if $\mathrm{d}\phi_p = 0$. A horizontally weakly conformal map with dilation $\lambda > 0$ (respectively, $\lambda = C > 0$, or $\lambda = 1$) is called a *conformal submersion* (respectively, a *homothetic*, or a *Riemannian submersion*).

Clearly, we have the following relationships among various conformal maps:

$$\{\text{Isometric immersions}\} \subset \{\text{Conformal immersions}\} \subset \{\text{Weakly conformal maps}\},$$

$$\{\text{Riemannian submersions}\} \subset \{\text{Conformal submersions}\} \subset \{\text{Horizontally weakly conformal maps}\}.$$

When the domain and the target manifolds have the same dimension, the weak conformality agrees with the horizontal weak conformality. In this case, conformal immersions are precisely conformal submersions, and we call them *conformal maps* in this note. Note that, unlike most other books, our definition of conformal

2010 *Mathematics Subject Classification.* Primary 58E20.

Key words and phrases. Biharmonic conformal maps, biharmonic conformal immersions, biharmonic Riemannian submersions, generalized harmonic morphisms.

This work was supported by a grant from the Simons Foundation (#427231, Ye-Lin Ou).

maps does not require conformal maps be diffeomorphisms, although they are local diffeomorphisms by the well known inverse mapping theorem.

A *harmonic map* is a map $\phi : (M^m, g) \to (N^n, h)$ between Riemannian manifolds whose tension field $\tau(\phi) = \mathrm{Trace}_g \nabla \mathrm{d}\phi$ vanishes identically. Harmonic maps include harmonic functions, geodesics, minimal submanifolds, and Riemannian submersions with minimal fibers as special cases. We refer the reader to [**17**], [**18**], [**19**] and [**43**] for the beautiful theory, important applications and interesting links of harmonic maps and the related topics.

A *biharmonic map* is a map $\phi : (M^m, g) \to (N^n, h)$ between Riemannian manifolds that is a critical point of the bienergy functional $E_2(\phi, g) = \frac{1}{2}\int_M |\tau(\phi)|^2 v_g$, where $\tau(\phi)$ is the tension field of ϕ. Biharmonic map equation is a system of the 4th order elliptic PDEs (see [**23**]):

$$\tau_2(\phi) := \mathrm{Trace}_g \left((\nabla^{\phi})^2 \tau(\phi) - R^N(\mathrm{d}\phi, \tau(\phi))\mathrm{d}\phi \right) = 0 \tag{1}$$

where R^N denotes the Riemannian curvature of the target manifold N.

A *biharmonic submanifold* is one whose defining isometric immersion is a biharmonic map. From the definition of biharmonic map and the well known fact that a submanifold is minimal if and only if its defining isometric immersion is harmonic, we have the following relations:

$$\begin{aligned} \{\text{Harmonic maps}\} &\subset \{\text{Biharmonic maps}\},\\ \{\text{Minimal submanifolds}\} &\subset \{\text{Biharmonic submanifolds}\}. \end{aligned}$$

For this reason, we use the name *proper biharmonic maps* for those biharmonic maps which are not harmonic, and *proper biharmonic submanifolds* for those biharmonic submanifolds which are not minimal.

By the nature of the biharmonic map equation in general, it is very difficult to get a solution, or try to get some geometric and/or topological information of the spaces which support such maps. However, some recent study have shown that studying biharmonic maps under some geometric constraints is both practical and fruitful. For example, by adding geometric condition of "being isometric immersion", we have

$$\{\text{Biharmonic submanifolds}\} = \{\text{Biharmonic maps}\} \cap \{\text{Isometric immersions}\},$$

which have been receiving a growing interest from many researchers and some interesting progress has been made since 2000. We refer the reader to a recent survey [**36**] for a progress report on the study of biharmonic submanifold.

Another natural geometric condition to impose is "being horizontally weakly conformal (HWC for short)", thus we study biharmonic HWC maps, which contains the following interesting subsets

$$\{\text{Bihar. Riemannian submersions}\} \subset \{\text{Bihar. Conformal submersions}\} \subset \{\text{Bihar. WHC maps}\},$$

Motivations to study biharmonic HWC maps include:

(i) Bihar. morphisms are special cases of bihar. HWC maps;
(ii) Generalized har. morphisms into $\mathbb{R}^n$ are special bihar. HWC maps;
(iii) Bihar. HWC maps are generalization of har. morphisms since

$$\{\text{Har. morphisms}\} = [\{\text{Har. maps}\} \cap \{\text{ WHC maps}\}] \subset [\{\text{Bihar. maps}\} \cap \{\text{WHC maps}\}].$$

Recall that a harmonic morphism is a map between Riemannian manifolds that preserves solutions of Laplace equation. Harmonic morphisms are characterized as HWC harmonic maps ([**20**], [**22**]). We refer the reader to the book [**9**] for a comprehensive study of the geometry of harmonic morphisms.

Recall also that a *biharmonic morphism* ([**30**]) is a map $\phi : (M, g) \to (N, h)$ between Riemannian manifolds which preserves solutions of bi-Laplace equation in the sense that $\Delta_N^2 f = 0$ implies $\Delta_M^2(f\circ\phi) = 0$ for any function $f : N \supset V \to \mathbb{R}$ with $\phi^{-1}(V)$ non-empty. It was proved in [**30**], [**25**] and [**26**] that any biharmonic morphism is a biharmonic HWC map.

A more general type of morphisms of harmonicity called *generalized harmonic morphisms* were recently studied in [**21**]. By definition, a generalized harmonic morphism is a map between Riemannian manifolds that pulls back germs of harmonic functions to germs of biharmonic functions. Among other things, it was proved in [**21**] that

(i) {Har. morphisms} $\subset$ {Generalized har. morphisms}
(ii) {Bihar. morphisms} $\subset$ {Generalized har. morphisms}
(iii) Generalized har. morphisms into $\mathbb{R}^n$ are special bihar. HWC maps;
(iv) There are many examples of proper biharmonic Riemannian submersions.

Biharmonic Riemannian submersions was first studied in [**28**]. For some recent work on this subject, see [**12**], [**26**], [**42**], [**21**], [**40**], and [**1**].

2. Biharmonic conformal maps between manifolds of the same dimension

2.1. Biharmonic conformal map equations. It is easily checked that the tension field of a conformal map $\phi : (M^m, g) \to (N^m, h)$ is $\tau(\phi) = -(m-2)d\phi(\nabla \ln\lambda)$. Thus, the picture of harmonic maps is very simple: every conformal map $\phi : (M^2, g) \to (N^2, h)$ is harmonic; and for $m \neq 2$, a conformal map is harmonic if and only if λ is a constant, i.e., ϕ is a homothety.

The following theorem shows that biharmonic conformal maps are far more complicated and hence have much more to study.

THEOREM 2.1. [**8**] *A conformal map $\phi : (M^m, g) \to (N^m, h)$ with $m \geq 3$ is biharmonic if and only if the conformal factor λ solves the PDE:*

$$\lambda\,\nabla\,\Delta\lambda - 3(\Delta\lambda)\nabla\lambda - \frac{m-4}{2}\nabla|\nabla\lambda|^2 + 2\lambda\,\mathrm{Ric}^M(\nabla\lambda) = 0, \tag{2}$$

where and in the rest of the paper, ∇, $|,|$, and Δ denote the gradient, the norm, and the Laplacian defined by the metric g, In particular, a conformal map from an Einstein 4-manifold with mathrmRicci^M $= ag$ is biharmonic if and only if

$$\Delta\lambda - a\lambda = A\lambda^3, \tag{3}$$

where A is a constant such taht

$$6A + \frac{2a}{\lambda^2} + R_h = 0 \tag{4}$$

holds for the scalar curvature R_h of the target manifold.

Interesting links to isoparametric functions and Yamabe type equations. (i) Biharmonicity and isoparametric functions.

The following interesting link was found in [**6**]

THEOREM 2.2 ([**5**], [**6**], see also [**8**]). *For $m > 2$ and $m \neq 4$, a conformal map $\phi : (M^m, g) \to (N^m, h)$ from an Einstein manifold is biharmonic if and only if the conformal factor λ is an isoparametric function.*

Here, an isoparametric function is a function $f : M \to R$ on a Riemannian manifold such that (i) $|\nabla f|^2 = \alpha(f)$, and (ii) $\Delta f = \beta(f)$ for some real variable functions α and β. Isoparametric functions on space forms have rich geometry related to hypersurfaces of constant principal curvatures which have been studied since 1918. For a good survey on this subject, see [**39**].

(ii) Biharmonic equation is a Yamabe type Equation. It is well known that the Yamabe problem is to find a metric h, within the conformal class $[g] = \{\sigma g|\, \sigma : M \to (0,\infty)\}$ of metrics on a closed Riemannian manifold (M^m, g), that has constant scalar curvature R_h. For $m \geq 3$ and $h = \lambda^{\frac{4}{m-2}} g$, the problem is equivalent to solving the Yamabe equation

$$-\frac{4(m-1)}{m-2}\Delta_g \lambda + R_g \lambda = R_h \lambda^{\frac{m+2}{m-2}}, \quad R_h = \text{constant}. \tag{5}$$

In particular, when $m = 4$, the Yamabe equation reads:

$$\Delta_g \lambda - \frac{1}{6} R_g \lambda = -\frac{1}{6} R_h \lambda^3, \quad R_h = \text{constant}. \tag{6}$$

More generally, the following *Yamabe type equation*

$$\Delta_g u - \alpha u = -\beta u^3, \quad \alpha \text{ and } \beta \text{ are functions}, \tag{7}$$

has been studied by many mathematicians (see, e.g., [**2**], [**3**], [**4**], and [**41**].

Theorem 2.1 shows that biharmonic equation for conformal maps from an Einstein 4-manifold is a Yamabe type equation.

(iii) Biharmonicity and Sobolev embedding Theorem.

The following link was found in [**8**]

COROLLARY 2.3. *For $m = 4$, the Euler-Lagrange equation for the extremal functions saturating the Sobolev inequality*

$$c\left(\int_{\mathbb{R}^m} |v|^{2m/(m-2)} dx\right)^{(m-2)/m} \leq \int_{\mathbb{R}^m} |\nabla v|^2 dx$$

is

$$\Delta v = -2v^3, \qquad \text{on } \mathbb{R}^4, \tag{8}$$

which is a special case of the biharmonic equation for conformal maps from Euclidean space $\phi : \mathbb{R}^4 \supseteq U \to (N^4, h)$.

We refer the reader to [**14**] for an explanation of the Euler-Lagrange equation for the extremal functions saturating the Sobolev inequality.

2.2. Some recent results. We would like to point out that it is the discovery of the link between biharmonic equation for conformal maps and the Yamabe type equation and the Euler-Lagrange equation for the extremal functions saturating the Sobolev inequality that provides us very useful tools in our classification of biharmonic Möbius transformations, understanding of biharmonic metrics, and proving some existence theorems about biharmonic conformal maps from a compact Einstein 4-manifold. The following is a list of some results from [**8**], we refer the reader to the original paper for all details and other results.

♣ A conformal map $\phi : (M^4, g) \to (N^4, h)$ with $\phi^* h = \lambda^2 g$ from an Einstein manifold with $\text{Ricci}^M = ag$ into a manifold with constant scalar curvature is biharmonic if and only if it is a homothety (hence a harmonic map), or

$$a = 0, \text{ and } \Delta\lambda = -\frac{R_h}{6}\lambda^3.$$

♣ The only possible case for proper biharmonic map to exist (even locally) between 4-dimensional space forms is: $\mathbb{R}^4 \supseteq U \to M^4(C)$, where $M^4(C)$ denotes a 4-dimensional space form. This confirms and extends a result in [**27**] which states that the only rotationally symmetric, proper biharmonic conformal diffeomorphisms between 4-dimensional space forms appear in the case of $\mathbb{R}^4 \supseteq U \to M^4(C)$, where $M^4(C)$.

♣ A conformal map $\phi : \mathbb{R}^4 \supseteq U \to \mathbb{R}^4$, or $\phi : \mathbb{R}^4 \supseteq U \to S^4$ is biharmonic if and only if it is a restriction of a Möbius transformation $\phi : \mathbb{R}^4 \supseteq U \to \mathbb{R}^4$ with

$$\phi(x) = a + \frac{\alpha A(x-b)}{|x-b|^{\varepsilon}} \quad a, b \in \mathbb{R}^4,\ \alpha \in \mathbb{R} \setminus \{0\},\ A \in O(4),\ \varepsilon \in \{0, 2\},$$

for all $x \in \mathbb{R}^4$. The first map is proper biharmonic if and only if $\varepsilon = 2$ whilst the seond map is always proper biharmonic.

♣ There exists a function λ on a closed Einstein 4-manifold with negative Ricci curvature so that $1_g : (M^4, g) \to (M^4, h = \lambda^2 g)$ is a proper biharmonic map.

♣ There exists no proper biharmonic conformal map from a close Ricci flat 4-manifold into any Riemannian 4-manifold;

♣ There exists an infinite family of conformal metrics $\lambda^2 g$ on the Euclidean 4-sphere so that the identity map $1_g : (S^4, g) \to (S^4, h = \lambda^2 g)$ is a proper biharmonic conformal diffeomorphism. Note that by the 2nd item of this list, no (not even local) proper biharmonic conformal map exists between the Euclidean 4-spheres.

Recall that a Riemannian metric $\bar{g}$ on a Riemannian manifold (M^m, g) is called a *harmonic metric* ([**16**]) (respectively, *biharmonic metric*) with respect to g if the identity map $1_g : (M^m, g) \to (M^m, \bar{g})$ is a harmonic map (respectively, *biharmonic map*).

COROLLARY 2.4 ([**11**],[**8**]). *The identity map $1_g : (M^m, g) \to (M^m, \bar{g} = \lambda^2 g)$ is biharmonic (i.e., the conformally related metric $\bar{g} = \lambda^2 g$ is biharmonic with respect to g) if and only if*

$$\lambda \nabla \Delta\lambda - 3(\Delta\lambda)\nabla\lambda - \frac{m-4}{2}\nabla|\nabla\lambda|^2 + 2\lambda\, \mathrm{Ric}^M(\nabla\lambda) = 0. \tag{9}$$

From this we have:

♣ A metric $h = \lambda^2 dx^2$ on an open set $U \subseteq \mathbb{R}^4$ is biharmonic if and only if λ is a solution of the Yamabe equation or, equivalently, $h = \lambda^2 dx^2$ has constant scalar curvature R_h and $\Delta\lambda = -\frac{R_h}{6}\lambda^3$ ([**8**]).

♣ The conformally flat metric $h = |x|^{2\alpha} \mathrm{d}x^2$ on $\mathbb{R}^4 \setminus \{0\}$ is biharmonic with respect to Euclidean metric $\mathrm{d}x^2$ if and only if $\alpha = -1$, or -2 ([**8**]).

EXAMPLE 1. The conformally flat metric $\lambda^2 \mathrm{d}x^2$ with $\lambda = |x|^{-1}, |x|^{-2}, \frac{2}{1 \pm |x|^2}$ is biharmonic with respect to Euclidean metric $\mathrm{d}x^2$ on an open set of $U \subset \mathbb{R}^4$. Equivalently, each of the following identity maps

$$1 : (\mathbb{R}^4 \setminus \{0\}, \mathrm{d}x^2) \to \left(\mathbb{R}^4 \setminus \{0\}, \frac{\mathrm{d}x^2}{|x|^2}\right), \quad 1 : (\mathbb{R}^4 \setminus \{0\}, \mathrm{d}x^2) \to \left(\mathbb{R}^4 \setminus \{0\}, \frac{\mathrm{d}x^2}{|x|^4}\right),$$

$$1 : (\mathbb{R}^4, \mathrm{d}x^2) \to \left(\mathbb{R}^4, \frac{4\mathrm{d}x^2}{(1+|x|^2)^2}\right), \qquad 1 : (\mathbb{B}^4, \mathrm{d}x^2) \to \left(\mathbb{B}^4, \frac{4\mathrm{d}x^2}{(1-|x|^2)^2}\right),$$

is a proper biharmonic map ([**8**] and [**26**]).

Note that several results (including the following one) on some obstructions on the existence of biharmonic metric on a compact Einstein manifold have been obtained in [**7**].

THEOREM 2.5. [**7**] *There exists no proper biharmonic metric on a compact Einstein manifold* (M^m, g) *with* $\mathrm{Ric}^g = a\,g$ *for either (i)* $a < 0$ *and* $m \geq 6$, *or (ii)* $a \leq 0$ *and* $m > 6$.

3. Biharmonic conformal immersions

In this section, we review some recent work on biharmonic conformal immersions done in [**32**], [**34**], [**35**], and [**37**].

3.1. Biharmonic conformal immersions. It is easily checked (see e.g., [**38**]) that the tension field of a conformal immersion $\phi : (M^m, g) \to (N^n, h)$ with $\phi^* h = \lambda^2 g$ is given by $\tau(\phi) = m\lambda^2\eta - (m-2)\mathrm{d}\phi(\nabla \ln \lambda)$, where η is the mean curvature vector field of the associated isometric immersion (submanifold) $\phi : (M^m, \bar{g} = \lambda^2 g) \to (N^n, h)$. Using this with $m = 2$, we have the well-known **Fact:** A conformal immersion $\phi : (M^2, g) \to (N^n, h)$ of a surface is harmonic if and only if its image $\phi(M^2) \subset (N^n, h)$ is a minimal surface.

It is this interesting link, together with the beautiful theory and important applications of minimal surfaces, and the relation

$$\{\text{Har. conformal immersions}\} \subset \{\text{Bihar. conformal immersions}\},$$

that motivates us to study biharmonic conformal immersions.

An idea here is to express the biharmonicity of a conformal immersion by using the data (including the Laplacian, the gradient, the shape operator and the mean curvature, etc) of its associated isometric immersion. A reason for this is not only that we can use some of the tools and results from submanifold theory but also that (as we hope) the study of biharmonic conformal immersions will help to understand the geometry of submanifolds.

The following transformations of the Jacobi operator and the bitension field under a conformal change of the domain metric for a generic map play an important role in our study of conformal immersions. Note that the idea of using a conformal change of the domain metric of a harmonic map to construct proper biharmonic maps or metrics had been successfully used in [**5**].

THEOREM 3.1. [**32**] *For a map* $\phi : (M^m, g) \to (N^n, h)$ *and a conformal change of the metric* $\bar{g} = \lambda^2 g$, *let* J^ϕ_g *and* $\tau_2(\phi, g)$ *(respectively,* $J^\phi_{\bar{g}}$ *and* $\tau_2(\phi, \bar{g})$*) denote the Jacobi operator and the bitension field of* ϕ *with respect to* g *(respectively,* $\bar{g}$*) respectively. Then, the following transformation hold*

$$\begin{aligned}
J^\phi_{\bar{g}}(X) &= \lambda^{-2}[J^\phi_g(X) - (m-2)\nabla^\phi_{\nabla \ln \lambda} X], \quad \textit{and} \\
\tau_2(\phi, \bar{g}) &= \lambda^{-4}\{\tau_2(\phi, g) - (m-2)J^\phi_g(\mathrm{d}\phi(\nabla \ln \lambda)) \\
&\quad -2(\Delta \ln \lambda + (m-4)\,|\nabla \ln \lambda|^2)\tau(\phi, g) + (m-6)\nabla^\phi_{\nabla \ln \lambda}\,\tau(\phi, g) \\
&\quad -2(m-2)(\Delta \ln \lambda + (m-4)\,|\nabla \ln \lambda|^2)\mathrm{d}\phi(\nabla \ln \lambda) \\
&\quad +(m-2)(m-6)\nabla^\phi_{\nabla \ln \lambda}\,\mathrm{d}\phi(\nabla \ln \lambda)\},
\end{aligned}$$

where the Jacobi operator is defined as

$$J^\phi_g(X) = -\{\mathrm{Trace}_g(\nabla^\phi\nabla^\phi - \nabla^\phi_{\nabla^M})X - \mathrm{Trace}_g\, R^N(\mathrm{d}\phi, X)\mathrm{d}\phi\} \tag{10}$$

for any vector field X *along the map* ϕ *so that* $J^\phi_g(\tau(\phi)) = -\tau_2(\phi)$.

PROPOSITION 3.2. [**32**] *A conformal immersion* $\phi : (M^m, g) \to (N^n, h)$ *with the conformal factor* λ *is biharmonic if and only if*

$$\lambda^4 \tau_2(\phi, \bar{g}) = -(m-2)J_g^\phi(\mathrm{d}\phi(\nabla \ln \lambda)) - m(\Delta\lambda^2)\eta + m(m-6)\lambda^2 \nabla^\phi_{\nabla \ln \lambda}\eta,$$

where $\tau_2(\phi, \bar{g})$ *and* η *are the bitension field and the mean curvature vector of the associated submanifold.*

Using this, we can check the following

EXAMPLE 2. The map $\phi : (\mathbb{R}^3 \times \mathbb{R}^+, g = \delta_{ij}) \to (H^5 = \mathbb{R}^4 \times \mathbb{R}^+, h = y_5^{-2}\delta_{\alpha\beta})$, with $\phi(x_1, \ldots, x_4) = (1, x_1, \ldots, x_4)$ gives is a proper biharmonic conformal immersions of the Euclidean space into the Hyperbolic space.

Note that a similar problem of studying biharmonicity of immersions via conformal changes on the target manifold was done in [**29**], where, among other things, the author proved that the equator of a Euclidean sphere (viewed as a totally geodesic submanifold) can be turned into a proper biharmonic submanifold via some particular conformal change of the Euclidean metric on the ambient sphere.

3.2. Biharmonic conformal immersions of surfaces. For a conformal immersion of a surface we have

THEOREM 3.3. [**37**] *A conformal immersion* $\phi : (M^2, \bar{g}) \to (N^3, h)$ *with conformal factor* λ *is biharmonic if and only if*

$$\begin{cases} \Delta(\lambda^2 H) - (\lambda^2 H)[|A|^2 - \mathrm{Ric}^N(\xi, \xi)] = 0, \\ A(\mathrm{grad}(\lambda^2 H)) + (\lambda^2 H)[\mathrm{grad}\, H - (\mathrm{Ric}^N(\xi))^\top] = 0. \end{cases} \tag{11}$$

where ξ, A, *and* H *are the unit normal vector field, the shape operator, and the mean curvature function of the surface* $\phi(M) \subset (N^3, h)$ *respectively, and the operators* Δ, grad *and* $|,|$ *are taken with respect to the induced metric* $g = \phi^* h = \lambda^2 \bar{g}$ *on the surface.*

It was shown in [**33**] that an isometric immersion (i.e., hypersurface) $M^m \to (N^{m+1}, h)$ is biharmonic if and only if

$$\begin{cases} \Delta H - H|A|^2 + H\mathrm{Ric}^N(\xi, \xi) = 0, \\ 2A\,(\mathrm{grad}\; H) + \frac{m}{2}\,\mathrm{grad}\; H^2 - 2\,H\,(\mathrm{Ric}^N\,(\xi))^\top = 0. \end{cases} \tag{12}$$

By comparing Equations (12) and (11), we see how biharmonic equation transforms under the change of the conformal factor from constant 1 to a function λ. The similar form of the first equation in the system allows us to use some of the tools used in and results from the biharmonic submanifold theory to study biharmonic conformal immersions.

In general, we have

$$\{\text{Isometric immersions}\} \subset \{\text{Conformal immersions}\} \text{ and hence}$$
$$\{\text{Bihar. isometric immersions}\} = \{\text{Bihar. submanifolds}\} \subset \{\text{Bihar. conformal immersions}\}$$

When the ambient space is a Euclidean space, a well known result, proved by Chen-Ishikawa [**15**] and Jiang [**24**] independently, shows that there exists no proper biharmonic isometric immersion from a surface into Euclidean space $\mathbb{R}^3$. However, the following result shows that there are infinitely many proper biharmonic conformal immersions of surfaces into $\mathbb{R}^3$.

PROPOSITION 3.4. [**32**] *For constants C_1 and C_2 with $C_2/C_1 < 0$, let $\lambda^2 = C_1e^z + C_2e^{-z}$ and $D = \{(\theta, z) \in (0, 2\pi) \times \mathbb{R} : z \neq \frac{1}{2}\ln(-C_2/C_1)\}$. Then, the map $\phi : (D, g = \lambda^{-2}(d\theta^2 + dz^2)) \to \mathbb{R}^3$ with $\phi(\theta, z) = (\cos\theta, \sin\theta, z)$ is a proper biharmonic conformal immersions of a cylinder of radius 1 into Euclidean space $\mathbb{R}^3$. In particular, for $D = \{(\theta, z) \in (0, 2\pi) \times \mathbb{R} : z \neq 0\}$, we have a proper biharmonic conformal immersion $\phi : (D, g = \sinh^{-1}(z)(d\theta^2 + dz^2)) \to \mathbb{R}^3$ with $\phi(\theta, z) = (\cos\theta, \sin\theta, z)$.*

The following theorem gives some obstructions to the existence of proper biharmonic conformal immersions.

THEOREM 3.5. [**37**] *Let $\phi : (M^2, \bar{g}) \to (\mathbb{R}^3, h_0)$ be a biharmonic conformal immersion into 3-dimensional Euclidean space with $\phi^*h_0 = \lambda^2\bar{g}$. Suppose the associated surface $\phi : (M^2, g = \lambda^2\bar{g}) \to (\mathbb{R}^3, h_0)$ is complete with the mean curvature function H satisfying $\int_M \lambda^4 H^2\, dv_g < \infty$. Then the biharmonic conformal immersion ϕ is a minimal (i.e., harmonic) immersion.*

A practical (also relatively easier) question to ask is: what submanifold admits a biharmonic conformal immersion into its ambient space? Here, a submanifold of (N^n, h) given by an isometric immersion $\phi : (M^m, g) \to (N^n, h)$ is said to *admit a biharmonic conformal immersion into its ambient space*, if there exists a function $\lambda : M^m \to (0, \infty)$ so that the conformal immersion $\phi : (M^m, \bar{g} = \lambda^{-2}g) \to (N^n, h)$ is a biharmonic map. When this happens, we also say that the submanifold $\phi : (M^m, g) \to (N^n, h)$ *can be biharmonically conformally immersed into its ambient space.*

We know that a surface $\phi : (M^2, g) \to (N^n, h)$ is minimal if and only if its defining isometric immersion ϕ a harmonic. In this case, it is well known that for any function $\lambda : M^2 \to \mathbb{R}^+$, the conformal immersion $\phi : (M^2, \bar{g} = \lambda^{-2}g) \to (N^n, h)$ is again harmonic and hence trivially biharmonic. So every minimal surface admits a trivial biharmonic conformal immersion into its ambient space.

An example of a surface that admits a proper biharmonic conformal immersion into its ambient space is the circular cylinder $\phi : \mathbb{R}^2 \to (\mathbb{R}^3, \delta_{ij})$, $\phi(x, y) = (R\cos\frac{x}{R}, R\sin\frac{x}{R}, y)$ in Euclidean 3-space. As we have seen in Proposition 3.4 that there exists $\lambda : \mathbb{R}^2 \to \mathbb{R}^+$ with $\lambda^{-2}(x, y) = e^{y/R}$ so that the conformal immersion $\phi : (\mathbb{R}^2, \lambda^{-2}(\mathrm{d}x^2 + \mathrm{d}y^2)) \to (\mathbb{R}^3, \delta_{ij})$ is a proper biharmonic map.

The following theorem shows that, locally, a circular cylinder is the only one among surfaces of constant mean curvature without umbilical point that admit a proper biharmonic conformal immersion into $\mathbb{R}^3$.

THEOREM 3.6. [**35**] *If a surface in $\mathbb{R}^3$ with non-zero constant mean curvature and no umbilical point can be biharmonically conformally immersed into $\mathbb{R}^3$, then, locally, it is an open portion of a circular cylinder.*

By a well-known results of [**13**], any proper biharmonbic isometric immersion of a surface into 3-sphere is given by the the standard isometric embedding $S^2(\frac{1}{\sqrt{2}}) \to S^3$ or a part of it. An interesting question to ask is whether there is a surface in Euclidean 3-sphere that admits a proper biharmonic conformal immersion into S^3. The following theorem gives a partial answer to the question.

THEOREM 3.7. [**37**] *If $\phi : (M^2, g = \phi^*h) \to (S^3, h)$ is a compact surface in Euclidean 3-sphere with $|A|^2 \geq 2$, and suppose that there is a function λ so that the* *conformal immersion* $\phi : (M^2, \bar{g} = \lambda^{-2}g) \to (S^3, h)$ *is biharmonic. Then, either* M^2 *is minimal, or* λ *is constant,* $M^2 = S^2(\frac{1}{2})$*, and* $\phi : M^2 \to S^3$ *is the standard embedding* $S^2(\frac{1}{\sqrt{2}}) \to S^3$.

Finally, we close this review with the following preliminary result about biharmonic conformal immersions of surfaces into a Euclidean space obtained from a consideration similar to Weierstrass representation of harmonic conformal immersions.

THEOREM 3.8. [**32**] *A map* $\phi : (M^2, g = \lambda^2(\mathrm{d}u^2 + \mathrm{d}v^2)) \to \mathbb{R}^n$ *into Euclidean space is a biharmonic conformal immersion if and only if the section* $\phi = \frac{\partial \varphi}{\partial z} = \frac{1}{2}(\varphi_u - i\varphi_v) = \phi^\alpha(z)\frac{\partial}{\partial y^\alpha}$ *solves the system*

$$\sum_{\alpha=1}^{n}(\phi^\alpha)^2 = 0, \quad \sum_{\alpha=1}^{n}|\phi^\alpha|^2 \neq 0, \tag{13}$$

$$\frac{\partial}{\partial \bar{z}}\frac{\partial}{\partial z}\left(\lambda^{-2}\frac{\partial \phi^\sigma}{\partial \bar{z}}\right) = 0, \quad \sigma = 1, 2, \ldots, n. \tag{14}$$

Acknowledgments

The author would like to thank Cezar Oniciuc and the referee for some comments and suggestions that help to improve the original manuscript.

References

[1] M. A. Akyol and Y.-L. Ou, *Biharmonic Riemannian submersions*, Ann. Mat. Pura Appl. (4) **198** (2019), no. 2, 559–570, DOI 10.1007/s10231-018-0789-x. MR3927170

[2] T. Aubin, *Problèmes isopérimétriques et espaces de Sobolev* (French), J. Differential Geometry **11** (1976), no. 4, 573–598. MR448404

[3] T. Aubin, *Équations différentielles non linéaires et problème de Yamabe concernant la courbure scalaire*, J. Math. Pures Appl. (9) **55** (1976), no. 3, 269–296. MR431287

[4] T. Aubin, *Some nonlinear problems in Riemannian geometry*, Springer Monographs in Mathematics, Springer-Verlag, Berlin, 1998. MR1636569

[5] P. Baird and D. Kamissoko, *On constructing biharmonic maps and metrics*, Ann. Global Anal. Geom. **23** (2003), no. 1, 65–75, DOI 10.1023/A:1021213930520. MR1952859

[6] P. Baird, A. Fardoun, and S. Ouakkas, *Conformal and semi-conformal biharmonic maps*, Ann. Global Anal. Geom. **34** (2008), no. 4, 403–414, DOI 10.1007/s10455-008-9118-8. MR2447908

[7] P. Baird, E. Loubeau, and C. Oniciuc, *Harmonic and biharmonic maps from surfaces*, Harmonic maps and differential geometry, Contemp. Math., vol. 542, Amer. Math. Soc., Providence, RI, 2011, pp. 223–230, DOI 10.1090/conm/542/10711. MR2796653

[8] P. Baird and Y.-L. Ou, *Biharmonic conformal maps in dimension four and equations of Yamabe-type*, J. Geom. Anal. **28** (2018), no. 4, 3892–3905, DOI 10.1007/s12220-018-0004-8. MR3881994

[9] P. Baird and J. C. Wood, *Harmonic morphisms between Riemannian manifolds*, London Mathematical Society Monographs. New Series, vol. 29, The Clarendon Press, Oxford University Press, Oxford, 2003. MR2044031

[10] P. Baird, E. Loubeau, and C. Oniciuc, *Harmonic and biharmonic maps from surfaces*, Harmonic maps and differential geometry, Contemp. Math., vol. 542, Amer. Math. Soc., Providence, RI, 2011, pp. 223–230, DOI 10.1090/conm/542/10711. MR2796653

[11] A. Balmuş, *Biharmonic properties and conformal changes*, An. Ştiinţ. Univ. Al. I. Cuza Iaşi. Mat. (N.S.) **50** (2004), no. 2, 361–372 (2005). MR2131943

[12] A. Balmuş, S. Montaldo, and C. Oniciuc, *Biharmonic maps between warped product manifolds*, J. Geom. Phys. **57** (2007), no. 2, 449–466, DOI 10.1016/j.geomphys.2006.03.012. MR2271198

[13] R. Caddeo, S. Montaldo, and C. Oniciuc, *Biharmonic submanifolds of S^3*, Internat. J. Math. **12** (2001), no. 8, 867–876, DOI 10.1142/S0129167X01001027. MR1863283

[14] S.-Y. A. Chang, *Conformal invariants and partial differential equations*, Bull. Amer. Math. Soc. (N.S.) **42** (2005), no. 3, 365–393, DOI 10.1090/S0273-0979-05-01058-X. MR2149088

[15] B.-Y. Chen and S. Ishikawa, *Biharmonic surfaces in pseudo-Euclidean spaces*, Mem. Fac. Sci. Kyushu Univ. Ser. A **45** (1991), no. 2, 323–347, DOI 10.2206/kyushumfs.45.323. MR1133117

[16] B.-Y. Chen and T. Nagano, *Harmonic metrics, harmonic tensors, and Gauss maps*, J. Math. Soc. Japan **36** (1984), no. 2, 295–313, DOI 10.2969/jmsj/03620295. MR740319

[17] J. Eells and L. Lemaire, *A report on harmonic maps*, Bull. London Math. Soc. **10** (1978), no. 1, 1–68, DOI 10.1112/blms/10.1.1. MR495450

[18] J. Eells and L. Lemaire, *Selected topics in harmonic maps*, CBMS Regional Conference Series in Mathematics, vol. 50, Published for the Conference Board of the Mathematical Sciences, Washington, DC; by the American Mathematical Society, Providence, RI, 1983. MR703510

[19] J. Eells and L. Lemaire, *Another report on harmonic maps*, Bull. London Math. Soc. **20** (1988), no. 5, 385–524, DOI 10.1112/blms/20.5.385. MR956352

[20] B. Fuglede, *Harmonic morphisms between Riemannian manifolds* (English, with French summary), Ann. Inst. Fourier (Grenoble) **28** (1978), no. 2, vi, 107–144. MR499588

[21] E. Ghandour and Y.-L. Ou, *Generalized harmonic morphisms and horizontally weakly conformal biharmonic maps*, J. Math. Anal. Appl. **464** (2018), no. 1, 924–938, DOI 10.1016/j.jmaa.2018.04.044. MR3794122

[22] T. Ishihara, *A mapping of Riemannian manifolds which preserves harmonic functions*, J. Math. Kyoto Univ. **19** (1979), no. 2, 215–229, DOI 10.1215/kjm/1250522428. MR545705

[23] G. Y. Jiang, *2-harmonic maps and their first and second variational formulas* (Chinese), Chinese Ann. Math. Ser. A **7** (1986), no. 4, 389–402. An English summary appears in Chinese Ann. Math. Ser. B **7** (1986), no. 4, 523. MR886529

[24] G. Y. Jiang, *Some nonexistence theorems on 2-harmonic and isometric immersions in Euclidean space* (Chinese), Chinese Ann. Math. Ser. A **8** (1987), no. 3, 377–383. An English summary appears in Chinese Ann. Math. Ser. B **8** (1987), no. 3, 389. MR924896

[25] E. Loubeau and Y.-L. Ou, *The characterization of biharmonic morphisms*, Differential geometry and its applications (Opava, 2001), Math. Publ., vol. 3, Silesian Univ. Opava, Opava, 2001, pp. 31–41. MR1978760

[26] E. Loubeau and Y.-L. Ou, *Biharmonic maps and morphisms from conformal mappings*, Tohoku Math. J. (2) **62** (2010), no. 1, 55–73, DOI 10.2748/tmj/1270041027. MR2654303

[27] S. Montaldo, C. Oniciuc, and A. Ratto, *Rotationally symmetric biharmonic maps between models*, J. Math. Anal. Appl. **431** (2015), no. 1, 494–508, DOI 10.1016/j.jmaa.2015.05.082. MR3357596

[28] C. Oniciuc, *Biharmonic maps between Riemannian manifolds*, An. Ştiinţ. Univ. Al. I. Cuza Iaşi. Mat. (N.S.) **48** (2002), no. 2, 237–248 (2003). MR2004799

[29] C. Oniciuc, *New examples of biharmonic maps in spheres*, Colloq. Math. **97** (2003), no. 1, 131–139, DOI 10.4064/cm97-1-12. MR2010548

[30] Y.-L. Ou, *Biharmonic morphisms between Riemannian manifolds*, Geometry and topology of submanifolds, X (Beijing/Berlin, 1999), World Sci. Publ., River Edge, NJ, 2000, pp. 231–239, DOI 10.1142/9789812792051_0018. MR1801915

[31] Y.-L. Ou, *p-harmonic morphisms, biharmonic morphisms, and nonharmonic biharmonic maps*, J. Geom. Phys. **56** (2006), no. 3, 358–374, DOI 10.1016/j.geomphys.2005.02.005. MR2171890

[32] Y.-L. Ou, *On conformal biharmonic immersions*, Ann. Global Anal. Geom. **36** (2009), no. 2, 133–142, DOI 10.1007/s10455-008-9153-5. MR2529469

[33] Y.-L. Ou, *Biharmonic hypersurfaces in Riemannian manifolds*, Pacific J. Math. **248** (2010), no. 1, 217–232, DOI 10.2140/pjm.2010.248.217. MR2734173

[34] Y.-L. Ou, *On f-biharmonic maps and f-biharmonic submanifolds*, Pacific J. Math. **271** (2014), no. 2, 461–477, DOI 10.2140/pjm.2014.271.461. MR3267537

[35] Y.-L. Ou, *Biharmonic conformal immersions into three-dimensional manifolds*, Mediterr. J. Math. **12** (2015), no. 2, 541–554, DOI 10.1007/s00009-014-0420-3. MR3350325

[36] Y.-L. Ou, *Some recent progress of biharmonic submanifolds*, Recent advances in the geometry of submanifolds—dedicated to the memory of Franki Dillen (1963–2013), Contemp. Math., vol. 674, Amer. Math. Soc., Providence, RI, 2016, pp. 127–139. MR3565250

[37] Y.-L. Ou, *f-biharmonic maps and f-biharmonic submanifolds II*, J. Math. Anal. Appl. **455** (2017), no. 2, 1285–1296, DOI 10.1016/j.jmaa.2017.06.033. MR3671226

[38] H. Takeuchi, *Some conformal properties of p-harmonic maps and a regularity for sphere-valued p-harmonic maps*, J. Math. Soc. Japan **46** (1994), no. 2, 217–234, DOI 10.2969/jmsj/04620217. MR1264939

[39] G. Thorbergsson, *A survey on isoparametric hypersurfaces and their generalizations*, Handbook of differential geometry, Vol. I, North-Holland, Amsterdam, 2000, pp. 963–995, DOI 10.1016/S1874-5741(00)80013-8. MR1736861

[40] H. Urakawa, *Harmonic maps and biharmonic maps on principal bundles and warped products*, J. Korean Math. Soc. **55** (2018), no. 3, 553–574. MR3800555

[41] J. Vétois and S. Wang, *Infinitely many solutions for cubic nonlinear Schrödinger equations in dimension four*, Adv. Nonlinear Anal. **8** (2019), no. 1, 715–724, DOI 10.1515/anona-2017-0085. MR3918401

[42] Z.-P. Wang and Y.-L. Ou, *Biharmonic Riemannian submersions from 3-manifolds*, Math. Z. **269** (2011), no. 3-4, 917–925, DOI 10.1007/s00209-010-0766-6. MR2860270

[43] S. T. Yau and R. Schoen, *Lectures on harmonic maps*, International Press Incorporated, Boston 385 Someville Ave, Someville, MA, U.S.A, 1997.

Department of Mathematics, Texas A & M University-Commerce, Commerce, Texas 75429

Email address: `yelin.ou@tamuc.edu`

Contemporary Mathematics
Volume **756**, 2020
https://doi.org/10.1090/conm/756/15210

On isoparametric linear Weingarten hypersurfaces in Riemannian and Lorentzian space forms

Cihan Özgür

ABSTRACT. Linear Weingarten hypersurfaces have been extensively studied by various aspects in the literature. In the present paper, we deal with the classifications of isoparametric linear Weingarten hypersurfaces in Riemannian and Lorentzian space forms.

1. Introduction

A hypersurface in a (semi)-Riemannian space form is said to be linear Weingarten if its normalized scalar curvature R and the mean curvature H satisfy

$$R = aH + b \tag{1.1}$$

for some constants $a, b \in \mathbb{R}$ [**16**]. A linear Weingarten hypersurface is a natural generalization of a hypersurface with constant scalar curvature or with constant mean curvature. In real space forms, linear Weingarten hypersurfaces have been studied by various authors in different ambient spaces, for example see [**3**], [**4**], [**7**], [**8**], [**13**], [**16**], [**17**], [**23**] and [**24**]. Furthermore, in [**2**], [**5**], [**9**], [**14**], [**15**] and [**18**], linear Weingarten hypersurfaces in Lorentzian space forms have been considered.

In [**21**], [**22**], P. J. Ryan classified isoparametric hypersurfaces in real space forms. In [**1**], Abe, Koike, Yamaguchi obtained the classifications of isoparametric hypersurfaces of semi-Riemannian space forms. In [**10**], Chen studied compact isoparametric hypersurfaces of $\mathbb{S}^{n+1}(1)$. In [**11**], Chen gave an open problem as "When is a finite type Dupin hypersurface of a hypersphere to be isoparametric?". The problem is still open.

The purpose of the present paper is to classify isoparametric linear Weingarten hypersurfaces in Riemannian and Lorentzian space forms. To obtain our results, we use P. J. Ryan's and Abe, Koike, Yamaguchi's classification results.

The paper is organized as follows: In Section 2, after we give P. J. Ryan's classification results, we state our results. In Section 3, after we give Abe, Koike, Yamaguchi's classification results, we present our results. As main results, we prove Theorem 2.1, Theorem 2.2, Theorem 2.3, Theorem 3.1, Theorem 3.2 and Theorem 3.3. To obtain the solutions for the inequalities and equations, we use Mathematica [**20**].

2010 *Mathematics Subject Classification.* Primary 53C40, 53C42, 53A07.

2. Isoparametric linear Weingarten hypersurfaces in Riemannian space forms

Let M be a hypersurface in a Riemannian space form $M^{n+1}(c)$ oriented by the unit normal vector field N. Denote the shape operator of M by A. Let $\lambda_1, \lambda_2, ..., \lambda_n$ be the principal curvatures of M. If M has constant principal curvatures then it is called an *isoparametric hypersurface* [**6**]. For the basic notions about hypersurfaces, we refer Chen's book [**12**].

The isoparametric hypersurfaces M of $\mathbb{R}^{n+1}$ were determined by Ryan in [**22**] as follows:

R1) $M = \{x \in \mathbb{R}^{n+1} : x_{n+1} = 0\}$ with $A = [0]$. In this case, M is isometric to $\mathbb{R}^n$.

R2) $M = \left\{x \in \mathbb{R}^{n+1} : \|x\|^2 = \frac{1}{c}\right\}$, with $A = \sqrt{c}I$. In this case, M is isometric to the sphere $\mathbb{S}^n(c)$.

R3) Cylinders over spheres: $M = \left\{x \in \mathbb{R}^{n+1} : (x_1)^2 + ... + (x_p)^2 = \frac{1}{c}\right\}$ with $A = \sqrt{c}I_p \oplus 0$, $n > p > 1$.

R4) Cylinders over complete plane curves. In this case, $A = \lambda I \oplus 0$ for some scalar function λ on the curve.

For the isoparametric linear Weingarten hypersurfaces in $\mathbb{R}^{n+1}$, we can state the following theorem.

THEOREM 2.1. *The isoparametric linear Weingarten hypersurfaces of* $\mathbb{R}^{n+1}$ *are classified as follows*:

i) M is isometric to $\mathbb{R}^n$. In this case, M satisfies $R = aH$ for all $a \in \mathbb{R}$.

ii) M is isometric to the sphere $\mathbb{S}^n\left(\frac{2b+a^2+|a|\sqrt{a^2+4b}}{2}\right)$, where

$$a \neq 0 \text{ and } -\frac{a^2}{4} \leq b \text{ or } a = 0 \text{ and } b > 0. \tag{2.1}$$

iii) M is isometric to the sphere $\mathbb{S}^n\left(\frac{2b+a^2-|a|\sqrt{a^2+4b}}{2}\right)$, where

$$a \in \mathbb{R} \text{ and } b > 0 \text{ or } a \in \mathbb{R} \text{ and } -\frac{a^2}{4} \leq b < 0. \tag{2.2}$$

iv) A cylinder over a sphere: $M = \left\{x \in \mathbb{R}^{n+1} : (x_1)^2 + ... + (x_p)^2 = \frac{1}{c}\right\}$, where

$$b = \frac{p(p-1)c - ap(n-1)\sqrt{c}}{n(n-1)} \tag{2.3}$$

$n > p > 1$ and $c > 0$.

v) M is a cylinder over a complete plane curve such that $A = -\frac{bn}{a}I \oplus 0$, $a \neq 0$.

PROOF. According to the Ryan's classification, we consider the following cases:

R1) If $A = [0]$, then we have $H = R = 0$. Hence M is isometric to $\mathbb{R}^n$ and satisfies $R = aH$ for all $a \in \mathbb{R}$.

R2) For $A = \sqrt{c}I$, we have $R = c$ and $H = \sqrt{c}$. So (1.1) gives us

$$c = \frac{a^2 + 2b + |a|\sqrt{a^2+4b}}{2} \text{ or } c = \frac{a^2 + 2b - |a|\sqrt{a^2+4b}}{2}.$$

If $c = \frac{a^2+2b+|a|\sqrt{a^2+4b}}{2} > 0$, then we have (2.1) and M is isometric to the sphere $\mathbb{S}^n\left(\frac{a^2+2b+|a|\sqrt{a^2+4b}}{2}\right)$. If $c = \frac{a^2+2b-|a|\sqrt{a^2+4b}}{2} > 0$, then we find (2.2) and M is isometric to the sphere $\mathbb{S}^n\left(\frac{a^2+2b-|a|\sqrt{a^2+4b}}{2}\right)$.

R3) When $A = \sqrt{c} I_p \oplus 0$, $n > p > 1$, we have $R = \frac{p(p-1)c}{n(n-1)}$ and $H = \frac{p\sqrt{c}}{n}$. So (1.1) gives (2.3) and $M = \left\{x \in \mathbb{R}^{n+1} : (x_1)^2 + ... + (x_p)^2 = \frac{1}{c}\right\}$ is a cylinder over a sphere.

R4) If $A = \lambda I \oplus 0$, then $R = 0$ and $H = \frac{\lambda}{n}$. So in view of (1.1), we have $a\lambda + bn = 0$. In this case, M is a cylinder over a complete plane curve such that $A = -\frac{bn}{a} I \oplus 0$.

Hence we get the result. □

The hyperbolic space in the upper half space model is defined by

$$\mathbb{H}^{n+1}(-1) = \left\{x \in \mathbb{R}^{n+2} : \left(x^1\right)^2 + ... + \left(x^{n+1}\right)^2 - \left(x^{n+2}\right)^2 = -1, \quad x^{n+2} > 0\right\}.$$

The isoparametric hypersurfaces M of $\mathbb{H}^{n+1} := \mathbb{H}^{n+1}(-1)$ were determined by Ryan in **[22]** as follows:

H1) $M = \left\{x \in \mathbb{H}^{n+1} : x_k = 0\right\}$, for $1 \le k \le n+1$ with $A = [0]$.

H2) $M = \left\{x \in \mathbb{H}^{n+1} : x_k = r > 0\right\}$, for $1 \le k \le n+1$ with $A = \sqrt{c+1} I$, where $c = -\frac{1}{r^2} \in (-1, 0)$; then $r \in (1, \infty)$. In this case, M is isometric to the hyperbolic space $\mathbb{H}^n(c)$.

H3) $M = \left\{x \in \mathbb{H}^{n+1} : x_{n+2} = x_{n+1} + 1\right\}$ with $A = I$. In this case, M is isometric to $\mathbb{R}^n$.

H4) $M = \left\{x \in \mathbb{H}^{n+1} : (x_1)^2 + ... + (x_{n+1})^2 = r^2\right\} = \mathbb{H}^{n+1} \cap \mathbb{S}^n(r)$ with $A = \sqrt{c+1} I$, where $c = \frac{1}{r^2} > 0$. In this case, M is isometric to the sphere $\mathbb{S}^n(r)$.

H5) $M = \left\{x \in \mathbb{H}^{n+1} : (x_1)^2 + ... + (x_{k+1})^2 = r^2\right\} = \mathbb{H}^{n+1} \cap \mathbb{S}^k(r)$ for $1 \le k \le n$ with $A = \lambda I_k \oplus \frac{1}{\lambda} I_{n-k}$, where $\lambda = \frac{\sqrt{r^2+1}}{r} > 0$.

For the isoparametric linear Weingarten hypersurfaces in $\mathbb{H}^{n+1}$, we can state the following theorem.

THEOREM 2.2. *The isoparametric linear Weingarten hypersurfaces of* $\mathbb{H}^{n+1}$ *are classified as follows:*

i) For $1 \le k \le n+1$, $M = \left\{x \in \mathbb{H}^{n+1} : x_k = 0\right\}$. *In this case,* M *satisfies* $R = aH - 1$ *for all* $a \in \mathbb{R}$.

ii) M *is isometric to the hyperbolic space* $\mathbb{H}^n\left(\frac{a^2+2b+|a|\sqrt{a^2+4b+4}}{2}\right)$, *where*

$$\begin{aligned} &-2 < b < -1 \text{ and } -|b| < a \le -\sqrt{-4-4b} \text{ or } \quad b = -1 \text{ and } -1 < a < 0 \\ &\text{or } -2 < b < -1 \text{ and } \sqrt{-4-4b} \le a < |b| \text{ or } \quad b = -1 \text{ and } 0 < a < 1 \\ &\text{or } -1 < b < 0 \text{ and } -|b| < a < |b|. \end{aligned} \tag{2.4}$$

iii) M *is isometric to hyperbolic space* $\mathbb{H}^n\left(\frac{a^2+2b-|a|\sqrt{a^2+4b+4}}{2}\right)$, *where*

$$\begin{aligned} &a \le -2 \text{ and } -|a| < b < -1 \text{ or } && a \le 2 \text{ and } -1 < b < |a| \\ &\text{or } -2 < a < 2 \text{ and } \tfrac{1}{4}\left(-4 - a^2\right) \le b < -1 \text{ or } && a \ge 2 \text{ and } -1 < b < |a| \\ &\text{or } a \ge 2 \text{ and } -|a| < b < -1 \text{ or } && -2 < a < 2 \text{ and } -1 < b < |a|. \end{aligned} \tag{2.5}$$

iv) M *is isometric to* $\mathbb{R}^n$, *in which case* $a = -b$.

v) M *is isometric to the sphere* $\mathbb{S}^n\left(\frac{1}{\sqrt{c}}\right)$, *where* $c = \frac{a^2+2b+|a|\sqrt{a^2+4b+4}}{2}$ *and*

$$\begin{aligned} &a < -2 \text{ and } b \ge \tfrac{1}{4}\left(-4 - a^2\right) \text{ or } \quad -2 \le a \le 2 \text{ and } b > -|a| \\ &\text{or } a > 2 \text{ and } b \ge \tfrac{1}{4}\left(-4 - a^2\right). \end{aligned} \tag{2.6}$$

vi) M is isometric to the sphere $\mathbb{S}^n\left(\frac{1}{\sqrt{c}}\right)$, *where* $c = \frac{a^2+2b-|a|\sqrt{a^2+4b+4}}{2}$ *and*

$$b < -2 \text{ and } -|b| < a \leq -\sqrt{-4-4b} \text{ or } \quad b > 0 \text{ and } -b < a < b \\ \text{or } b < -2 \text{ and } \sqrt{-4-4b} \leq a < |b|. \tag{2.7}$$

vii) $M = \left\{x \in \mathbb{H}^{n+1} : (x_1)^2 + ... + (x_{k+1})^2 = r^2\right\} = \mathbb{H}^{n+1} \cap \mathbb{S}^k(r)$ *for* $1 \leq k \leq n$, *where*

$$\begin{aligned} k(k-1)(\lambda^2-1) + (n-k)(n-k-1)\left(\tfrac{1}{\lambda^2}-1\right) \\ = a(n-1)\left[\lambda k + \tfrac{n-k}{\lambda}\right] + n(n-1)b \end{aligned} \tag{2.8}$$

and $\lambda = \frac{\sqrt{r^2+1}}{r} > 0$.

PROOF. According to the Ryan's classification, we consider the following cases:

H1) When $A = [0]$ we have $H = 0$ and $R = -1$. In this case, M satisfies $R = aH - 1$ for all $a \in \mathbb{R}$.

H2) For $A = \sqrt{c+1}I$, where $c = -\frac{1}{r^2} \in (-1, 0)$, we have $R = c$ and $H = \sqrt{c+1}$. Then (1.1) gives

$$c = \frac{a^2 + 2b + |a|\sqrt{a^2+4b+4}}{2} \text{ or } c = \frac{a^2 + 2b - |a|\sqrt{a^2+4b+4}}{2}.$$

If $c = \frac{a^2+2b+|a|\sqrt{a^2+4b+4}}{2} \in (-1,0)$, then we have (2.4) and M is isometric to the hyperbolic space $\mathbb{H}^n\left(\frac{a^2+2b+|a|\sqrt{a^2+4b+4}}{2}\right)$. If $c = \frac{a^2+2b-|a|\sqrt{a^2+4b+4}}{2} \in (-1,0)$, then we have (2.5) and M is isometric to the hyperbolic space $\mathbb{H}^n\left(\frac{a^2+2b-|a|\sqrt{a^2+4b+4}}{2}\right)$.

H3) If $A = I$, we have $R = 0$ and $H = 1$. So (1.1) implies $a = -b$. In this case, M is isometric to $\mathbb{R}^n$.

H4) In the case $A = \sqrt{c+1}I$, where $c = \frac{1}{r^2} > 0$, we have $R = c$ and $H = \sqrt{c+1}$. Then in view of (1.1), we have

$$c = \frac{2b + a^2 + |a|\sqrt{a^2+4b+4}}{2} \text{ or } c = \frac{2b + a^2 - |a|\sqrt{a^2+4b+4}}{2}.$$

If $c = \frac{a^2+2b+|a|\sqrt{a^2+4b+4}}{2} > 0$, then we have (2.6) and M is isometric to $\mathbb{S}^n\left(\frac{1}{\sqrt{c}}\right)$.
If $c = \frac{a^2+2b-|a|\sqrt{a^2+4b+4}}{2} > 0$, then we have (2.7) and M is isometric to $\mathbb{S}^n\left(\frac{1}{\sqrt{c}}\right)$.

H5) For $A = \lambda I_k \oplus \frac{1}{\lambda} I_{n-k}$, where $\lambda = \frac{\sqrt{r^2+1}}{r} > 0$, we have

$$R = \frac{k\left[(k-1)(\lambda^2-1) + n - k)\right] + (n-k)(n-k-1)\left(\frac{1}{\lambda^2}-1\right)}{n(n-1)}$$

and

$$H = \frac{\lambda k + \frac{n-k}{\lambda}}{n}.$$

Then (1.1) gives us (2.8). In this case, $M = \mathbb{H}^{n+1} \cap \mathbb{S}^k(r)$ for $1 \leq k \leq n$.

This proves the theorem. □

The isoparametric hypersurfaces M of $\mathbb{S}^{n+1} := \mathbb{S}^{n+1}(1)$, which have at most two distinct principal curvatures, were determined by Ryan in [**21**] and [**22**] as follows:

S1) M is umbilical and $M = \left\{x \in \mathbb{S}^{n+1} : x^{n+2} = \sqrt{1-r^2}\right\}$ for $r \in (0,1)$ with $A = \frac{\sqrt{1-r^2}}{r}I$. In this case, M is isometric to $S^n(r)$.

S2) M is the generalized Clifford torus $M = \mathbb{S}^p(c_1) \times \mathbb{S}^{n-p}(c_2)$ with

$$A = \frac{1}{\sqrt{c_1 + c_2}} \left(c_1 I_p \oplus (-c_2) I_{n-p}\right), \quad \frac{1}{c_1} + \frac{1}{c_2} = 1$$

and $1 < p < n - 1$.

For the isoparametric linear Weingarten hypersurfaces in $\mathbb{S}^{n+1}$, we can state the following theorem.

THEOREM 2.3. *The isoparametric linear Weingarten hypersurfaces, which have at most two distinct principal curvatures, of* $\mathbb{S}^{n+1}$ *are classified as follows:*

i) M is umbilical and $M = \{x \in \mathbb{S}^{n+1} : x^{n+2} = \sqrt{1 - r^2}\}$, where $a = \frac{1-br^2}{r\sqrt{1-r^2}}$ and $r \in (0, 1)$. In this case, M is isometric to $\mathbb{S}^n(r)$.

ii) M is the Clifford torus $\mathbb{S}^p(c_1) \times \mathbb{S}^{n-p}(c_2)$ satisfying

$$\begin{aligned} &p\left[(p-1)\left(1+c_1^2\right)+(n-p)\left(1-c_1c_2\right)\right] \\ &+(n-p)\left[p\left(1-c_1c_2\right)+(n-p-1)\left(1+c_2^2\right)\right] \\ &= a\,(n-1)\left(pc_1-(n-p)\,c_2\right)+n(n-1)b\left(\sqrt{c_1+c_2}\right), \end{aligned} \tag{2.9}$$

where $\frac{1}{c_1} + \frac{1}{c_2} = 1$ and $1 \le p < n - 1$.

PROOF. According to the Ryan's classification, we consider the following cases:

S1) For $A = \frac{\sqrt{1-r^2}}{r} I$, we have $R = \frac{1}{r^2}$ and $H = \frac{\sqrt{1-r^2}}{r}$. Then (1.1) implies $a = \frac{1-br^2}{r\sqrt{1-r^2}}$. In this case, M is isometric to $\mathbb{S}^n(r)$, where $r \in (0, 1)$.

S2) If $A = \frac{1}{\sqrt{c_1+c_2}} \left(r_1 I_p \oplus (-r_2) I_{n-p}\right)$, $\frac{1}{r_1} + \frac{1}{r_2} = 1$ and $1 \le p < n - 1$, then we have

$$\begin{aligned} R = &\frac{1}{\sqrt{c_1+c_2}} \frac{p\left[(p-1)\left(1+c_1^2\right)+(n-p)\left(1-c_1c_2\right)\right]}{n(n-1)} \\ &+\frac{1}{\sqrt{c_1+c_2}} \frac{(n-p)\left[p\left(1-c_1c_2\right)+(n-p-1)\left(1+c_2^2\right)\right]}{n(n-1)} \end{aligned}$$

and $H = \frac{1}{\sqrt{c_1+c_2}} \frac{pc_1-(n-p)c_2}{n}$. Hence using (1.1), we have (2.9). In this case, M is the Clifford torus $\mathbb{S}^p(c_1) \times \mathbb{S}^{n-p}(c_2)$. This completes the proof □

3. Isoparametric linear Weingarten hypersurfaces in Lorentzian space forms

Let M be a hypersurface of a Lorentzian space form $M_1^{n+1}(c)$ and for a certain normal vector field N, let $A = A_N$ be the shape operator. Let $\lambda_1, \lambda_2, ..., \lambda_n$ be the principal curvatures of M. If M has constant principal curvatures and its shape operator is diagonalized, then it is called an *isoparametric hypersurface* [**19**].

In [**1**], Abe, Koike and Yamaguchi showed that the isoparametric hypersurfaces in the Minkowski space $\mathbb{R}_1^{n+1}$ have the following cases:

RL1) $\mathbb{R}^n = \{x = (x_1, ..., x_{n+1}) \in \mathbb{R}_1^{n+1} : x_1 = 0\}$, $A = [0]$,

RL2) $\mathbb{H}^n(c) = \{x \in \mathbb{R}_1^{n+1} : -x_1^2 + \sum_{i=2}^{n+1} x_i^2 = \frac{1}{c},\ c < 0\}$, $A = \pm\sqrt{-c}I$,

RL3) $\mathbb{R}^r \times \mathbb{H}^{n-r} = \{x \in \mathbb{R}_1^{n+1} : -x_1^2 + \sum_{i=r+2}^{n+1} x_i^2 = \frac{1}{c},\ c < 0\}$, $A = \pm\left(0_r \oplus \sqrt{-c}I_{n-r}\right)$,

RL4) $\mathbb{R}_1^n = \{x \in \mathbb{R}_1^{n+1} : x_{n+1} = 0\}$, $A = [0]$,

RL5) $\mathbb{S}_1^n(c) = \{x \in \mathbb{R}_1^{n+1} : -x_1^2 + \sum_{i=2}^{n+1} x_i^2 = \frac{1}{c},\ c > 0\}$, $A = \pm\sqrt{c}I$,

RL6) $\mathbb{R}^r \times \mathbb{S}_1^{n-r} = \{x \in \mathbb{R}_1^{n+1} : -x_1^2 + \sum_{i=r+2}^{n+1} x_i^2 = \frac{1}{c},\ c > 0\}$, $A = \pm(0_r \oplus \sqrt{c} I_{n-r})$.

Here (RL1)-(RL3) are spacelike hypersurfaces and (RL4)-(RL6) are Lorentzian hypersurfaces in $\mathbb{R}_1^{n+1}$.

For the isoparametric linear Weingarten hypersurfaces in the Minkowski space $\mathbb{R}_1^{n+1}$, we can state the following theorem.

THEOREM 3.1. *The isoparametric linear Weingarten hypersurfaces of the Minkowski space $\mathbb{R}_1^{n+1}$ are classified as follows:*

i) M is isometric to $\mathbb{R}^n$. In this case, M satisfies $R = aH$ for all $a \in \mathbb{R}$.

ii) M is isometric to the hyperbolic space $\mathbb{H}^n\left(\frac{-a^2+2b+|a|\sqrt{a^2-4b}}{2}\right)$, where

$$a \in \mathbb{R} \text{ and } b < 0 \text{ or } a \in \mathbb{R} \text{ and } 0 < b \leq \frac{a^2}{4}. \tag{3.1}$$

iii) M is isometric to the hyperbolic space $\mathbb{H}^n\left(\frac{-a^2+2b-|a|\sqrt{a^2-4b}}{2}\right)$, where

$$a \neq 0 \text{ and } b \leq \frac{a^2}{4} \text{ or } a = 0 \text{ and } b < 0. \tag{3.2}$$

iv) $M = \mathbb{R}^r \times \mathbb{H}^{n-r} = \{x \in \mathbb{R}_1^{n+1} : -x_1^2 + \sum_{i=r+2}^{n+1} x_i^2 = \frac{1}{c}, c < 0\}$, where

$$b = \frac{-c(n-r-1)(n-r) \mp a\sqrt{-c}(n-1)(n-r)}{n(n-1)}. \tag{3.3}$$

v) M is isometric to the Minkowski space $\mathbb{R}_1^n$. In this case, M satisfies $R = aH$ for all $a \in \mathbb{R}$.

vi) M is isometric to the de Sitter space $\mathbb{S}_1^n\left(\frac{a^2+2b+|a|\sqrt{a^2+4b}}{2}\right)$, where

$$a \neq 0 \text{ and } -\frac{a^2}{4} \leq b \text{ or } a = 0 \text{ and } b > 0. \tag{3.4}$$

vii) M is isometric to the de Sitter space $\mathbb{S}_1^n\left(\frac{a^2+2b-|a|\sqrt{a^2+4b}}{2}\right)$, where

$$a \in \mathbb{R} \text{ and } b > 0 \text{ or } a \in \mathbb{R} \text{ and } -\frac{a^2}{4} \leq b < 0. \tag{3.5}$$

viii) $M = \mathbb{R}^r \times \mathbb{S}_1^{n-r} = \{x \in \mathbb{R}_1^{n+1} : -x_1^2 + \sum_{i=r+2}^{n+1} x_i^2 = \frac{1}{c},\ c > 0\}$, where

$$b = \frac{c(n-r-1)(n-r) \mp a\sqrt{c}(n-1)(n-r)}{n(n-1)}. \tag{3.6}$$

PROOF. According to the Abe, Koike and Yamaguchi's classification, we consider the following cases:

RL1) For $A = [0]$, we have $H = R = 0$. Hence M is isometric to $\mathbb{R}^n$ and satisfies $R = aH$ for all $a \in \mathbb{R}$.

RL2) In the case $A = \pm\sqrt{-c}I$, we have $R = c$ and $H = \pm\sqrt{-c}$. So, in view of (1.1), we have

$$c = \frac{-a^2 + 2b + |a|\sqrt{a^2-4b}}{2} \text{ or } c = \frac{-a^2 + 2b - |a|\sqrt{a^2-4b}}{2}.$$

If $c = \frac{-a^2+2b+|a|\sqrt{a^2-4b}}{2} < 0$, then we have (3.1) and M is isometric to the hyperbola $\mathbb{H}^n(\frac{-a^2+2b+|a|\sqrt{a^2-4b}}{2})$. If $c = \frac{-a^2+2b-|a|\sqrt{a^2-4b}}{2} < 0$, then we have (3.2) and M is isometric to the hyperbola $\mathbb{H}^n\left(\frac{-a^2+2b-|a|\sqrt{a^2-4b}}{2}\right)$.

RL3) When $A = \pm\left(0_r \oplus \sqrt{-c}I_{n-r}\right)$, we have $R = -\frac{(n-r)(n-r-1)}{n(n-1)}c$ and $H = \pm\frac{(n-r)\sqrt{-c}}{n}$. So (1.1) implies (3.3). In this case, $M = \mathbb{R}^r \times \mathbb{H}^{n-r}$.

RL4) If $A = [0]$, then we have $H = R = 0$. Hence M is isometric to the Minkowski space $\mathbb{R}_1^n$ and satisfies $R = aH$ for all $a \in \mathbb{R}$.

RL5) For $A = \pm\sqrt{c}I$, we have $R = c$ and $H = \pm\sqrt{c}$. So using (1.1), we get

$$c = \frac{a^2 + 2b + |a|\sqrt{a^2+4b}}{2} \text{ or } c = \frac{a^2 + 2b - |a|\sqrt{a^2+4b}}{2}.$$

If $c = \frac{a^2+2b+|a|\sqrt{a^2+4b}}{2} > 0$, then we have (3.4) and M is isometric to the de Sitter space $\mathbb{S}_1^n\left(\frac{a^2+2b+|a|\sqrt{a^2+4b}}{2}\right)$. If $c = \frac{a^2+2b-|a|\sqrt{a^2+4b}}{2} > 0$, then we have (3.5) and M is isometric to the de Sitter space $\mathbb{S}_1^n\left(\frac{a^2+2b-|a|\sqrt{a^2+4b}}{2}\right)$.

RL6) When $A = \pm\left(0_r \oplus \sqrt{c}I_{n-r}\right)$, we have $R = \frac{(n-r)(n-r-1)}{n(n-1)}c$ and $H = \pm\frac{(n-r)\sqrt{c}}{n}$. So (1.1) implies (3.6). In this case, $M = \mathbb{R}^r \times \mathbb{S}_1^{n-r}$.

Then, we complete the proof. □

Abe, Koike and Yamaguchi [**1**] classified isoparametric hypersurfaces of anti-de Sitter space $\mathbb{H}_1^{n+1}(-1)$ as follows:

HL1) $\mathbb{H}^n(c) = \{x \in \mathbb{H}_1^{n+1}(-1) \subset \mathbb{R}_2^{n+2} : x_1 = \sqrt{\frac{1}{c}+1},\, c \leq -1\}$, $A = \pm\sqrt{-1-c}I$,

HL2) $\mathbb{H}^r(c_1) \times \mathbb{H}^{n-r}(c_2) = \{x \in \mathbb{H}_1^{n+1}(-1) \subset \mathbb{R}_2^{n+2} : -x_1^2 + \sum_{i=3}^{r+2} x_i^2 = \frac{1}{c_1},\ -x_2^2 + \sum_{i=r+3}^{n+2} x_i^2 = \frac{1}{c_2}\}$, $(\frac{1}{c_1}+\frac{1}{c_2} = -1,\, c_1 < 0,\, c_2 < 0)$, $A = \pm\left(\sqrt{-1-c_1}I_r \oplus \left(-\sqrt{-1-c_2}\right)I_{n-r}\right)$,

HL3) $\mathbb{R}_1^n = \{x \in \mathbb{H}_1^{n+1}(-1) \subset \mathbb{R}_2^{n+2} : x_1 = x_{n+2} + t_0,\, t_0 > 0\}$, $A = \pm I$,

HL4) $\mathbb{S}_1^n(c) = \{x \in \mathbb{H}_1^{n+1}(-1) \subset \mathbb{R}_2^{n+2} : x_1 = \sqrt{\frac{1}{c}+1},\, c > 0\}$, $A = \pm\sqrt{1+c}I$,

HL5) $\mathbb{H}_1^n(c) = \{x \in \mathbb{H}_1^{n+1}(-1) \subset \mathbb{R}_2^{n+2} : x_{n+2} = \sqrt{-1-\frac{1}{c}},\, -1 < c < 0\}$, $A = \pm\sqrt{1+c}I$,

HL6) $\mathbb{S}_1^r(c_1) \times \mathbb{H}^{n-r}(c_2) = \{x \in \mathbb{H}_1^{n+1}(-1) \subset \mathbb{R}_2^{n+2} : -x_1^2 + \sum_{i=3}^{r+2} x_i^2 = \frac{1}{c_1},\ -x_2^2 + \sum_{i=r+3}^{n+2} x_i^2 = \frac{1}{c_2}\}$, $(\frac{1}{c_1}+\frac{1}{c_2} = 1,\, c_1 > 0,\, c_2 < 0)$, $A = \pm\left(\sqrt{1+c_1}I_r \oplus \sqrt{1+c_2}I_{n-r}\right)$. Here (HL1), (HL2) are spacelike hypersurfaces and (HL3), (HL4), (HL5), (HL6) are Lorentzian hypersurfaces in $\mathbb{H}_1^{n+1} := \mathbb{H}_1^{n+1}(-1)$.

For the isoparametric linear Weingarten hypersurfaces in $\mathbb{H}_1^{n+1}$, we have the following theorem.

THEOREM 3.2. *The isoparametric linear Weingarten hypersurfaces of anti-de Sitter space $\mathbb{H}_1^{n+1}$ are classified as follows:*

i) M is isometric to hyperbolic space $\mathbb{H}^n\left(\frac{-a^2+2b+|a|\sqrt{a^2-4b-4}}{2}\right)$ or $\mathbb{H}^n\left(\frac{-a^2+2b-|a|\sqrt{a^2-4b-4}}{2}\right)$, where

$$a \in \mathbb{R} \text{ and } b \leq \frac{1}{4}(-4+a^2). \tag{3.7}$$

ii) M is isometric to $\mathbb{H}^r(c_1) \times \mathbb{H}^{n-r}(c_2)$, *where*

$$\begin{aligned}&\left[(r-1)c_1 + (n-r)\left(-1+\sqrt{-1-c_1}\sqrt{-1-c_2}\right)\right]r\\ &+(n-r)\left[\left(-1+\sqrt{-1-c_1}\sqrt{-1-c_2}\right)r+(n-r-1)c_2\right]\\ &=\pm a(n-1)\left(r\sqrt{-1-c_1}-(n-r)\sqrt{-1-c_2}\right)+bn(n-1)\end{aligned} \tag{3.8}$$

and $\frac{1}{c_1}+\frac{1}{c_2}=-1$, $c_1<0$, $c_2<0$.

iii) M is isometric to the Minkowski space $\mathbb{R}^n_1$. *In this case,* $b=\mp a$.

iv) M is isometric to the de Sitter space $\mathbb{S}^n_1\left(\frac{a^2+2b+|a|\sqrt{a^2+4b+4}}{2}\right)$, *where*

$$\begin{aligned}&a<-2 \text{ and } b\geq \tfrac{1}{4}(-4-a^2) \text{ or } -2\leq a\leq 2 \text{ and } b>-|a|\\ &\text{or } a>2 \text{ and } b\geq \tfrac{1}{4}(-4-a^2).\end{aligned} \tag{3.9}$$

v) M is isometric to the de Sitter space $\mathbb{S}^n_1\left(\frac{a^2+2b-|a|\sqrt{a^2+4b+4}}{2}\right)$, *where*

$$\begin{aligned}&a<-2 \text{ and } \tfrac{1}{4}(-4-a^2)\leq b<-|a| \text{ or } a<-2 \text{ and } b>|a|\\ &\text{or } -2\leq a\leq 2 \text{ and } b>|a| \text{ or } a>2 \text{ and } b>a\\ &\text{or } a>2 \text{ and } \tfrac{1}{4}(-4-a^2)\leq b<-|a|.\end{aligned} \tag{3.10}$$

vi) M is isometric to the anti-de Sitter space $\mathbb{H}^n_1\left(\frac{a^2+2b+|a|\sqrt{a^2+4b+4}}{2}\right)$, *where*

$$-2<a<2 \text{ and } \frac{1}{4}\left(-4-a^2\right)\leq b<-|a|. \tag{3.11}$$

vii) M is isometric to the anti-de Sitter space $\mathbb{H}^n_1\left(\frac{a^2+2b-|a|\sqrt{a^2+4b+4}}{2}\right)$, *where*

$$a\leq -2 \text{ and } -|a|<b<|a| \text{ or } -2<a<2 \text{ and } \tfrac{1}{4}\left(-4-a^2\right)\leq b<|a| \text{ or}$$

$$a\geq 2 \text{ and } -a<b<a. \tag{3.12}$$

viii) M is isometric to $\mathbb{S}^r_1(c_1)\times\mathbb{H}^{n-r}(c_2)$, *where*

$$\begin{aligned}&\left[(r-1)c_1+(n-r)\left(-1+\sqrt{1+c_1}\sqrt{1+c_2}\right)\right]r\\ &+(n-r)\left[\left(-1+\sqrt{1+c_1}\sqrt{1+c_2}\right)r+(n-r-1)c_2\right]\\ &=\pm a(n-1)\left(r\sqrt{1+c_1}+(n-r)\sqrt{1+c_2}\right)+bn(n-1),\end{aligned} \tag{3.13}$$

$(\frac{1}{c_1}+\frac{1}{c_2}=1$, $c_1>0$, $c_2<0)$.

Proof. HL1) For $A=\pm\sqrt{-1-c}I$, we have $R=c$ and $H=\pm\sqrt{-1-c}$. Then using (1.1), we get

$$c=\frac{-a^2+2b+|a|\sqrt{a^2-4b-4}}{2} \text{ or } c=\frac{-a^2+2b-|a|\sqrt{a^2-4b-4}}{2}.$$

If $c=\frac{-a^2+2b+|a|\sqrt{a^2-4b-4}}{2}\leq -1$ or $c=\frac{-a^2+2b-|a|\sqrt{a^2-4b-4}}{2}\leq -1$, then we have (3.7). In this case, M is isometric to $\mathbb{H}^n\left(\frac{-a^2+2b+a\sqrt{a^2-4b-4}}{2}\right)$.

HL2) If $A=\pm\left(\sqrt{-1-c_1}I_r\oplus\left(-\sqrt{-1-c_2}\right)I_{n-r}\right)$, then we have

$$\begin{aligned}R=&\frac{\left[(r-1)c_1+(n-r)\left(-1+\sqrt{-1-c_1}\sqrt{-1-c_2}\right)\right]r}{n(n-1)}\\ &+\frac{(n-r)\left[\left(-1+\sqrt{-1-c_1}\sqrt{-1-c_2}\right)r+(n-r-1)c_2\right]}{n(n-1)}\end{aligned}$$

and $H=\pm\frac{r\sqrt{-1-c_1}-(n-r)\sqrt{-1-c_2}}{n}$. So (1.1) implies (3.8). In this case, M is isometric to $\mathbb{H}^r(c_1)\times\mathbb{H}^{n-r}(c_2)$.

HL3) In the case $A = \pm I$, we have $R = 0$ and $H = \pm 1$. Then (1.1) gives us $b = \mp a$. In this case, M is isometric to $\mathbb{R}^n_1$.

HL4) When $A = \pm\sqrt{1+c}I$, we have $R = c$ and $H = \pm\sqrt{1+c}$, where $c > 0$. Then using (1.1), we get

$$c = \frac{a^2 + 2b + |a|\sqrt{a^2 + 4b + 4}}{2} \text{ or } c = \frac{a^2 + 2b - |a|\sqrt{a^2 + 4b + 4}}{2}. \tag{3.14}$$

If $c = \frac{a^2+2b+|a|\sqrt{a^2+4b+4}}{2} > 0$, then we have (3.9) and M is isometric to $\mathbb{S}^n_1\left(\frac{a^2+2b+|a|\sqrt{a^2+4b+4}}{2}\right)$. If $c = \frac{a^2+2b-|a|\sqrt{a^2+4b+4}}{2} > 0$, then we have (3.10) and M is isometric to $\mathbb{S}^n_1\left(\frac{a^2+2b-|a|\sqrt{a^2+4b+4}}{2}\right)$.

HL5) For $A = \pm\sqrt{1+c}I$, we have $R = c$ and $H = \pm\sqrt{1+c}$, where $-1 \le c < 0$. Then we have the same solutions (3.14) for c.

If $-1 \le c = \frac{a^2+2b+|a|\sqrt{a^2+4b+4}}{2} < 0$, then we have (3.11) and M is isometric to $\mathbb{H}^n_1\left(\frac{a^2+2b+|a|\sqrt{a^2+4b+4}}{2}\right)$. If $-1 \le c = \frac{a^2+2b-|a|\sqrt{a^2+4b+4}}{2} < 0$, then we have (3.12) and M is isometric to $\mathbb{H}^n_1\left(\frac{a^2+2b-|a|\sqrt{a^2+4b+4}}{2}\right)$.

HL6) If $A = \pm\left(\sqrt{1+c_1}I_r \oplus \sqrt{1+c_2}I_{n-r}\right)$, then we have

$$R = \frac{\left[(r-1)c_1 + (n-r)\left(-1 + \sqrt{1+c_1}\sqrt{1+c_2}\right)\right]r}{n(n-1)} + \frac{(n-r)\left[\left(-1 + \sqrt{1+c_1}\sqrt{1+c_2}\right)r + (n-r-1)c_2\right]}{n(n-1)}$$

and $H = \pm\frac{r\sqrt{1+c_1}+(n-r)\sqrt{1+c_2}}{n}$. Then in view of (1.1), we obtain (3.13).

Then, the proof is completed. □

The isoparametric hypersurfaces of de Sitter space $\mathbb{S}^{n+1}_1 := \mathbb{S}^{n+1}_1(1)$ were determined by Abe, Koike and Yamaguchi in [**1**] as follows:

SL1) $\mathbb{R}^n = \{x \in \mathbb{S}^{n+1}_1(1) \subset \mathbb{R}^{n+2}_1 : x_1 = x_{n+2} + t_0,\ t_0 > 0\}$, $A = \pm I$,

SL2) $\mathbb{S}^n(c) = \{x \in \mathbb{S}^{n+1}_1(1) \subset \mathbb{R}^{n+2}_1 : x_1 = \sqrt{\frac{1}{c} - 1},\ 0 < c \le 1\}$, $A = \pm\sqrt{1-c}I$,

SL3) $\mathbb{H}^n(c) = \{x \in \mathbb{S}^{n+1}_1(1) \subset \mathbb{R}^{n+2}_1 : x_{n+2} = \sqrt{1 - \frac{1}{c}},\ c < 0\}$, $A = \pm\sqrt{1-c}I$,

SL4) $\mathbb{S}^r(c_1) \times \mathbb{H}^{n-r}(c_2) = \{x \in \mathbb{S}^{n+1}_1(1) \subset \mathbb{R}^{n+2}_1 : \sum_{i=1}^{r+2} x_i^2 = \frac{1}{c_1},\ -x_1^2 + \sum_{i=r+3}^{n+2} x_i^2 = \frac{1}{c_2}\}$, $(\frac{1}{c_1} + \frac{1}{c_2} = 1,\ c_1 > 0,\ c_2 < 0)$, $A = \pm\left(\sqrt{1-c_1}I_r \oplus \sqrt{1-c_2}I_{n-r}\right)$,

SL5) $\mathbb{S}^n_1(c) = \{x \in \mathbb{S}^{n+1}_1(1) \subset \mathbb{R}^{n+2}_1 : x_{n+2} = \sqrt{1 - \frac{1}{c}},\ c \ge 1\}$, $A = \pm\sqrt{c-1}I$,

SL6) $\mathbb{S}^r(c_1) \times \mathbb{S}^{n-r}_1(c_2) = \{x \in \mathbb{S}^{n+1}_1(1) \subset \mathbb{R}^{n+2}_1 : \sum_{i=2}^{r+2} x_i^2 = \frac{1}{c_1},\ -x_1^2 + \sum_{i=r+3}^{n+2} x_i^2 = \frac{1}{c_2}\}$, $(\frac{1}{c_1} + \frac{1}{c_2} = 1,\ c_1 > 0,\ c_2 > 0)$, $A = \pm\left(\sqrt{c_1 - 1}I_r \oplus \left(-\sqrt{c_2 - 1}I_{n-r}\right)\right)$.

Here (SL1)-(SL4) are spacelike hypersurfaces and (SL5), (SL6) are Lorentzian hypersurfaces of $\mathbb{S}^{n+1}_1(1)$.

For the isoparametric linear Weingarten hypersurfaces in $\mathbb{S}^{n+1}_1$, we have the following theorem.

THEOREM 3.3. *The isoparametric linear Weingarten hypersurfaces of the de Sitter space $\mathbb{S}^{n+1}_1$ are classified as follows:*

i) M is isometric to $\mathbb{R}^n$ and $b = \mp a$.

ii) M is isometric to the sphere $\mathbb{S}^n\left(\frac{-a^2+2b+|a|\sqrt{a^2-4b+4}}{2}\right)$, *where*

$$\begin{array}{c} a\leq -2 \text{ and } -|a|<b<|a| \text{ or } \quad -2<a<2 \text{ and } -|a|<b\leq \frac{1}{4}(4+a^2) \\ \text{or } a\geq 2 \text{ and } -a<b<a. \end{array} \tag{3.15}$$

iii) M is isometric to the sphere $\mathbb{S}^n\left(\frac{-a^2+2b-|a|\sqrt{a^2-4b+4}}{2}\right)$, *where*

$$-2<a<2 \text{ and } |a|<b\leq \frac{1}{4}(4+a^2). \tag{3.16}$$

iv) M is isometric to the hyperbolic space $\mathbb{H}^n\left(\frac{-a^2+2b+|a|\sqrt{a^2-4b+4}}{2}\right)$, *where*

$$\begin{array}{ll} a<-2 \text{ and } b<-|a| \text{ or } & a<-2 \text{ and } |a|<b\leq \frac{1}{4}(4+a^2) \\ \text{or } -2\leq a\leq 2 \text{ and } b<-|a| \text{ or } & a>2 \text{ and } b<-a \\ \text{or } a>2 \text{ and } a<b\leq \frac{1}{4}(4+a^2). & \end{array} \tag{3.17}$$

v) M is isometric to the hyperbolic space $\mathbb{H}^n\left(\frac{-a^2+2b-|a|\sqrt{a^2-4b+4}}{2}\right)$, *where*

$$\begin{array}{c} a<-2 \text{ and } b\leq \frac{1}{4}(4+a^2) \text{ or } \quad -2\leq a\leq 2 \text{ and } b<|a| \text{ or } \\ a>2 \text{ and } b\leq \frac{1}{4}(4+a^2). \end{array} \tag{3.18}$$

vi) M is isometric to $\mathbb{S}^r(c_1)\times\mathbb{H}^{n-r}(c_2)$, *where*

$$\begin{aligned} &\left[(r-1)c_1+(n-r)\left(1-\sqrt{1-c_1}\sqrt{1-c_2}\right)\right]r \\ &+(n-r)\left[\left(1-\sqrt{1-c_1}\sqrt{1-c_2}\right)r+(n-r-1)c_2\right] \\ &=\pm a(n-1)\left(r\sqrt{1-c_1}+(n-r)\sqrt{1-c_2}\right)+bn(n-1) \end{aligned} \tag{3.19}$$

and $\frac{1}{c_1}+\frac{1}{c_2}=1$, $c_1>0$, $c_2<0$, $A=\pm\left(\sqrt{1-c_1}I_r\oplus\sqrt{1-c_2}I_{n-r}\right)$.

vii) M is isometric to the de Sitter space $\mathbb{S}_1^n\left(\frac{a^2+2b+|a|\sqrt{a^2+4b-4}}{2}\right)$ *or* $\mathbb{S}_1^n\left(\frac{a^2+2b-|a|\sqrt{a^2+4b-4}}{2}\right)$, *where*

$$a\in\mathbb{R} \text{ and } b\geq \frac{1}{4}(4-a^2). \tag{3.20}$$

viii) M is isometric to $\mathbb{S}^r(c_1)\times\mathbb{S}_1^{n-r}(c_2)$, *where*

$$\begin{aligned} &\left[(r-1)c_1+(n-r)\left(1-\sqrt{c_1-1}\sqrt{c_2-1}\right)\right]r \\ &+(n-r)\left[\left(1-\sqrt{c_1-1}\sqrt{c_2-1}\right)r+(n-r-1)c_2\right] \\ &=\pm a(n-1)\left(r\sqrt{c_1-1}-(n-r)\sqrt{c_2-1}\right)+bn(n-1) \end{aligned} \tag{3.21}$$

and $\frac{1}{c_1}+\frac{1}{c_2}=1$, $c_1>0$, $c_2>0$.

PROOF. SL1) If $A=\pm I$, then we have $R=0$ and $H=\pm 1$. So (1.1) implies $b=\mp a$.

SL2) For $A=\pm\sqrt{1-c}I$, we have $R=c$ and $H=\pm\sqrt{1-c}$. Then by using of (1.1), we find

$$c=\frac{-a^2+2b+|a|\sqrt{a^2-4b+4}}{2} \text{ or } c=\frac{-a^2+2b-|a|\sqrt{a^2-4b+4}}{2}. \tag{3.22}$$

If $c=\frac{-a^2+2b+|a|\sqrt{a^2-4b+4}}{2}$ and $0<c\leq 1$, then we have (3.15) and M is isometric to the sphere $\mathbb{S}^n\left(\frac{-a^2+2b+|a|\sqrt{a^2-4b+4}}{2}\right)$. If $c=\frac{-a^2+2b-|a|\sqrt{a^2-4b+4}}{2}$ and $0<c\leq 1$, then we have (3.16) and M is isometric to the sphere $\mathbb{S}^n\left(\frac{-a^2+2b-|a|\sqrt{a^2-4b+4}}{2}\right)$.

SL3) In the case $A=\pm\sqrt{1-c}I$, we have the same solutions (3.22) for c.

If $c = \frac{-a^2+2b+|a|\sqrt{a^2-4b+4}}{2} < 0$, then we have (3.17) and M is isometric to the hyperbolic space $\mathbb{H}^n\left(\frac{-a^2+2b+|a|\sqrt{a^2-4b+4}}{2}\right)$. If $c = \frac{-a^2+2b-|a|\sqrt{a^2-4b+4}}{2} < 0$, then we have (3.18) and M is isometric to the hyperbolic space $\mathbb{H}^n\left(\frac{-a^2+2b-|a|\sqrt{a^2-4b+4}}{2}\right)$.

SL4) When $A = \pm\left(\sqrt{1-c_1}I_r \oplus \sqrt{1-c_2}I_{n-r}\right)$, we have

$$R = \frac{\left[(r-1)c_1 + (n-r)\left(1-\sqrt{1-c_1}\sqrt{1-c_2}\right)\right]r}{n(n-1)} + \frac{(n-r)\left[\left(1-\sqrt{1-c_1}\sqrt{1-c_2}\right)r + (n-r-1)c_2\right]}{n(n-1)}$$

and $H = \pm\frac{r\sqrt{1-c_1}+(n-r)\sqrt{1-c_2}}{n}$. So (1.1) implies (3.19).

SL5) For $A = \pm\sqrt{c-1}I$, we have $R = c$ and $H = \pm\sqrt{c-1}$. Then in view of (1.1), we find

$$c = \frac{a^2+2b+|a|\sqrt{a^2+4b-4}}{2} \text{ or } c = \frac{a^2+2b-|a|\sqrt{a^2+4b-4}}{2}.$$

If $c = \frac{a^2+2b+|a|\sqrt{a^2+4b-4}}{2} \geq 1$ or $c = \frac{a^2+2b-a\sqrt{a^2+4b-4}}{2} \geq 1$, then we have (3.20). In these cases, M is isometric to the de Sitter space $\mathbb{S}_1^n\left(\frac{a^2+2b+|a|\sqrt{a^2+4b-4}}{2}\right)$ or $\mathbb{S}_1^n\left(\frac{a^2+2b-|a|\sqrt{a^2+4b-4}}{2}\right)$.

SL6) For $A = \pm\left(\sqrt{c_1-1}I_r \oplus \left(-\sqrt{c_2-1}I_{n-r}\right)\right)$, we have

$$R = \frac{\left[(r-1)c_1 + (n-r)\left(1-\sqrt{c_1-1}\sqrt{c_2-1}\right)\right]r}{n(n-1)} + \frac{(n-r)\left[\left(1-\sqrt{c_1-1}\sqrt{c_2-1}\right)r + (n-r-1)c_2\right]}{n(n-1)}$$

and $H = \pm\frac{r\sqrt{c_1-1}-(n-r)\sqrt{c_2-1}}{n}$. So (1.1) gives us (3.21). Hence we complete the proof of the theorem. □

References

[1] N. Abe, N. Koike, and S. Yamaguchi, *Congruence theorems for proper semi-Riemannian hypersurfaces in a real space form*, Yokohama Math. J. **35** (1987), no. 1-2, 123–136. MR928379

[2] L. J. Alías, H. F. de Lima, and F. R. dos Santos, *New characterizations of linear Weingarten spacelike hypersurfaces in the de Sitter space*, Pacific J. Math. **292** (2018), no. 1, 1–19, DOI 10.2140/pjm.2018.292.1. MR3708256

[3] C. P. Aquino, H. F. de Lima, and M. A. L. Velásquez, *A new characterization of complete linear Weingarten hypersurfaces in real space forms*, Pacific J. Math. **261** (2013), no. 1, 33–43, DOI 10.2140/pjm.2013.261.33. MR3037557

[4] C. P. Aquino, H. F. de Lima, and M. A. L. Velásquez, *Generalized maximum principles and the characterization of linear Weingarten hypersurfaces in space forms*, Michigan Math. J. **63** (2014), no. 1, 27–40, DOI 10.1307/mmj/1395234357. MR3189466

[5] C. P. Aquino, H. F. de Lima, and M. A. L. Velásquez, *Characterizations of linear Weingarten spacelike hypersurfaces in Lorentz space forms*, Rocky Mountain J. Math. **45** (2015), no. 1, 13–27, DOI 10.1216/RMJ-2015-45-1-13. MR3334203

[6] T. E. Cecil, *Isoparametric and Dupin hypersurfaces*, SIGMA Symmetry Integrability Geom. Methods Appl. **4** (2008), Paper 062, 28, DOI 10.3842/SIGMA.2008.062. MR2434936

[7] X. Chao and P. Wang, *Linear Weingarten hypersurfaces in Riemannian space forms*, Bull. Korean Math. Soc. **51** (2014), no. 2, 567–577, DOI 10.4134/BKMS.2014.51.2.567. MR3194399

[8] X. Chao and P. Wang, *Linear Weingarten hypersurfaces in locally symmetric manifolds*, Hokkaido Math. J. **46** (2017), no. 1, 29–40, DOI 10.14492/hokmj/1498788095. MR3677874

[9] X. Chao and Y. Lv, *Linear Weingarten spacelike hypersurfaces in Lorentz space forms with prescribed Gauss map*, Kodai Math. J. **41** (2018), no. 1, 98–114, DOI 10.2996/kmj/1521424826. MR3777389

[10] B.-Y. Chen, *2-type submanifolds and their applications*, Chinese J. Math. **14** (1986), no. 1, 1–14. MR861185

[11] B.-Y. Chen, *Some open problems and conjectures on submanifolds of finite type*, Soochow J. Math. **17** (1991), no. 2, 169–188. MR1143504

[12] B.-Y. Chen, *Total mean curvature and submanifolds of finite type*, 2nd ed., Series in Pure Mathematics, vol. 27, World Scientific Publishing Co. Pte. Ltd., Hackensack, NJ, 2015. With a foreword by Leopold Verstraelen. MR3362186

[13] H. Chen and X. Wang, *Stability and eigenvalue estimates of linear Weingarten hypersurfaces in a sphere*, J. Math. Anal. Appl. **397** (2013), no. 2, 658–670, DOI 10.1016/j.jmaa.2012.08.003. MR2979602

[14] J. N. Gomes, H. F. de Lima, F. R. dos Santos, and M. A. L. Velásquez, *On the complete linear Weingarten spacelike hypersurfaces with two distinct principal curvatures in Lorentzian space forms*, J. Math. Anal. Appl. **418** (2014), no. 1, 248–263, DOI 10.1016/j.jmaa.2014.03.090. MR3198877

[15] Z. H. Hou and D. Yang, *Linear Weingarten spacelike hypersurfaces in de Sitter space*, Bull. Belg. Math. Soc. Simon Stevin **17** (2010), no. 5, 769–780. MR2777769

[16] H. Li, Y. J. Suh, and G. Wei, *Linear Weingarten hypersurfaces in a unit sphere*, Bull. Korean Math. Soc. **46** (2009), no. 2, 321–329, DOI 10.4134/BKMS.2009.46.2.321. MR2502796

[17] X. X. Li and Q. Q. Li, *Rigidity theorems of linear Weingarten hypersurfaces in* $\mathbb{R}^{n+1}$ (Chinese, with English and Chinese summaries), Chinese Ann. Math. Ser. A **38** (2017), no. 3, 295–302; English transl., Chinese J. Contemp. Math. **38** (2017), no. 3, 237–244. MR3727728

[18] H. F. de Lima and M. A. L. Velásquez, *On the geometry of linear Weingarten spacelike hypersurfaces in the de Sitter space*, Bull. Braz. Math. Soc. (N.S.) **44** (2013), no. 1, 49–65, DOI 10.1007/s00574-013-0003-0. MR3077634

[19] M. A. Magid, *Lorentzian isoparametric hypersurfaces*, Pacific J. Math. **118** (1985), no. 1, 165–197. MR783023

[20] Wolfram Research, Inc., Mathematica, Version 10.1, Champaign, IL, 2015

[21] P. J. Ryan, *Homogeneity and some curvature conditions for hypersurfaces*, Tohoku Math. J. (2) **21** (1969), 363–388, DOI 10.2748/tmj/1178242949. MR253243

[22] P. J. Ryan, *Hypersurfaces with parallel Ricci tensor*, Osaka Math. J. **8** (1971), 251–259. MR296859

[23] J. F. da Silva, H. F. de Lima, and M. A. L. Velásquez, *Stability of generalized linear Weingarten hypersurfaces immersed in the Euclidean space*, Publ. Mat. **62** (2018), no. 1, 95–111, DOI 10.5565/PUBLMAT6211805. MR3738184

[24] S. Shu, *Linear Weingarten hypersurfaces in a real space form*, Glasg. Math. J. **52** (2010), no. 3, 635–648, DOI 10.1017/S0017089510000480. MR2679920

Department of Mathematics, Balikesir University, 10145, Çağış, Balikesir, TURKEY
Email address: cozgur@balikesir.edu.tr

Contemporary Mathematics
Volume **756**, 2020
https://doi.org/10.1090/conm/756/15211

Lagrangian submanifolds of the nearly Kähler 6-sphere and Chen's equality

Ramesh Sharma

ABSTRACT. We obtain the following results: (i) If the normal curvature operator of a Lagrangian submanifold M of the nearly Kähler S^6 (with almost complex structure J) annihilates a vector field V normal to M, then the shape operator with respect to V vanishes, and M is quasi-Einstein and is Sasakian if JV is Killing on M. (ii) If a Lagrangian submanifold M of the nearly Kähler S^6 satisfies the Chen's equality and is conformally flat, then it is the totally geodesic S^3. Finally we prove the result: (iii) If the almost contact metric structure induced by a unit tangent vector field on a Lagrangian submanifold of the nearly Kähler S^6 is normal, then it is Sasakian.

1. Introduction

The Cayley multiplication can be used to define a vector cross product on the purely imaginary octonions R^7 by the formula:

$$u \times v = \frac{1}{2}(uv - vu)$$

and the standard inner product on R^7 is given by

$$(u, v) = -\frac{1}{2}(uv + vu).$$

It turns out that

$$u \times (v \times w) + (u \times v) \times w = 2(u, w)v - (u, v)w - (w, v)u.$$

The Cayley multiplication on the octonions $\mathcal{O}$ is thus given by

$$(r + u)(s + v) = rs - (u, v) + rv + su + (u \times v)$$

for $r, s \in Re(\mathcal{O})$ and $u, v \in Im(\mathcal{O})$. It is also known that the group G_2 of automorphisms of $\mathcal{O}$ is precisely the group of isometries of R^7 preserving the vector cross product.

An almost complex structure J on the unit 6-sphere $S^6 = \{x \in R^7 : ||x|| = 1\}$ is defined by $Ju = x \times u$, where $x \in S^6$ and $u \in T_x S^6$. It turns out that the Riemannian metric g on S^6 induced from R^7 is a Hermitian metric with respect to J, i.e. $g(JX, JY) = g(X, Y)$ for arbitrary $X, Y \in \mathfrak{X}(S^6)$. It is also known

2010 *Mathematics Subject Classification.* Primary 53B25; Secondary 53C25.
Key words and phrases. Nearly Kähler unit 6-sphere, Lagrangian submanifold, Chen's equality, Sasakian, Quasi-Einstein, Normal curvature operator.

that the Riemannian connection $\bar{\nabla}$ of (S^6, g) satisfies the condition: $(\bar{\nabla}_X J)(Y) + (\bar{\nabla}_Y J)(X) = 0$. An almost Hermitian manifold satisfying this condition is known as a nearly Kähler (or Tachibana) manifold. Thus, S^6 is nearly Kähler, but not Kähler. For details, we refer to Dillen and Vrancken [**DV**]. Due to this special structure, S^6 has attracted the attention of several researchers. A submanifold M of the nearly Kähler S^6 is said to be Lagrangian if J maps the tangent bundle of M onto the normal bundle, and the normal bundle onto the tangent bundle. We denote the metrics of S^6 and M by the same letter g. In [**E**], Ejiri showed that a Lagrangian submanifold of the nearly Kähler S^6 is orientable and minimal. Deshmukh and Hadi [**DH**] proved the following result: "Let M be a compact 3-dimensional Lagrangian submanifold of S^6 with nearly Kähler structure. Then there exists a global unit vector field ξ on M, and if $J\xi$ is parallel in the normal bundle, then M is a Sasakian manifold with ξ as the Reeb vector field." Intrigued by this result, Vrancken [**V**] showed that the second fundamental form of a Sasakian Lagrangian submanifold M of the nearly Kähler S^6 annihilates the Reeb vector field, and provided a complete classification of such submanifolds. In [**SDA**], Sharma, Deshmukh and Al-Solamy showed that a global unit tangent vector field ξ on a Lagrangian submanifold M of the nearly Kähler S^6 induces an almost contact metric structure on M, which is Sasakian if and only if $J\xi$ is parallel in the normal bundle. Thus, for a unit vector field ξ tangent to a Lagrangian submanifold M of nearly Kähler S^6, the condition that $J\xi$ is parallel in the normal bundle implies that the shape operator of M along any normal vector field vanishes.

In [**SD**], Sharma and Deshmukh proved that, if the normal connection of a Lagrangian submanifold M of the nearly Kähler S^6 is flat, then M is the totally geodesic S^3. In this paper, weakening the flatness condition of the normal connection we establish the following result.

THEOREM 1.1. *If the normal curvature operator of a Lagrangian submanifold M of the nearly Kähler S^6 annihilates a vector field V normal to M, then the shape operator with respect to V vanishes and M is quasi-Einstein. Further, M is Sasakian if in addition, JV is Killing on M.*

For a minimal submanifold M of a real space form of constant curvature c, the Chen's inequality [**C**]

$$\delta_M \leq \frac{1}{2}(n+1)(n-2)$$

holds, where the invariant δ_M is given at a point p by $\tau(p) - inf(K)(p)$, $inf(K)$ is the function assigning to each point $p \in M$ the infimum of $K(\pi)$, and π runs over all planes in T_pM. The equality holds at a point p if and only if the index of relative nullity, i.e. the dimension of the subspace of T_pM annihilated by the second fundamental form σ of M is at least $n-2$ at that point. As the Lagrangian submanifold of the nearly Kähler S^6 is 3-dimensional and minimal, the Chen's equality $\delta_M = 2$ holds at a point if and only if the index of relative nullity is at least 1 at that point. So, if the Chen's equality holds at each point of M, then there exists a unit vector field ξ tangent to M such that $\sigma(\xi, X) = 0$ for an arbitrary tangent vector field X on M, and M is quasi-Einstein (i.e. the two eigenvalues of the Ricci tensor with respect to the eigenspace orthogonal to ξ are equal). For details we refer to Deszcz, Dillen, Verstraelen and Vrancken [**DDVV**]. As pointed out earlier, for a Sasakian Lagrangian submaifold of nearly Kähler S^6, $\sigma(\xi, X) = 0$, and hence Chen's equality is evidently satisfied [**DV**].

Now let us recall the following result of Ejiri [**E**]: "If a Lagrangian submanifold M of the nearly Kähler S^6 has constant curvature c, then either $c = 1$ (i.e. M is totally geodesic) or $c = \frac{1}{16}$". We note that if a manifold has constant curvature, then it is conformally flat, however the converse need not be true. We generalize Ejiri's result and obtain the following.

THEOREM 1.2. *If a Lagrangian submanifold M of the nearly Kähler S^6 satisfies Chen's equality and is conformally flat, then it is the totally geodesic S^3.*

REMARK 1.3. This result holds also when "conformally flat" is replaced by either "locally symmetric" (because a 3-dimensional locally symmetric manifold is conformally flat), or "parallel second fundamental form" (this implies, by Gauss equation, that M is locally symmetric).

Our next result is

PROPOSITION 1.4. *Let M be a 3-dimensional submanifold (not necessarily Lagrangian) of S^6, and ξ be a unit vector field tangent to M belonging to the relative null space everywhere on M. Then M is quasi-Einstein. Further, (i) if ξ is Killing on M then M is Sasakian, (ii) if M is conformally flat then it is the totally geodesic S^3.*

Finally, we note the fact that a Sasakian structure is normal almost contact, but the converse need not be true. We examine the converse problem in the present context, and prove the following result.

THEOREM 1.5. *If the almost contact metric structure induced by a unit tangent vector field on a Lagrangian submanifold of the nearly Kähler S^6 is normal, then it is Sasakian.*

REMARK 1.6. Theorem 1.5 provides an extrinsic characterization of a 3-dimensional Sasakian structure.

2. Preliminaries

On a Lagrangian submanifold M of the nearly Kähler S^6, the tensor field G of type (1,2) defined on S^6 by $G(X,Y) = (\bar{\nabla}_X J)(Y)$ satisfies the following properties (see [**DOVV**] and [**E**]):

$$G(X,Y) = -G(Y,X) \tag{2.1}$$

$$G(X,JY) = -JG(X,Y) \tag{2.2}$$

$$g(G(X,Y),Z) = -g(G(X,Z),Y) \tag{2.3}$$

$$\begin{aligned} g(G(X,Y),G(Z,W)) \;&-\; g(X,Z)g(Y,W) - g(X,W)g(Y,Z) \\ &+\; g(JX,Z)g(Y,JW) - g(JX,W)g(Y,JZ) \end{aligned} \tag{2.4}$$

$$(\bar{\nabla}_X G)(Y,Z) = g(Y,JZ)X + g(X,Z)JY - g(X,Y)JZ \tag{2.5}$$

for arbitrary $X,Y,Z \in \mathfrak{X}(S^6)$. If ∇ and $\nabla^{\perp}$ denote the Riemannian connection of a submanifold M of a Riemannian manifold $\bar{M}$, and the connection in the normal bundle, respectively, then we have the Gauss and Weingarten formulas:

$$\bar{\nabla}_X Y = \nabla_X Y + \sigma(X,Y), \quad \bar{\nabla}_X N = -A_N X + \nabla^{\perp}_X N$$

for arbitrary $X, Y \in \mathfrak{X}(M)$, where A_N denotes the shape operator of M with respect to a normal vector field N, defined by $g(A_N X, Y) = g(\sigma(X,Y), N)$, and $\nabla^\perp$ denotes the normal connection. Let R and $\bar{R}$ denote the curvature tensors of M and $\bar{M}$ respectively, X, Y, Z, W any vector fields tangent to M and U, V any vector fields normal to M. The curvature tensor $R^\perp$ of the normal bundle (normal curvature operator) is defined by

$$R^\perp(X,Y)U = \nabla^\perp_X \nabla^\perp_Y U - \nabla^\perp_Y \nabla^\perp_X U - \nabla^\perp_{[X,Y]} U. \tag{2.6}$$

It may be noted that the normal curvature operator is conformally invariant (Chen [**C1**]), i.e. if $^*g = \Omega^2 g$ for a smooth function Ω on M, and $^*R^\perp$ is the normal curvature operator corresponding to $^*g$, then $^*R^\perp(X,Y) = R^\perp(X,Y)$ for any $X, Y \in \mathfrak{X}(M)$. First we recall the following equations of Gauss, Codazzi and Ricci, respectively, for a general submanifold M of a Riemannian manifold $\bar{M}$:

$$\begin{aligned} g(\bar{R}(X,Y)Z,W) &= g(R(X,Y)Z,W) \\ &- g(\sigma(X,W),\sigma(Y,Z)) + g(\sigma(Y,W),\sigma(X,Z)) \end{aligned} \tag{2.7}$$

$$(\bar{R}(X,Y)Z)^\perp = (\nabla_X \sigma)(Y,Z) - (\nabla_Y \sigma)(X,Z) \tag{2.8}$$

$$g(\bar{R}(X,Y)U,V) = g(R^\perp(X,Y)U,V) + g(A_V A_U X - A_U A_V X, Y). \tag{2.9}$$

For a Lagrangian submanifold of the nearly Kähler S^6 we have the following formulas ([**E**], Schafer and K. Smoczyk [**SS**]):

$$\nabla^\perp_X JY = J\nabla_X Y + G(X,Y) \tag{2.10}$$

$$\sigma(X,Y) = JA_{JY}X \tag{2.11}$$

$$JG(X, JG(Y,Z)) = g(X,Z)Y - g(X,Y)Z \tag{2.12}$$

$$-\sigma(X, JG(Y,Z)) + JG(\sigma(X,Y),Z) + JG(Y,\sigma(Z,X)) = 0. \tag{2.13}$$

We also know [**E**] that $G(X,Y)$ is normal to M for any $X, Y \in \mathfrak{X}(M)$.

3. Almost contact metric structure

A $(2n+1)$-dimensional smooth manifold M is said to be an almost contact metric manifold if carries a global 1-form η, a unit vector field ξ, a (1,1)-tensor field φ, and a Riemannian metric g satisfying

$$\varphi^2 = -I + \eta \otimes \xi, \quad \eta(\xi) = 1, \tag{3.1}$$

$$g(\varphi X, \varphi Y) = g(X,Y) - \eta(X)\eta(Y).$$

Obviously, $\varphi\xi = 0$, $\eta \circ \varphi = 0$, $g(X,\xi) = \eta(X)$. An almost contact metric is called Sasakian if

$$(\nabla_X \varphi)Y = g(X,Y)\xi - \eta(Y)X \tag{3.2}$$

where ∇ is the Riemannian connection of g. For details we refer to Blair [**B**]. It is known that a Riemannian manifold is Sasakian if and only if it has a unit Killing vector field ξ such that

$$R(X,Y)\xi = g(\xi,Y)X - g(\xi,X)Y. \tag{3.3}$$

4. Proofs of the results

PROOF OF THEOREM 1.1. The Ricci equation (2.9) can be expressed as

$$\begin{aligned} g(\bar{R}(X,Y)JZ, JW) &= g(R^{\perp}(X,Y)JZ, JW) \\ &+ g(A_{JW}A_{JZ}X, Y) - g(A_{JZ}A_{JW}X, Y) \end{aligned} \tag{4.1}$$

because any normal vector field is the image of a tangent vector field under J. As the ambient space is S^6, the left side of (4.1) vanishes. By hypothesis, there exists a unit tangent vector field ξ on M such that $R^{\perp}(X,Y)J\xi = 0$ for all $X, Y \in \mathfrak{X}(M)$. Setting $Z = \xi$ in (4.1) we find that

$$g(A_{JW}A_{J\xi}X - A_{J\xi}A_{JW}X, Y) = 0.$$

The use of (2.11) and the skew-symmetry of J transforms the above equation as

$$g(\sigma(J\sigma(X,\xi), W) - \sigma(J\sigma(W,X),\xi), JY) = 0. \tag{4.2}$$

Next, using the formula [**DV**]:

$$g(\sigma(X,Y), JZ) = g(\sigma(X,Z), JY) \tag{4.3}$$

for a Lagrangian submanifold of nearly Kähler S^6 in (4.2), gives

$$g(\sigma(W,Y), \sigma(X,\xi)) = g(\sigma(\xi,Y), \sigma(W,X)). \tag{4.4}$$

On the other hand, substituting ξ for Z in the Gauss equation (2.7) and using (4.4) we obtain

$$R(X,Y)\xi = g(\xi,Y)X - g(\xi,X)Y. \tag{4.5}$$

Contracting it at Y provides

$$Ric(X,\xi) = 2g(X,\xi) \tag{4.6}$$

and substituting ξ for X in the above equation we get

$$Ric(\xi,\xi) = 2. \tag{4.7}$$

Going back to Gauss equation (2.7), contracting it at X and W with respect to a local orthonormal frame (e_i), and using the minimality of M we find that

$$Ric(Y,Z) = 2g(Y,Z) - \sum_{i=1}^{3} g(\sigma(e_i,Y), \sigma(e_i,Z)).$$

Substituting ξ for Y and Z in the foregoing equation and using (4.6) yields the relation

$$\sum_{i=1}^{3} g(\sigma(e_i,\xi), \sigma(e_i,\xi) = 0.$$

The last equation shows that $\sigma(X,\xi) = 0$ for any $X \in \mathfrak{X}(M)$, i.e. $A_{J\xi} = 0$. As M is 3-dimensional, its Weyl tensor vanishes, and hence using (4.5) and (4.6) we find that M is quasi-Einstein, more specifically

$$Ric = (\frac{r}{2} - 1)g + (3 - \frac{r}{2})\eta \otimes \eta. \tag{4.8}$$

where r is the scalar curvature of (M,g) and η is the dual 1-form of ξ, i.e $\eta(X) = g(\xi, X)$ for all $X \in \mathfrak{X}(M)$. The last part follows from the following characterization of a Sasakian 3-manifold (Prop. 7.5 in Blair [**B**], also Stromenger [**ST**]):"A

Riemannian manifold (M,g) is Sasakian if and only if it admits a unit Killing field ξ such that $R(X,Y)\xi = g(\xi,Y)X - g(X,\xi)Y$." This completes the proof. □

PROOF OF THEOREM 1.2. By hypothesis, Chen's equality holds on M, and hence there exists a unit vector field ξ tangent to M such that $\sigma(\xi,X) = 0$, i.e. $A_{J\xi} = 0$. Taking into account that the ambient space is S^6, the Gauss equation (2.7) reduces to

$$\begin{aligned} g(R(X,Y)Z,W) &= g(Y,Z)g(X,W) - g(X,Z)g(Y,W) \\ &+ g(\sigma(X,W),\sigma(Y,Z)) - g(\sigma(Y,W),\sigma(X,Z)). \end{aligned} \tag{4.9}$$

Taking its trace with respect to X and W we get

$$Ric(Y,Z) = 2g(Y,Z) - \sum_{i=1}^{3} g(\sigma(e_i,Y),\sigma(e_i,Z)). \tag{4.10}$$

Setting $Z = \xi$ and using $\sigma(X,\xi) = 0$ in (4.10) provides

$$Ric(Y,\xi) = 2\eta(Y). \tag{4.11}$$

Now, as M is 3-dimensional, its Weyl tensor vanishes identically, and hence

$$\begin{aligned} g(R(X,Y)Z,W) &= g(Y,Z)Ric(X,W) - g(X,Z)Ric(Y,W) \\ &+ g(X,W)Ric(Y,Z) - g(Y,W)Ric(X,Z) \\ &- \frac{1}{2}[g(Y,Z)g(X,W) - g(Y,W),g(X,Z)]. \end{aligned} \tag{4.12}$$

Substituting ξ for X and W in (4.9) and using $\sigma(X,\xi) = 0$ we immediately get

$$g(R(\xi,Y)Z,\xi) = g(Y,Z) - \eta(Y)\eta(Z).$$

This equation used in conjunction with (4.12) for $X = W = \xi$, and equation (4.11) shows that

$$Ric(Y,Z) = (\frac{r}{2} - 1)g(Y,Z) + (3 - \frac{r}{2})\eta(Y)\eta(Z). \tag{4.13}$$

At this point, we use conformal flatness of M, i.e. the Cotton tensor vanishes:

$$(\nabla_X Ric)(Y,Z) - (\nabla_Y Ric)(X,Z) - \frac{1}{4}[(Xr)g(Y,Z) - (Yr)g(X,Z)] = 0.$$

Substituting the expression for the Ricci tensor from (4.13) in the preceding equation gives

$$\begin{aligned} [(Xr)g(Y,Z) - (Yr)g(X,Z)] &= 2[(Xr)\eta(Y)\eta(Z) - (Yr)\eta(X)\eta(Z)] \\ &- 2(6-r)[((\nabla_X\eta)Y - (\nabla_Y\eta)X)\eta(Z) \\ &+ \eta(Y)(\nabla_X\eta)Z - \eta(X)(\nabla_Y\eta)Z]. \end{aligned} \tag{4.14}$$

Now first substituting ξ for Z in the above equation we have

$$(Xr)\eta(Y) - (Yr)\eta(X) = 2(6-r)[(\nabla_X\eta)Y - (\nabla_Y\eta(X]. \tag{4.15}$$

Then substituting ξ for Z in the above equation shows that

$$2(6-r)\nabla_\xi\xi = \xi r - Dr. \tag{4.16}$$

Next, contracting (4.14) with respect to Y and Z provides

$$(\xi r)\xi = (6-r)[\nabla_\xi\xi + (div\xi)\xi]. \tag{4.17}$$

Its inner product with ξ yields the relation:

$$\xi r = (6 - r) div \xi. \tag{4.18}$$

Two cases arise: (i) $r = 6$ on M in which case equations (4.13) and (4.10) imply that M is totally geodesic S^3, or (ii) $r \neq 6$ on some open dense subset $\mathcal{U}$ of M. We rule out the case (ii) by working on $\mathcal{U}$ as follows. From equations (4.17) and (4.18) we get $\nabla_\xi \xi = 0$ and hence (4.16) reduces to $Dr = (\xi r)\xi$ on $\mathcal{U}$. Consequently, it follows from (4.15) that $d\eta = 0$ on $\mathcal{U}$. Using these consequences back in (4.14), and substituting $Y = \xi$ we obtain

$$\nabla_X \xi = \frac{\xi r}{4}[X - \eta(X)\xi]. \tag{4.19}$$

Its g-trace is $2(div\xi) = \xi r$. Comparing it with (4.18) we conclude that $(r-4)\xi r = 0$, and so $\xi r = 0$ on $\mathcal{U}$. This reduces the equation (4.19) to $\nabla \xi = 0$ which, in turn, shows that $Ric(Y, \xi) = 0$ contradicting (4.11). Hence M is the totally geodesic S^3, completing the proof. □

Proof of Proposition 1.4. Here, M is a 3-dimensional submanifold (not necessarily Lagrangian) of S^6. As the index of relative nullity of M is 1 everywhere, we have $\sigma(X, \xi) = 0$ on M. Contracting the Gauss equation (4.9) at X and W, and substituting ξ for Z we obtain

$$Ric(Y, \xi) = 2\eta(Y). \tag{4.20}$$

Going back to (4.9) and substituting $Z = \xi$ we also get

$$R(X, Y)\xi = \eta(Y)X - \eta(X)Y. \tag{4.21}$$

As M is 3-dimensional, its Weyl tensor is zero, and hence the two preceding equations imply

$$Ric = (\frac{r}{2} - 1)g + (3 - \frac{r}{2})\eta \otimes \eta, \tag{4.22}$$

i.e. M is quasi-Einstein. Now, if ξ is Killing on (M, g), then by the result in [**B**], stated at the end of the proof of Theorem 1, M becomes Sasakian. If M is conformally flat, then proceeding as in the proof of Theorem 2, we can show that M is totally geodesic S^3, completing the proof.

□

Proof of Theorem 1.5. As shown in [**SDA**], a unit tangent vector field ξ on a Lagrangian submanifold M of the nearly Kähler S^6 induces an almost contact metric structure defined by

$$\varphi X = G(X, J\xi) \tag{4.23}$$

for an arbitrary $X \in \mathfrak{X}(M)$. This definition makes sense because $G(X, Y)$ is normal to M for any two vector fields X, Y tangent to M. Defining a (1,1) tensor F on M by $\nabla^\perp_X J\xi = JFX$, one can show [**SDA**] that

$$(\nabla_X \varphi)Y = g(X, Y)\xi - \eta(Y)X + G(Y, JFX), \tag{4.24}$$

and substituting $Y = \xi$ in the above equation gives

$$\nabla_X \xi = -\varphi X + FX. \tag{4.25}$$

Now, since M is a 3-dimensional normal almost contact metric manifold, we have the following formulas: (Olszak [**O**])

$$\begin{aligned}(\nabla_X\varphi)Y &= a[g(X,Y)\xi - \eta(Y)X] \\ &+ b[g(\varphi X,Y)\xi - \eta(Y)\varphi X]\end{aligned} \tag{4.26}$$

$$\nabla_X\xi = -a\varphi X + b[X - \eta(X)\xi], \tag{4.27}$$

where a and b are smooth functions on M. Comparing the equations (4.24), (4.25) with equations (4.26), (4.27) provides

$$\begin{aligned}G(Y,JFX) &= (a-1)[g(X,Y)\xi - \eta(Y)X] \\ &+ b[g(\varphi X,Y) - \eta(Y)\varphi X].\end{aligned} \tag{4.28}$$

From (2.2), $G(Y,JFX) = -JG(Y,FX)$. Also, equations (4.25) and (4.27) yield

$$FX = (1-a)\varphi X + b[X - \eta(X)\xi].$$

Using the last two equations along with (4.23) and a straightforward computation we get

$$\begin{aligned}(a-1)JG(Y,\varphi X) &- bJG(Y,X) \\ &= (a-1)[g(X,Y)\xi - \eta(Y)X] \\ &+ b[g(\varphi X,Y)\xi - \eta(Y)\varphi X + \eta(X)\varphi Y].\end{aligned} \tag{4.29}$$

At this point, we substitute φX for X in the above equation in order to get

$$\begin{aligned}-(a-1)JG(Y,X) &- bJG(Y,\varphi X) \\ &= (a-1)[g(\varphi X,Y)\xi - \eta(Y)\varphi X + \eta(X)\varphi Y] \\ &+ b[-g(X,Y)\xi + \eta(Y)X].\end{aligned}$$

Eliminating $JG(Y,\varphi X)$ between the above two equations, and then applying J we obtain

$$[b^2 + (a-1)^2][G(X,Y) + g(\varphi X,Y)J\xi - \eta(Y)J\varphi X + \eta(X)J\varphi Y] = 0.$$

Hence, either (i) $a = 1, b = 0$ on M, or (ii) there is an open dense subset $\mathcal{U}$ of M on which

$$G(X,Y) = \eta(Y)J\varphi X - \eta(X)J\varphi Y + g(X,\varphi Y)J\xi.$$

Let us rule out the possibility (ii). By definition,

$$\begin{aligned}G(X,Y) = (\bar\nabla_X J)(Y) &= \bar\nabla_X JY - J\bar\nabla_X Y \\ &- A_{JY}X - J\nabla_X Y - J\sigma(X,Y) = -J\nabla_X Y.\end{aligned} \tag{4.30}$$

Comparing the above set of equations with the equation preceding it, and then operating by J shows

$$\nabla_X Y = \eta(X)\varphi Y - \eta(Y)\varphi X - g(X,\varphi Y)\xi.$$

Taking its inner product with Y provides $g(\nabla_X Y, Y) = 0$, and which, in turn, shows that $||Y||_g$ is constant. But Y is an arbitrary tangent vector field on $\mathcal{U}$, and hence we arrive at a contradiction, ruling out the possibility (ii). Hence (i) holds on M, i.e. M is Sasakian. This completes the proof. □

5. Concluding remarks

1. As pointed out in Section 3, a unit tangent vector field ξ on a Lagrangian submanifold M of the nearly Kähler S^6 induces an almost contact metric structure on M. Let us see an interplay between this structure and the nullity condition $\sigma(X,\xi) = 0$, i.e. $A_{J\xi} = 0$ which is equivalent to the Chen's equality. Using this condition in equations (2.2), (2.11) and (2.13) and a straightforward calculation one obtains $A_{JX}\varphi = -\varphi A_{JX}$, i.e. $A_V\varphi = -\varphi A_V$ for any vector field V normal to M. Hence the shape operator of M with respect to any normal vector field, anti-commutes with the almost contact structure φ.

2. In [**DV**], Lagrangian submanifolds of the nearly Kähler S^6 satisfying Chen's equality were classified in terms of complex curves in $CP(4)$ lifted via the Hopf fibration $S^5(1) \to CP(4)$ when M is contained in some 5-dimensional totally geodesic $S^5(1)$ and as tubes of radius $\pi/2$ in the direction of the second normal space over an almost complex curve in $S^6(1)$ when M is linearly full in $S^6(1)$.

References

[B] D. E. Blair, *Riemannian geometry of contact and symplectic manifolds*, 2nd ed., Progress in Mathematics, vol. 203, Birkhäuser Boston, Inc., Boston, MA, 2010. MR2682326

[C] B.-Y. Chen, *Pseudo-Riemannian geometry, δ-invariants and applications*, World Scientific Publishing Co. Pte. Ltd., Hackensack, NJ, 2011. With a foreword by Leopold Verstraelen. MR2799371

[C1] F. J. E. Dillen and L. C. A. Verstraelen (eds.), *Handbook of differential geometry. Vol. I*, North-Holland, Amsterdam, 2000. MR1736851

[DDVV] R. Deszcz, F. Dillen, L. Verstraelen, and L. Vrancken, *Quasi-Einstein totally real submanifolds of the nearly Kähler 6-sphere*, Tohoku Math. J. (2) **51** (1999), no. 4, 461–478, DOI 10.2748/tmj/1178224715. MR1725622

[DH] S. Deshmukh and K. El Hadi, *A Sasakian structure on a 3-dimensional totally real submanifold of the nearly Kaehler 6-sphere*, Panamer. Math. J. **2** (1992), no. 3, 43–52. MR1173148

[DOVV] F. Dillen, B. Opozda, L. Verstraelen, and L. Vrancken, *On totally real 3-dimensional submanifolds of the nearly Kaehler 6-sphere*, Proc. Amer. Math. Soc. **99** (1987), no. 4, 741–749, DOI 10.2307/2046486. MR877050

[DV] F. Dillen and L. Vrancken, *Totally real submanifolds in $S^6(1)$ satisfying Chen's equality*, Trans. Amer. Math. Soc. **348** (1996), no. 4, 1633–1646, DOI 10.1090/S0002-9947-96-01626-1. MR1355070

[E] N. Ejiri, *Totally real submanifolds in a 6-sphere*, Proc. Amer. Math. Soc. **83** (1981), no. 4, 759–763, DOI 10.2307/2044249. MR630028

[O] Z. Olszak, *Normal almost contact metric manifolds of dimension three*, Ann. Polon. Math. **47** (1986), no. 1, 41–50, DOI 10.4064/ap-47-1-41-50. MR859423

[SD] R. Sharma and S. Deshmukh, *On Lagrangian submanifolds of the nearly Kaehler 6-sphere*, Recent advances in the geometry of submanifolds—dedicated to the memory of Franki Dillen (1963–2013), Contemp. Math., vol. 674, Amer. Math. Soc., Providence, RI, 2016, pp. 153–160. MR3565252

[SDA] R. Sharma, S. Deshmukh, and F. Al-Solamy, *Almost contact Lagrangian submanifolds of nearly Kaehler 6-sphere*, Results Math. **65** (2014), no. 1-2, 143–153, DOI 10.1007/s00025-013-0335-5. MR3162434

[SS] L. Schäfer and K. Smoczyk, *Decomposition and minimality of Lagrangian submanifolds in nearly Kähler manifolds*, Ann. Global Anal. Geom. **37** (2010), no. 3, 221–240, DOI 10.1007/s10455-009-9181-9. MR2595677

[ST] C. Stromenger, *Sasakian manifolds: Differential forms, Curvature and Conformal Killing forms*, Ph.D. Thesis, University of Cologne, 2010.

[V] L. Vrancken, *Special Lagrangian submanifolds of the nearly Kaehler 6-sphere*, Glasg. Math. J. **45** (2003), no. 3, 415–426, DOI 10.1017/S0017089503001356. MR2005344

Department of Mathematics, University of New Haven, West Haven, Connecticut 06516

Email address: `rsharma@newhaven.edu`

Contemporary Mathematics
Volume **756**, 2020
https://doi.org/10.1090/conm/756/15212

Sesquilinear forms and symmetric spaces

Gudlaugur Thorbergsson

Dedicated to Professor Bang-Yen Chen on the occasion of his 75th birthday

ABSTRACT. Let f be a sesquilinear form on $\mathbb{F}^n$ with positive Witt index r where $\mathbb{F}$ is $\mathbb{R}$, $\mathbb{C}$, or $\mathbb{H}$. Let $N_i(\mathbb{F}^n, f)$ denote the space of i-dimensional totally isotropic subspaces of $\mathbb{F}^n$ with respect to f where $i \leq r$. Then our main result will be the observation that $N_i(\mathbb{F}^n, f)$ is a symmetric space if and only if $n = 2i$. This gives us seven series of compact symmetric spaces. If we add the three series of Grassmannians $G_i(\mathbb{F}^n)$ over $\mathbb{F}$, we get all ten series of classical compact symmetric spaces.

1. Introduction

1.1. Symmetric spaces. A *symmetric space* is a Riemannian manifold M with the property that for every p in M there is an isometry σ_p of M to itself fixing p and reversing the orientation of the geodesics passing through p. It easily follows from the definition that a symmetric space M is homogeneous and can hence be written as a coset space K/L where K is the isometry group of M; see [**4**], Ch. IV, §3.

We will phrase our main result in terms of symmetric pairs. Let (K, L) be a pair consisting of a Lie group K and a compact subgroup L. We say that (K, L) is a *symmetric pair* if there is an involutive automorphism σ of K with

$$K_0^\sigma \subset L \subset K^\sigma$$

where K^σ is the fixed point group of σ and K_0^σ denotes the identity component of K^σ.

If (K, L) is a symmetric pair, then every K-invariant Riemannian metric on K/L makes it into a symmetric space. A K-invariant Riemannian metric on K/L can always be found by averaging.

If K is the isometry group of a symmetric space M and L the isotropy group of some point in M, then (K, L) is a symmetric pair.

We will say that $M = K/L$ is a *classical* symmetric space and (K, L) a *classical* symmetric pair if K and L are classical matrix groups or finite quotients or coverings of such groups.

1.2. Sesquilinear forms. We let $\mathbb{F}$ denote the reals $\mathbb{R}$, the complex numbers $\mathbb{C}$, or the quaternions $\mathbb{H}$. We will call $\mathbb{F}$ a field in spite of the noncommutativity of $\mathbb{H}$. We will consider the set $\mathbb{F}^n$ of n-tuples of elements in $\mathbb{F}$ (which we will call

2010 *Mathematics Subject Classification.* Primary 53C35, 11E39.

vectors) as a right vector space over $\mathbb{F}$, i.e., we will multiply a vector in $\mathbb{F}^n$ by a scalar from $\mathbb{F}$ on the right hand side. It is the noncommutativity of the quaternions that makes it necessary to decide on which side we multiply the scalars.

We let $\sigma : \mathbb{F} \to \mathbb{F}$ denote an *anti-automorphism* of $\mathbb{F}$, i.e.,

$$\sigma(\alpha\beta) = \sigma(\beta)\sigma(\alpha),$$

which we will assume to be continuous.[1] These anti-automorphisms are the identities on $\mathbb{R}$ and $\mathbb{C}$, the conjugation on $\mathbb{C}$, and the conjugation composed with an inner automorphism on $\mathbb{H}$. For our purposes, we can disregard the inner automorphism in the case of $\mathbb{H}$ and assume that an anti-automorphism is the conjugation. We let ϵ denote either 1 or -1. A *sesquilinear form* is by definition a map

$$f : \mathbb{F}^n \times \mathbb{F}^n \to \mathbb{F}$$

that is additive in both arguments and satisfies

$$f(x\alpha, y\beta) = \sigma(\alpha) f(x, y)\beta$$

for all x and y in $\mathbb{F}^n$ and all α and β in $\mathbb{F}$; see [**1**], §3 and especially [**8**], §8.1. A sesquilinear form f is said to be (σ, ϵ)-*Hermitian* if

$$f(x, y) = \epsilon\sigma(f(y, x))$$

for all x and y in $\mathbb{F}^n$.

A (σ, ϵ)-Hermitian form f is said to be *nondegenerate* if $f(x, y) = 0$ for all y in $\mathbb{F}^n$ implies $x = 0$. In the following, we will assume all forms to be nondegenerate.

1.3. Totally isotropic subspaces. Let f be a (σ, ϵ)-Hermitian form on $\mathbb{F}^n$. We will say that a subspace of $\mathbb{F}^n$ is *totally isotropic with respect to* f if the restriction of f to it is identically zero. The *Witt index of* f is defined to be the maximal dimension r of a totally isotropic subspace of $\mathbb{F}^n$ with respect to f. Note that r is at most half the dimension of $\mathbb{F}^n$, i.e., $2r \leq n$.

The automorphism group $\mathrm{Aut}(\mathbb{F}^n, f)$ is clearly a closed subgroup of $\mathrm{GL}(n, \mathbb{F})$ and hence a Lie group. Let W_1 and W_2 be subspaces of $\mathbb{F}^n$ that are isometric with respect to the restrictions of f. Then, by a theorem of Witt, there is an element of $\mathrm{Aut}(\mathbb{F}^n, f)$ that maps W_1 to W_2; see [**1**], p. 71.

Let $N_i(\mathbb{F}^n, f)$ denote the space of i-dimensional totally isotropic subspaces of $(\mathbb{F}^n, f)$ for $i \leq r$. Then it is clear from the theorem of Witt we just quoted that $\mathrm{Aut}(\mathbb{F}^n, f)$ acts transitively on $N_i(\mathbb{F}^n, f)$. The isotropy group of the action of $\mathrm{Aut}(\mathbb{F}^n, f)$ on $N_i(\mathbb{F}^n, f)$ at some given element is a closed subgroup. It follows that we can write $N_i(\mathbb{F}^n, f)$ as a quotient space of $\mathrm{Aut}(\mathbb{F}^n, f)$ and a closed subgroup and hence that $N_i(\mathbb{F}^n, f)$ is a differentiable manifold, which is in fact a closed submanifold of the Grassmannian $G_i(\mathbb{F}^n)$ and hence compact.

In the following, we will simplify the notation by setting $G = \mathrm{Aut}(\mathbb{F}^n, f)$. We will say that a Lie group, which we do not assume to be connected, is *reductive* if its Lie algebra is the direct sum of an Abelian and a semisimple Lie algebra.

We will need the following lemma on the structure of G.

LEMMA 1.1. *Assume the Witt index r of $(\mathbb{F}^n, f)$ is positive. Then G is a noncompact, possibly disconnected, reductive Lie group. If K is a maximal compact subgroup of G, then K acts transitively on $N_i(\mathbb{F}^n, f)$ for all $i \leq r$.*

[1] (Anti)-automorphisms of $\mathbb{R}$ and $\mathbb{H}$ are continuous, but $\mathbb{C}$ has discontinuous ones.

We will prove the above lemma and the following theorem in Section 2 by looking at the different possibilities for a (σ, ϵ)-Hermitian form f on $\mathbb{F}^n$. The group G will in all cases turn out to be a well-known classical Lie group, which we know to be reductive. The center of G is in all cases either zero- or one-dimensional. It is also well known that any two maximal compact subgroups in these reductive groups G are conjugate to each other.

We can now formulate our main result.

THEOREM 1.2. *Assume the Witt index r of $(\mathbb{F}^n, f)$ is positive and that L is the isotropy group of the action of K on $N_i(\mathbb{F}^n, f)$ at some given totally isotropic i-plane. We assume furthermore that $r \geq 2$ if $\mathbb{F} = \mathbb{R}$ and $\epsilon = 1$.*

Then (K, L) is a symmetric pair if and only if i is equal to the Witt index r and $n = 2r$.

If $r = 1$ and $\epsilon = 1$, then K/L is diffeomorphic to a sphere for all three fields, but only for the real field is (K, L) a symmetric pair; see the remark at the end of 2.1.1 for details.

Finally, we would like to point out that it is assumed in the definition of symmetric pairs on p. 209 in [**4**] that the group here denoted by K is connected. We have decided not to follow this since the spaces $N_k(\mathbb{F}^{2k}, f)$ are not always connected. This is of course not essential for our results and could be modified; see also the remarks in Section 3.2 at the end of the paper.

2. The proofs of Lemma 1.1 and Theorem 1.2

It is an easy task to list the possible types of (σ, ϵ)-Hermitian forms by going through the possibilities for $\mathbb{F}$, σ, and ϵ . We bring the result in the following table. In the last column, we explain how the forms are usually referred to.

$\mathbb{R}$	$\sigma =$ identity	$\epsilon = +1$	symmetric forms
		$\epsilon = -1$	symplectic forms
$\mathbb{C}$	$\sigma =$ identity	$\epsilon = +1$	symmetric forms
		$\epsilon = -1$	symplectic forms
	$\sigma =$ conjugation	$\epsilon = +1$	Hermitian forms
		$\epsilon = -1$	skew-Hermitian forms
$\mathbb{H}$	$\sigma =$ conjugation	$\epsilon = +1$	Hermitian forms
		$\epsilon = -1$	skew-Hermitian forms

There are eight lines in the table. Note that a Hermitian form over $\mathbb{C}$ becomes skew-Hermitian if we multiply it by i and vice versa. This will make these two cases equivalent from our point of view since the automorphism groups and the spaces $N_i(\mathbb{F}^n, f)$ do not change when we multiply f by i. In the quaternionic case, the Hermitian and the skew-Hermitian forms are genuinely different.

We are therefore left with *seven different types of (σ, ϵ)-Hermitian forms.* It turns out that each type has a *normal form*, which will lead to a complete classification of these sesquilinear forms. These normal forms are well known from Linear Algebra with the possible exception of the skew-Hermitian forms on $\mathbb{H}^n$. All cases can be found in [**2**], §12. We will use some elementary theory of symmetric spaces and Lie groups in the proofs, but no use will be made of the classifications of symmetric spaces and simple Lie groups.

It will be convenient to group the seven types of normal forms as follows:

(1) The symmetric forms over $\mathbb{R}$ and the Hermitian forms over $\mathbb{C}$ and $\mathbb{H}$, which we will refer to as *inner products on* $\mathbb{F}^n$.
(2) The symplectic forms over $\mathbb{R}$ and $\mathbb{C}$
(3) The symmetric forms over $\mathbb{C}$ and the skew-Hermitian forms over $\mathbb{H}$.

We will consider the seven cases separately and prove the lemma and the theorem for each case.

2.1. Inner products. Let f be a form on $\mathbb{F}^n$ that is symmetric if $\mathbb{F} = \mathbb{R}$ and Hermitian if $\mathbb{F} = \mathbb{C}$ or $\mathbb{H}$. Then there is a basis of $\mathbb{F}^n$ such that f can be written as

$$f(x, y) = \sum_{i=1}^{k} \bar{x}_i y_i - \sum_{i=k+1}^{n} \bar{x}_i y_i$$

where we disregard the bars in the formula in the real case. The pair $(k, n-k)$ is called the *signature of* f. It is easy to see that the Witt index r of f is equal to $\min\{k, n-k\}$. We will mostly assume that $k \leq n-k$ and hence that $r = k$.

2.1.1. *Inner products on* $\mathbb{R}^n$. The automorphism group $G = \mathrm{Aut}(\mathbb{R}^n, f)$ is denoted by $\mathrm{O}(k, n-k)$ and called the *orthogonal group of type* $(k, n-k)$. The group $G = \mathrm{O}(k, n-k)$ has four connected components when $k \geq 1$. The group $G = \mathrm{O}(k, n-k)$ is simple except in some exceptional cases where it is semi-simple.[2] We set $\mathrm{O}(n) = \mathrm{O}(n, 0)$. The group $\mathrm{O}(n)$ is compact and has two components. The group $\mathrm{O}(k) \times \mathrm{O}(n-k)$ is a maximal compact subgroup in $\mathrm{O}(k, n-k)$. It is easy to see that the action of $\mathrm{O}(k) \times \mathrm{O}(n-k)$ on $N_i(\mathbb{R}^n, f)$ for all $i \leq r = k$ is transitive.

We would now like to prove Theorem 1.2 in this case. Let P be the i-plane in $N_i(\mathbb{R}^n, f)$ spanned by $e_1 + e_{k+1}, \ldots, e_i + e_{k+i}$ where $e_1, \ldots, e_n$ is the standard basis of $\mathbb{R}^n$. We denote by L the isotropy group of P under the action of $K = \mathrm{O}(k) \times \mathrm{O}(n-k)$. Our goal is to show that (K, L) is a symmetric pair if and only if $n = 2i$. Now it is easy to see that a block matrix (A, B) in $K = \mathrm{O}(k) \times \mathrm{O}(n-k)$ is contained in L if and only if A is a block matrix of the form (C, D) in $\mathrm{O}(k)$ and B is a block matrix of the form (C, E) in $\mathrm{O}(n-k)$ where $C \in \mathrm{O}(i)$, $D \in \mathrm{O}(k-i)$, and $E \in \mathrm{O}(n-k-i)$. In particular, L is equal to the diagonal group $\Delta(\mathrm{O}(k))$ in K if $i = k = n-i$. This proves that (K, L) is a symmetric pair when $i = k$ and $2k = n$ since then $\Delta(\mathrm{O}(k))$ is the fixed point group of the involution of $K = \mathrm{O}(k) \times \mathrm{O}(k)$ that switches the two factors. The corresponding symmetric space $\mathrm{O}(k) \times \mathrm{O}(k)/\Delta(\mathrm{O}(k))$ can easily be seen to coincide with

$$\mathrm{O}(k).$$

Note that the symmetric space $N_k(\mathbb{R}^{2k}, f)$ has two components.

It is left to prove that (K, L) is not a symmetric pair when $1 < i < k$ or $2k < n$. We give an indirect proof and assume that $i < k$ or $2k < n$ and that L is open and closed in the fixed point group of an involution σ on K. Note that (the identity component of) K can consist of two, three, or four simple factors. We first assume that K has two factors. Now we know from [**4**], Theorem 5.3 on p. 379, that the involution σ either leaves the simple factors of K invariant or interchanges two such factors. The fact that L cannot be written as a product of subgroups of $\mathrm{O}(k)$ and $\mathrm{O}(n-k)$ shows that σ must interchange the factors which is not possible if $2k < n$, and L would have smaller dimension than the fixed point group if $i < k$ and $2k = n$.

[2] It will not be important for us to know when $\mathrm{O}(k, n-k)$ is simple, but one finds necessary and sufficient conditions for this in [**5**] on p. 362.

Similar arguments can be used when K has more than two simple factors. This finishes the proof of both Lemma 1.1 and Theorem 1.2 in this case.

It is easy to see that (K, L) is a symmetric pair and K/L a sphere when $i = r = 1$. In fact, if f is an inner product with signature $(n, 1)$ on $\mathbb{R}^{n+1}$, then $G = \mathrm{O}(n, 1)$ and (K, L) is the symmetric pair $(\mathrm{O}(n) \times \mathrm{O}(1), \mathrm{O}(n-1) \times \mathrm{O}(1))$ which has the more familiar symmetric pair $(\mathrm{SO}(n), \mathrm{SO}(n-1))$ as its identity component.[3]

2.1.2. *Inner products on* $\mathbb{C}^n$. This case is very similar to the case of inner products on $\mathbb{R}^n$, see 2.1.1, so we only sketch it.

The automorphism group $G = \mathrm{Aut}(\mathbb{C}^n, f)$ is denoted by $\mathrm{U}(k, n-k)$ and called the *unitary group of type* $(k, n-k)$. Note that G has a one-dimensional center that consists of multiples of the identity by complex numbers with absolute value one. The group G modulo the center is isomorphic to the kernel of the map

$$\det : G = \mathrm{U}(k, n-k) \to \mathbb{C},$$

which is denoted by $\mathrm{SU}(k, n-k)$ and called the *special unitary group of type* $(k, n-k)$. The group $\mathrm{SU}(k, n-k)$ is simple which implies that G is reductive. Furthermore, $K = \mathrm{U}(k) \times \mathrm{U}(n-k)$ is a maximal compact subgroup of G where $\mathrm{U}(k) = \mathrm{U}(k, 0)$ and $\mathrm{U}(n-k) = \mathrm{U}(n-k, 0)$. It is easy to see that the action of $\mathrm{U}(k) \times \mathrm{U}(n-k)$ on $N_i(\mathbb{C}^n, f)$ for $i \leq \min\{k, n-k\}$ is transitive. The group K is connected and its center is two-dimensional.

The proof Theorem 1.2 in this case is very similar to the arguments in 2.1.1. The symmetric space $N_k(\mathbb{C}^{2k}, f)$ coincides with

$$\mathrm{U}(k).$$

Note that $N_k(\mathbb{C}^{2k}, f)$ has a one-dimensional Euclidean factor.

2.1.3. *Inner products on* $\mathbb{H}^n$. We will be very brief since this case is very similar to the previous two cases; see 2.1.1 and 2.1.2.

The automorphism group $G = \mathrm{Aut}(\mathbb{H}^n, f)$ is denoted by $\mathrm{Sp}(k, n-k)$ and is called the *quaternionic unitary group of type* $(k, n-k)$. The group $\mathrm{Sp}(k, n-k)$ is simple and connected and has $\mathrm{Sp}(k) \times \mathrm{Sp}(n-k)$ as a maximal compact subgroup acting transitively on $N_i(\mathbb{H}^n, f)$ for $i \leq \min\{k, n-k\}$ where $\mathrm{Sp}(k) = \mathrm{Sp}(k, 0)$ and $\mathrm{Sp}(n-k) = \mathrm{Sp}(n-k, 0)$.

Again, we can argue as in 2.1.1 to prove Theorem 1.2 in this case. Here there is a slight simplification since K has precisely two simple factors. The symmetric space $N_k(\mathbb{H}^{2k}, f)$ coincides with

$$\mathrm{Sp}(k),$$

which is irreducible.

2.2. The symplectic forms. Let f be a symplectic form on $\mathbb{F}^n$. It follows that n must be even, i.e., $n = 2k$, and $\mathbb{F}$ is either $\mathbb{R}$ or $\mathbb{C}$, and there is a basis of $\mathbb{F}^n$ such that f can be written as

$$f(x, y) = \sum_{i=1}^{k} (x_i y_{k+i} - x_{k+i} y_i).$$

The Witt index of f is equal to k, which is half the dimension n.

The maximal totally isotropic subspaces in $(\mathbb{F}^n, f)$ are usually called *Lagrangian subspaces* and $N_k(\mathbb{F}^n, f)$ is called a *Lagrangian Grassmannian (over* $\mathbb{F}$*)*.

[3]Putting an S in front of the symbol for a group will always mean passing to the subgroup of elements with determinant equal to one.

2.2.1. *Symplectic forms on* $\mathbb{R}^{2k}$. The automorphism group $G = \mathrm{Aut}(\mathbb{R}^{2k}, f)$ is denoted by $\mathrm{Sp}(2k, \mathbb{R})$ and called the *symplectic group over* $\mathbb{R}$.[4] It is well known that the group $\mathrm{Sp}(2k, \mathbb{R})$ is simple and connected. The form f is the imaginary part of the standard inner product on $\mathbb{C}^k$ that we assume to be identified with $\mathbb{R}^{2k}$. It now follows easily that $\mathrm{U}(k)$ is a maximal compact group in $\mathrm{Sp}(2k, \mathbb{R})$. To see that $\mathrm{U}(k)$ acts transitively on $N_i(\mathbb{R}^{2k}, f)$ for $i \leq k$, let P be a plane in $N_i(\mathbb{R}^{2k}, f)$ with an orthonormal basis $\hat{e}_1, \dots \hat{e}_i$. Then $\hat{e}_1, \dots \hat{e}_i$ is clearly also orthonormal with respect to the unitary scalar product on $\mathbb{C}^k$ and it follows at once that $\mathrm{U}(k)$ acts transitively on $N_i(\mathbb{R}^{2k}, f)$.

We would now like to write $N_k(\mathbb{R}^{2k}, f)$ as a coset space. We let P be the subspace of $\mathbb{R}^{2k}$ spanned by the first k standard basis vectors. It is clear that P is a Lagrangian subspace and that an element of $\mathrm{U}(k)$ leaves it invariant if and only if it is contained in $\mathrm{O}(k)$. The involution σ on $\mathrm{U}(k)$ that sends A to the conjugated matrix $\bar{A}$ has $\mathrm{O}(k)$ as a fixed point group. It follows that $(\mathrm{U}(k), \mathrm{O}(k))$ is a symmetric pair and the symmetric space $N_k(\mathbb{R}^{2k}, f)$ coincides with

$$\mathrm{U}(k)/\,\mathrm{O}(k).$$

Note that $N_k(\mathbb{R}^{2k}, f)$ has a one-dimensional Euclidean factor.

It is left to prove that $N_i(\mathbb{R}^{2k}, f)$ is not a symmetric space for $i < k$. Assume it is a symmetric space and let P be the plane in $N_i(\mathbb{R}^{2k}, f)$ spanned by $e_1, \dots, e_i$. Then the isotropy group at P of the action of $\mathrm{U}(k)$ on $N_i(\mathbb{R}^{2k}, f)$ is $\mathrm{O}(i) \times \mathrm{U}(k-i)$. Let σ be an involution of $\mathrm{U}(k)$ corresponding to the symmetric pair $(\mathrm{U}(k), \mathrm{O}(i) \times \mathrm{U}(k-i))$. Let $\mathfrak{k}$ and $\mathfrak{p}$ be the $+1$- and -1-eigenspaces of $d\sigma_e$, respectively. A straightforward calculation now shows that $[\mathfrak{p}, \mathfrak{p}]$ is not a subspace of $\mathfrak{k}$ as it should be since $d\sigma_e$ is a Lie algebra automorphism. This contradiction shows that $N_i(\mathbb{R}^{2k}, f)$ is not a symmetric space for $i < k$ and finishes the proof of Theorem 1.2 in this case.

2.2.2. *Symplectic forms on* $\mathbb{C}^{2k}$. The automorphism group $G = \mathrm{Aut}(\mathbb{C}^{2k}, f)$ is denoted by $\mathrm{Sp}(2k, \mathbb{C})$ and called the *symplectic group over* $\mathbb{C}$.[5] It is well known that the group $\mathrm{Sp}(2n, \mathbb{C})$ is simple and connected. Similar to the real case in 2.2.1, we identify $\mathbb{C}^{2k}$ with $\mathbb{H}^k$ and use the fact that f is the j-part of the standard quaternionic scalar product on $\mathbb{H}^k$; see [**3**], Ch. I, §VIII. It follows that $\mathrm{Sp}(k)$ is a maximal compact subgroup of $\mathrm{Sp}(2k, \mathbb{C})$. It now follows as in 2.2.1 that the action of $\mathrm{Sp}(k)$ on $N_i(\mathbb{R}^{2k}, f)$ is transitive for $i \leq k$. We let P be the subspace of $\mathbb{C}^{2k}$ spanned by the first k standard basis vectors. Then we see with arguments as in 2.2.1 that $\mathrm{U}(k)$ is the isotropy subgroup at P under the action of $\mathrm{Sp}(k)$. We would now like to define an involution σ on $\mathrm{Sp}(k)$ having $\mathrm{U}(k)$ as a fixed point group. We write the quaternionic matrix $A \in \mathrm{Sp}(k)$ as $A = C + jD$, where C and D are matrices with complex entries. We set $\sigma(C + jD) = \bar{C} + j\bar{D}$ and calculate directly that σ is the inner group automorphism that we get by conjugating by $I_k j$. The fixed point group of σ consists of those elements of $\mathrm{Sp}(k)$ that leave $\mathbb{R}^k + j\mathbb{R}^k$ invariant. By the definition of $\mathrm{Sp}(k)$, the fixed point group can be described as the subgroup that leaves the inner product restricted to $\mathbb{R}^k + j\mathbb{R}^k$ invariant, i.e., it is $\mathrm{U}(k)$. It follows that $(\mathrm{Sp}(k), \mathrm{U}(k))$ is a symmetric pair and the symmetric space $N_k(\mathbb{C}^{2k}, f)$ coincides with

$$\mathrm{Sp}(k)/\,\mathrm{U}(k),$$

which is irreducible.

[4] This group is frequently denoted by $\mathrm{Sp}(k, \mathbb{R})$.

[5] This group is frequently denoted by $\mathrm{Sp}(k, \mathbb{C})$.

It is left to prove that $N_i(\mathbb{C}^{2k}, f)$ is not a symmetric space for $i < k$. Here we can use arguments similar to those in the previous case; see 2.2.1.

2.3. Symmetric forms on $\mathbb{C}^n$ and skew-Hermitian forms on $\mathbb{H}^n$. Here we are considering two cases that turn out to have rather similar normal forms.

2.3.1. *Symmetric forms on $\mathbb{C}^n$.* Let f be a symmetric form on $\mathbb{C}^n$. Then there is a basis of $\mathbb{C}^n$ such that f can be written as

$$f(x, y) = \sum_{i=1}^{n} x_i y_i.$$

The automorphism group of $(\mathbb{C}^n, f)$ is called the *complex orthogonal group* and is denoted by $\mathrm{O}(n, \mathbb{C})$. The group $\mathrm{O}(n, \mathbb{C})$ has two connected components that coincide with the preimages of 1 and -1 under the determinant. Furthermore, it has $\mathrm{O}(n)$ as a maximal compact subgroup. The Witt index r of f is equal to $\left[\frac{n}{2}\right]$.

We will now show that the action of $\mathrm{O}(n)$ is transitive on $N_i(\mathbb{C}^n, f)$ for all $i \leq \left[\frac{n}{2}\right]$. To do this we split f into real and imaginary parts. Let $u + iv$ and $w + iz$ be elements in $\mathbb{C}^n = \mathbb{R}^n + i\mathbb{R}^n$. Then

$$f(u + iv, w + iz) = \sum_{j=1}^{n}(u_j w_j - v_j z_j) + i \sum_{j=1}^{n}(u_j z_j + v_j w_j).$$

Let P be an element in $N_i(\mathbb{C}^n, f)$. We choose an orthogonal basis $z_1 = u_1 + iv_1, \ldots, z_i = u_i + iv_i$ in P with respect to the standard Hermitian scalar product in $\mathbb{C}^n$ where u_j and v_j are elements in $\mathbb{R}^n$ and assume that $\|z_j\|^2 = 2$ for all i. Then the equations $\|z_j\|^2 = 2$ and $f(z_j, z_j) = 0$ imply

$$\|u_j\|^2 = \|v_j\|^2 = 1 \text{ and } \langle u_j, v_j \rangle = 0.$$

Furthermore, the equations $\langle z_h, z_j \rangle = 0$ and $f(z_h, z_j) = 0$ for $h \neq j$ imply

$$\langle u_h, u_j \rangle = \langle v_h, v_j \rangle = 0 \text{ and } \langle u_h, v_j \rangle = 0.$$

As a consequence, we see that $u_1, \ldots, u_i, v_1, \ldots, v_i$ is an orthonormal set in $\mathbb{R}^n$. Conversely, we see that every such orthonormal set $u_1^*, \ldots, u_i^*, v_1^*, \ldots, v_i^*$ of $\mathbb{R}^n$ gives rise to a subspace in $\mathbb{C}^n$ spanned by $z_1^* = u_1^* + iv_1^*, \ldots, z_i^* = u_i^* + iv_i^*$ that is contained in $N_i(\mathbb{C}^n, f)$. Note that $2i \leq n$ since i is less than or equal to the Witt index, which is equal to $\left[\frac{n}{2}\right]$. It follows that the maximal compact subgroup $\mathrm{O}(n)$ of $\mathrm{O}(n, \mathbb{C})$ acts transitively on $N_i(\mathbb{C}^n, f)$ for all $i \leq \left[\frac{n}{2}\right]$.

Now assume that $n = 2k$ and let $P \in N_k(\mathbb{C}^{2k}, f)$. Then the above considerations make clear that the isotropy group of the action of $\mathrm{O}(2k)$ at P is $\mathrm{U}(k)$. Let σ be the involution of $\mathrm{O}(2k)$ defined as the conjugation by

$$J_k = \begin{pmatrix} 0 & I_k \\ -I_k & 0 \end{pmatrix}.$$

Then $\mathrm{U}(k)$ is the fixed point group of σ and it follows that $(\mathrm{O}(2k), \mathrm{U}(k))$ is a symmetric pair. We have therefore proved that $N_k(\mathbb{C}^{2k}, f)$ is a symmetric space that coincides with

$$\mathrm{O}(2k)/\,\mathrm{U}(k).$$

We see that $N_k(\mathbb{C}^{2k}, f)$ has two connected components. One frequently refers to one to the components as an *orthogonal Grassmannian.*

It is left to prove that $N_i(\mathbb{C}^n, f)$ is not a symmetric space if $i < \left[\frac{n}{2}\right]$. To do this, we use methods analogous to those for the symplectic forms in 2.2; see the end of 2.2.1.

2.3.2. *Skew-Hermitian forms on* $\mathbb{H}^n$. Let f be a skew-Hermitian form on $\mathbb{H}^n$. Then there is a basis of $\mathbb{H}^n$ such that f can be written as

$$f(x, y) = \sum_{i=1}^{n} \bar{x}_i j y_i;$$

see [**2**], Satz on p. 434.[6] Another normal form will turn out to be more important for us. Analogous to the splitting of f into real and imaginary parts in 2.3.1, we will write $\mathbb{H}^n = \mathbb{C}^n + j\mathbb{C}^n$. We consider $u + jv$ and $w + jz$ in $\mathbb{H}^n = \mathbb{C}^n + j\mathbb{C}^n$. As in [**7**], we consider the form f on $\mathbb{H}^n$ defined by setting

$$f(u + jv, w + jz) = i\sum_{k=1}^{n}(\bar{u}_k w_k - \bar{v}_k z_k) + j\sum_{k=1}^{n}(u_k z_k + v_k w_k).$$

and verify that it is both nondegenerate and skew-Hermitian over $\mathbb{H}$. This form is not equal to the normal form above, but they are of course equal up to a basis change. The first sum in the definition of f is a nondegenerate Hermitian form with Witt index n on $\mathbb{C}^{2n} = \mathbb{C}^n + \mathbb{C}^n$ and the second sum is a nondegenerate symmetric form on $\mathbb{C}^{2n} = \mathbb{C}^n + \mathbb{C}^n$.

The automorphism group of $(\mathbb{H}^n, f)$ is called the *quaternionic anti-unitary group* and is denoted by $\mathrm{U}_\alpha(n, \mathbb{H})$.[7] The group $\mathrm{U}_\alpha(n, \mathbb{H})$ is connected and simple, and has $\mathrm{U}(n)$ as a maximal compact subgroup where we have embedded $\mathrm{U}(n)$ into $\mathrm{U}_\alpha(n, \mathbb{H})$ by letting $A \in \mathrm{U}(n)$ send $u + jv$ to $Au + j\bar{A}v$. The Witt index r of f is equal to $\left[\frac{n}{2}\right]$.

We will now show that the action of $\mathrm{U}(n)$ is transitive on $N_i(\mathbb{H}^n, f)$ for all $i \leq \left[\frac{n}{2}\right]$. The arguments will be very similar to those we saw in 2.3.1, so we will only sketch them. Let P be an element in $N_i(\mathbb{H}^n, f)$ for some $i \leq \left[\frac{n}{2}\right]$ and let $z_1 = u_1 + jv_1, \ldots, z_i = u_n + jv_i$ be a basis in P such that

$$\langle z_h, z_j \rangle = 2\delta_{hj}$$

where $\langle z_h, z_j \rangle$ denotes the standard quaternionic inner product in $\mathbb{H}^n$. The equation $\langle z_h, z_h \rangle = 2$ is equivalent to

$$(u_h, u_h) + (v_h, v_h) = 2$$

where (u, v) is the standard Hermitian scalar product in $\mathbb{C}^n$.

On the other hand, $f(z_h, z_h) = 0$ is equivalent to

$$(u_h, u_h) - (v_h, v_h) = 0 \text{ and } \phi(u_h, v_h) = 0$$

---

[6]There is no proof of this normal form in [**2**]. A very short and elementary proof that any two nondegenerate skew-Hermitian forms on $\mathbb{H}^n$ are equivalent can be found in [**6**].

[7]Our notation is similar to the one on p. 435 in [**2**], but one should note that many authors denote this group by $\mathrm{SO}^*(2n)$. The reason for this notation is that $\mathrm{U}_\alpha(n, \mathbb{H})$ is isomorphic to $\mathrm{SO}(2n, \mathbb{C}) \cap \mathrm{SU}(n, n)$. One sees this with help of the second normal form above. First one observes that an automorphism of $\mathbb{C}^{2n}$ that comes from an automorphism of $\mathbb{H}^n$ has real determinant, which must be positive since $\mathrm{GL}(n, \mathbb{H})$ is connected. Such an endomorphism of $\mathbb{C}^{2n}$ preserving the first sum in the normal form belongs to $\mathrm{SU}(n, n)$ and to $\mathrm{SO}(2n, \mathbb{C})$ if it preserves the second sum. Finally one shows, that $\mathrm{SO}(2n, \mathbb{C}) \cap \mathrm{SU}(n, n)$ belongs to the image of $\mathrm{GL}(n, \mathbb{H})$ in $\mathrm{GL}(2n, \mathbb{C})$.

One can find a definition of a *special quaternionic anti-unitary group* $\mathrm{SU}_\alpha(n, \mathbb{H})$ in the literature, but our remarks show that this group actually coincides with $\mathrm{U}_\alpha(n, \mathbb{H})$.

where ϕ is the standard symmetric form $\phi(u,v) = \sum_{k=1}^{n} u_k v_k$ on $\mathbb{C}^n$. Note that $\phi(u,v) = 0$ is equivalent to $(u,\bar{v}) = 0$. Hence we get

$$(u_h, u_h) = (v_h, v_h) = 1 \text{ and } (u_h, \bar{v}_h) = 0.$$

Furthermore, $\langle z_h, z_j \rangle = 0$ für $h \neq j$ is equivalent to

$$(u_h, u_j) + (v_h, v_j) = 0 \text{ and } (u_h, \bar{v}_j) - (v_h, \bar{u}_j) = 0$$

and $f(z_h, z_j) = 0$ is equivalent to

$$(u_h, u_j) - (v_h, v_j) = 0 \text{ and } (u_h, \bar{v}_j) + (v_h, \bar{u}_j) = 0.$$

As a consequence of these considerations, we see that $u_1, \ldots, u_i, \bar{v}_1, \ldots, \bar{v}_i$ is an orthonormal set in $\mathbb{C}^n$. Note that $i \leq \left[\frac{n}{2}\right]$. Conversely, every such orthonormal set in $\mathbb{C}^n$ gives rise to a subspace P' in $N_i(\mathbb{H}^n, f)$. It is now clear that the action of $\mathrm{U}(n)$ is transitive on $N_i(\mathbb{H}^n, f)$ since we are assuming by definition that $A \in \mathrm{U}(n)$ sends $u + jv$ to $Au + j\bar{A}v$.

We now assume that $n = 2k$. We would like to show that $N_k(\mathbb{H}^{2k}, f)$ is a symmetric space. To see this we determine the isotropy group of the action of $\mathrm{U}(2k)$ on $N_k(\mathbb{H}^{2k}, f)$ at the k-plane P spanned by the quaternionic orthogonal basis $z_1 = e_1 + je_{k+1}, \ldots, z_k = e_k + je_{2k}$, where the norms of the elements are all equal to $\sqrt{2}$. Assume that $A \in \mathrm{U}(2k)$ leaves P invariant. Then A sends the orthogonal basis into $Az_1 = Ae_1 + j\bar{A}e_{k+1}, \ldots, Az_k = Ae_k + j\bar{A}e_{2k}$, which is an another quaternionic orthogonal basis of P. This shows that the isotropy group at P is isomorphic to $\mathrm{Sp}(k)$.

We would now like to prove that $(\mathrm{U}(2k), \mathrm{Sp}(k))$ is a symmetric pair. Note that the group $\mathrm{Sp}(k))$ is embedded into $\mathrm{U}(2k)$ as the subgroup of matrices of the form

$$\begin{pmatrix} A & B \\ -\bar{B} & \bar{A} \end{pmatrix}.$$

where A and B are complex $k \times k$-matrices. This subgroup is the fixed point group of the automorphism σ of $\mathrm{U}(2k)$ defined by $\sigma(C) = J_k \bar{C} J_k^{-1}$ where J_k is defined as in 2.3.1. This shows that $(\mathrm{U}(2k), \mathrm{Sp}(k))$ is a symmetric pair and we have proved that $N_k(\mathbb{H}^{2k}, f)$ is a symmetric pair that coincides with

$$\mathrm{U}(2k)/\,\mathrm{Sp}(k),$$

which has a one-dimensional Euclidean factor. It is called the *quaternionic orthogonal Grassmannian.*

It is left to prove that $N_i(\mathbb{H}^n, f)$ is not a symmetric space if $i < \left[\frac{n}{2}\right]$. This we do with arguments analogous to those we used in 2.2.1.

3. Final Remarks

3.1. The classification of compact symmetric spaces. If we go through the seven cases of (σ, ϵ)-Hermitian forms in the proof in Section 2, then we see that the different $N_n(\mathbb{F}^{2n}, f)$ are the symmetric spaces in the following table.

inner product on $\mathbb{R}^{2n}$	$\mathrm{O}(n)$
inner product on $\mathbb{C}^{2n}$	$\mathrm{U}(n)$
inner product on $\mathbb{H}^{2n}$	$\mathrm{Sp}(n)$
symplectic form on $\mathbb{R}^{2n}$	$\mathrm{U}(n)/\mathrm{O}(n)$
symplectic form on $\mathbb{C}^{2n}$	$\mathrm{Sp}(n)/\mathrm{U}(n)$
symmetric form on $\mathbb{C}^{2n}$	$\mathrm{O}(2n)/\mathrm{U}(n)$
skew-Hermitian form on $\mathbb{H}^{2n}$	$\mathrm{U}(2n)/\mathrm{Sp}(n)$

Note that $\mathrm{O}(n)$ and $\mathrm{O}(2n)/\mathrm{U}(n)$ both have two connected components, and that $\mathrm{U}(n)$, $\mathrm{U}(n)/\mathrm{O}(n)$, and $\mathrm{U}(2n)/\mathrm{Sp}(n)$ all have a one-dimensional Euclidean factor. If we consider a connected component or split off a one-dimensional Euclidean factor where needed, the table above leads to the following seven irreducible compact symmetric spaces:

$$\mathrm{SO}(n), \mathrm{SU}(n), \mathrm{Sp}(n), \mathrm{SU}(n)/\mathrm{SO}(n), \mathrm{Sp}(n)/\mathrm{U}(n), \mathrm{SO}(2n)/\mathrm{U}(n), \mathrm{SU}(2n)/\mathrm{Sp}(n).$$

We would now like to compare these spaces with É. Cartan's classification of *classical* symmetric spaces, which can be found in [**4**], pp. 516 and 518. Cartan divides the class of irreducible compact symmetric spaces into type I and type II. The classical spaces in type I consist of the three series of Grassmann manifolds $G_k(\mathbb{F}^n)$ of k-planes in $\mathbb{F}^n$ and the four series

$$\mathrm{SU}(n)/\mathrm{SO}(n), \mathrm{Sp}(n)/\mathrm{U}(n), \mathrm{SO}(2n)/\mathrm{U}(n), \mathrm{SU}(2n)/\mathrm{Sp}(n),$$

which are all in our list. Type II consists of the classical compact groups

$$\mathrm{SO}(n), \mathrm{SU}(n), \mathrm{Sp}(n).$$

(Cartan divides the series $\mathrm{SO}(n)$ into two series according to their root systems, but that is not of importance to us).

Summarizing, the ten series of classical compact symmetric spaces in Cartan's classification are all related to the seven series $N_n(\mathbb{F}^{2n}, f)$ except the three series of Grassmannians $G_k(\mathbb{F}^n)$.

3.2. Classical R-spaces. The *connected components* of the spaces $N_i(\mathbb{F}^n, f)$ for $i \leq r$ that we have been considering are all examples of the R-spaces that were introduced by J. Tits in the 1950's in a study that would later lead to the notion of a spherical building; see [**8**].

Let G be a noncompact connected semisimple Lie group and P a parabolic subgroup. Then the quotient G/P is called an *R-space*. Let K be a maximal compact subgroup of G. Then the action of K on G/P is transitive and we can write $G/P = K/L$. If (K, L) is a symmetric pair, we call $G/P = K/L$ a *symmetric R-space*. It is now common to refer to R-spaces as *(real) generalized flag manifolds*.

The automorphism groups $\mathrm{Aut}(\mathbb{F}^n, f)$ we have been studying so far are in general neither connected nor semisimple, but reductive. If we take the identity components of these reductive groups, one can still define R-spaces and symmetric R-spaces as above, but it would not lead to anything new since these spaces coincide for a reductive group and that same reductive group modulo center. In the two cases $\mathrm{O}(k, n-k)$ and $\mathrm{O}(n, \mathbb{C})$, the automorphism groups are not connected. If we had considered instead the identity components $\mathrm{SO}_0(k, k)$ and $\mathrm{SO}(2k, \mathbb{C})$, we would have been led to one of the two components of $N_k(\mathbb{F}^{2k}, f)$.

It turns out that the classical symmetric R-spaces are precisely the three series of Grassmannians and the connected components of the spaces $N_n(\mathbb{F}^{2n}, f)$. We do not go into further detail since we have discussed these spaces in detail in [**7**].

References

[1] N. Bourbaki, *Éléments de mathématique. Première partie: Les structures fondamentales de l'analyse. Livre II: Algèbre. Chapitre 9: Formes sesquilinéaires et formes quadratiques* (French), Actualités Sci. Ind. no. 1272, Hermann, Paris, 1959. MR0107661

[2] E. Brieskorn, *Lineare Algebra und analytische Geometrie. II.* Friedr. Vieweg & Sohn, Braunschweig, 1985.

[3] C. Chevalley, *Theory of Lie Groups. I*, Princeton Mathematical Series, vol. 8, Princeton University Press, Princeton, N. J., 1946. MR0015396

[4] S. Helgason, *Differential geometry, Lie groups, and symmetric spaces*, Pure and Applied Mathematics, vol. 80, Academic Press, Inc. [Harcourt Brace Jovanovich, Publishers], New York-London, 1978. MR514561

[5] A. W. Knapp, *Lie groups beyond an introduction*, Progress in Mathematics, vol. 140, Birkhäuser Boston, Inc., Boston, MA, 1996. MR1399083

[6] D. W. Lewis, *Congruence of quaternionic skew-Hermitian matrices*, Irish Math. Soc. Bull. **39** (1997), 6–9. MR1492336

[7] G. Thorbergsson, *Classical symmetric R-spaces*, Rend. Semin. Mat. Univ. Politec. Torino **74** (2016), no. 1, 329–354. MR3772594

[8] J. Tits, *Buildings of spherical type and finite BN-pairs*, Lecture Notes in Mathematics, Vol. 386, Springer-Verlag, Berlin-New York, 1974. MR0470099

Mathematisches Institut, Universität zu Köln, Weyertal 86–90, 50931 Köln, Germany

Email address: gthorber@mi.uni-koeln.de

Contemporary Mathematics
Volume **756**, 2020
https://doi.org/10.1090/conm/756/15213

Lagrangian submanifolds of the complex quadric as Gauss maps of hypersurfaces of spheres

Joeri Van der Veken and Anne Wijffels

ABSTRACT. The Gauss map of a hypersurface of a unit sphere $S^{n+1}(1)$ is a Lagrangian immersion into the complex quadric Q^n and, conversely, every Lagrangian submanifold of Q^n is locally the image under the Gauss map of several hypersurfaces of $S^{n+1}(1)$. In this paper, we give explicit constructions for these correspondences and we prove a relation between the principal curvatures of a hypersurface of $S^{n+1}(1)$ and the local angle functions of the corresponding Lagrangian submanifold of Q^n. The existence of such a relation is remarkable since the definition of the angle functions depends on the choice of an almost product structure on Q^n and since several hypersurfaces of $S^{n+1}(1)$, with different principal curvatures, correspond to the same Lagrangian submanifold of Q^n.

1. The geometry of the complex quadric

Let $\mathbb{C}P^{n+1}(4)$ be the complex projective space of complex dimension $n+1$ equipped with the Fubini-Study metric g_{FS} of constant holomorphic sectional curvature 4. Then the Hopf fibration $\pi : S^{2n+3}(1) \subseteq \mathbb{C}^{n+2} \to \mathbb{C}P^{n+1}(4) : z \mapsto [z]$ is a Riemannian submersion from the unit sphere of real dimension $2n+3$ to $\mathbb{C}P^{n+1}(4)$. Remark that for any $z \in S^{2n+3}(1)$ we have $\pi^{-1}\{[z]\} = \{e^{it}z \mid t \in \mathbb{R}\}$ and $\ker(d\pi)_z = \operatorname{span}\{iz\}$. The complex structure J on $\mathbb{C}P^{n+1}(4)$ is induced from multiplication by i on $TS^{2n+3}(1)$ and it is well-known that $(\mathbb{C}P^{n+1}(4), g_{FS}, J)$ is a Kähler manifold.

We define the *complex quadric* of complex dimension n as the following complex hypersurface of $\mathbb{C}P^{n+1}(4)$:

$$Q^n = \{[(z_0, \ldots, z_{n+1})] \in \mathbb{C}P^{n+1}(4) \mid z_0^2 + \ldots + z_{n+1}^2 = 0\}.$$

If Q^n is equipped with the induced metric $g_{FS}|_{Q^n}$, which we will denote by g, and the induced almost complex structure $J|_{Q^n}$, which we will again denote by J, then (Q^n, g, J) is of course a Kähler manifold itself. The inverse image of Q^n under the Hopf fibration is the $(2n+1)$-dimensional Stiefel manifold

$$V^{2n+1} = \left\{ u + iv \;\middle|\; u, v \in \mathbb{R}^{n+2},\ \langle u, u\rangle = \langle v, v\rangle = \frac{1}{2},\ \langle u, v\rangle = 0 \right\} \subseteq S^{2n+3}(1),$$

2010 *Mathematics Subject Classification.* Primary 53C42; Secondary 53D12, 53B25.

The first author is supported by the Excellence Of Science project G0H4518N of the Belgian government and both authors are supported by project 3E160361 of the KU Leuven Research Fund.

where $\langle \cdot, \cdot \rangle$ denotes the Euclidean inner product on $\mathbb{R}^{n+2}$. From this perspective, it is easy to see that Q^n can be identified with the Grassmannian of oriented 2-planes in $\mathbb{R}^{n+2}$ and hence, as a homogeneous space, is

$$Q^n = \frac{\mathrm{SO}(n+2)}{\mathrm{SO}(n) \times \mathrm{SO}(2)}.$$

Denote by $\mathcal{A}$ the set of all shape operators of Q^n in $\mathbb{C}P^{n+1}(4)$ associated with unit normal vector fields. Since we need it in the next sections, we allow for elements of $\mathcal{A}$ to be defined only on a subset of Q^n. One can deduce the following (see for example [**8**] or [**10**]).

LEMMA 1.1. *Any $A \in \mathcal{A}$ is involutive, symmetric and anti-commutes with J.*

This implies in particular that $\mathcal{A}$ is a family of almost product structures. However, these almost product structures are not integrable. In fact, we have the following equalities, which can be found in [**10**].

LEMMA 1.2. *Let ζ be a unit normal vector field along Q^n in $\mathbb{C}P^{n+1}(4)$ with corresponding shape operator A. Then there exists a non-zero one-form s such that $\nabla_X^{\mathbb{C}P^{n+1}(4)}\zeta = -AX + s(X)J\zeta$ and $\nabla_X^{Q^n} A = s(X)JA$ for all X tangent to Q^n, where $\nabla^{\mathbb{C}P^{n+1}(4)}$ and ∇^{Q^n} are the Levi Civita connections of $\mathbb{C}P^{n+1}(4)$ and (Q^n, g) respectively.*

The equation of Gauss for Q^n as a submanifold of $\mathbb{C}P^{n+1}(4)$ yields the following expression for the Riemann-Christoffel curvature tensor of Q^n:

$$\begin{aligned} R^{Q^n}(X,Y)Z &= g(Y,Z)X - g(X,Z)Y \\ &+ g(JY,Z)JX - g(JX,Z)JY - 2g(JX,Y)JZ \\ &+ g(AY,Z)AX - g(AX,Z)AY + g(JAY,Z)JAX - g(JAX,Z)JAY, \end{aligned} \tag{1.1}$$

where A is any element of $\mathcal{A}$. It follows directly from (1.1) that Q^n is Einstein.

REMARK 1.3. Although the almost product structures in $\mathcal{A}$ are non-integrable, the complex quadric of complex dimension 2 is in fact a Riemannian product. Indeed, it was proven in [**1**] that a homogeneous Einstein manifold of real dimension 4 must have either constant sectional curvature or constant holomorphic sectional curvature, or must be a Riemannian product of two surfaces of equal constant Gaussian curvature $S^2(c) \times S^2(c)$ or $H^2(c) \times H^2(c)$. It follows from (1.1) that Q^2 does not have constant (holomorphic) sectional curvature and from computing the maximal sectional curvature, we see that $Q^2 = S^2(4) \times S^2(4)$.

2. Lagrangian submanifolds of the complex quadric

An isometric immersion $f : M^n \to Q^n$ of a manifold of real dimension n into Q^n is said to be *Lagrangian* if J maps the tangent space to M^n at any point into the normal space to M^n at that point and vice versa. If $f : M^n \to Q^n$ is such a Lagrangian submanifold and $A \in \mathcal{A}$ is defined at least along $f(M^n)$, it was proven in [**2**] that, in a neighborhood of any point of M^n, there exist an orthonormal frame $\{e_1, \ldots, e_n\}$ on M^n and local angle functions $\theta_1, \ldots, \theta_n$ such that

$$A(df)e_j = \cos(2\theta_j)(df)e_j - \sin(2\theta_j)J(df)e_j \tag{2.1}$$

for all $j = 1, \ldots, n$. Clearly, the angle functions are only defined up to addition with an integer multiple of π and they depend on the choice of A.

REMARK 2.1 (Choice of A along a Lagrangian submanifold of Q^n). Assume that, apart from a Lagrangian immersion $f : M^n \to Q^n$, also a horizontal lift $\hat{f} : M^n \to V^{2n+1}$ of f is given. It follows from [**7**] that any Lagrangian immersion into Q^n locally allows such a horizontal lift. If M^n is simply connected, the horizontal lift can be defined globally. Since the normal space to V^{2n+1} in $S^{2n+3}(1) \subseteq \mathbb{C}^{n+2}$ at a point z is the complex span of $\bar{z}$, one can take ζ, defined by $\zeta_{f(p)} = (d\pi)_{\hat{f}(p)}\left(\overline{\hat{f}(p)}\right)$, as a unit normal vector field to Q^n in $\mathbb{C}P^{n+1}(4)$ along the image of f and the corresponding shape operator is given by $AX = -(d\pi)_{\hat{f}(p)}\left(\overline{\hat{X}}\right)$, where X is any vector tangent to Q^n at a point $f(p)$ and $\hat{X}$ is its horizontal lift to $\hat{f}(p)$. In the special case that v is tangent to M^n at a point p, we have

$$A(df)_p(v) = -(d\pi)_{\hat{f}(p)}\left(\overline{d\hat{f}}\right)_p(v). \tag{2.2}$$

This A can be extended to an element of $\mathcal{A}$, defined in a neighborhood of $f(M^n)$.

3. Lagrangian submanifolds of the complex quadric as Gauss maps

Several possible definitions for the Gauss map of a hypersurface of a round sphere can be found in the literature. We consider here a definition which was studied in [**6**]. Let $a : M^n \to S^{n+1}(1)$ be a hypersurface of a unit sphere and denote by b a unit normal to the hypersurface, tangent to the sphere. Then the Gauss map of a is the following map from M^n to the complex quadric Q^n:

$$G : M^n \to Q^n : p \mapsto [a(p) + ib(p)].$$

Looking at $a(p)$ and $b(p)$ as vectors in $\mathbb{R}^{n+2}$, one has $(a(p) + ib(p))/\sqrt{2} \in V^{2n+1}$, such that $[a(p) + ib(p)] = [(a(p) + ib(p))/\sqrt{2}]$ is indeed an element of Q^n.

An interesting property that this Gauss map shares with the classical Gauss map of a hypersurface of a Euclidean space is that *parallel hypersurfaces have the same Gauss maps.* Indeed, a parallel hypersurface to a given hypersurface a is obtained by, starting at any point $a(p)$ of the hypersurface, traveling over a distance t along a geodesic of the ambient space with the unit normal $b(p)$ as initial velocity. If the ambient space is $S^{n+1}(1)$, it is easy to see that any parallel hypersurface to $a : M^n \to S^{n+1}(1)$ is given by $a_t : M^n \to S^{n+1}(1) : p \mapsto (\cos t)a(p) + (\sin t)b(p)$ for some $t \in \mathbb{R}$. If $|t|$ is small enough, a_t will, at least locally, be an immersion. A straightforward computation shows that $b_t = -(\sin t)a + (\cos t)b$ is a unit normal to a_t such that b_0 equals the original b. We conclude that the Gauss map of a_t is given by $G_t = [a_t + ib_t] = [e^{-it}(a + ib)] = [a + ib] = G$.

This Gauss map has attracted quite some attention in recent years, especially in the case of isoparametric hypersurfaces of spheres, i.e., hypersurfaces for which all principal curvatures are constant. We mention for example the works [**3**], [**4**] and [**5**]. In [**9**], a study of this Gauss map was proposed as a structural approach to gain a better understanding of the notorious family of isoparametric hypersurfaces of spheres. In [**2**] a correspondence between the principal curvatures of an isoparametric hypersurface of a sphere and the angle functions of its Gauss map in the sense of (2.1) was given. Indeed, it turns out that the Gauss map of a hypersurface of a sphere is a Lagrangian immersion into Q^n. The following theorem includes this statement and, more importantly, generalizes the result from [**2**] to arbitrary hypersurfaces of spheres. As mentioned in the abstract, this result is remarkable since both the principal curvatures and the angle functions depend on certain choices.

THEOREM 3.1. *Let $a : M^n \to S^{n+1}(1)$ be a hypersurface with unit normal b. Then the Gauss map $G : M^n \to Q^n : p \mapsto [a(p)+ib(p)]$ is a Lagrangian immersion. Moreover, if A is chosen as in Remark 2.1 using the canonical horizontal lift*

$$\hat{G} : M^n \to V^{2n+1} : p \mapsto \frac{1}{\sqrt{2}}(a(p) + ib(p)), \tag{3.1}$$

then the relation between the principal curvatures $\lambda_1, \ldots, \lambda_n$ of a, with respect to the shape operator associated to b, and the angle functions $\theta_1, \ldots, \theta_n$ of G is

$$\lambda_j = \cot \theta_j \tag{3.2}$$

for $j = 1, \ldots, n$.

Conversely, if $f : M^n \to Q^n$ is a Lagrangian immersion, then for every point of M^n there exist an open neighborhood U of that point in M^n and an immersion $a : U \to S^{n+1}(1)$ with Gauss map $f|_U$. This immersion is not unique, nor are its principal curvature functions. However, for any choice of a, a local frame of principal directions for a is adapted to f in the sense that (2.1) holds for any choice of A and the principal curvature functions $\lambda_1, \ldots, \lambda_n$ of a are related to the corresponding local angle functions $\theta_1, \ldots, \theta_n$ by

$$\cot(\theta_j - \theta_k) = \pm\frac{\lambda_j\lambda_k + 1}{\lambda_j - \lambda_k} \tag{3.3}$$

for $j, k = 1, \ldots, n$ in points where $\lambda_j \neq \lambda_k$.

PROOF. Let $\{e_1, \ldots, e_n\}$ be a local orthonormal frame of principal directions on M^n for the immersion $a : M^n \to S^{2n+1}(1)$, say $Se_j = \lambda_j e_j$, where S is the shape operator associated to b. It follows directly from (3.1) that

$$(d\hat{G})e_j = \frac{1}{\sqrt{2}}(1 - i\lambda_j)e_j. \tag{3.4}$$

These vector fields are orthogonal to $i\hat{G}$ and hence $\hat{G}$ is indeed horizontal. Moreover, G is Lagrangian since $(d\hat{G})e_j$ and $i(d\hat{G})e_k$ are orthogonal for all $j, k = 1, \ldots, n$.

For the choice of A given in Remark 2.1, we have, using (2.2) and (3.4),

$$\begin{aligned} A(dG)e_j &= -(d\pi)\left(\overline{d\hat{G}}\right)e_j = -(d\pi)\left(\overline{(d\hat{G})e_j}\right) = -(d\pi)\left(\frac{1}{\sqrt{2}}(1 + i\lambda_j)e_j\right) \\ &= -(d\pi)\left(\frac{1-\lambda_j^2}{1+\lambda_j^2}(d\hat{G})e_j + \frac{2\lambda_j}{1+\lambda_j^2}i(d\hat{G})e_j\right) = \frac{\lambda_j^2-1}{\lambda_j^2+1}(dG)e_j - \frac{2\lambda_j}{\lambda_j^2+1}J(dG)e_j. \end{aligned}$$

Comparing this to (2.1) implies that the angle functions associated to A are determined by

$$\cos(2\theta_j) = \frac{\lambda_j^2 - 1}{\lambda_j^2 + 1}, \qquad \sin(2\theta_j) = \frac{2\lambda_j}{\lambda_j^2 + 1}$$

and hence $\lambda_j = \cot \theta_j$.

Conversely, let $f : M^n \to Q^n$ be a Lagrangian immersion and fix a point $p_0 \in M^n$. If, for some open neighborhood U of p_0 in M^n, the restriction $f|_U$ is the Gauss map of a hypersurface $a : U \to S^{n+1(1)}$ with unit normal b, the first part of the proof implies that $U \to V^{2n+1} : p \mapsto (a(p) + ib(p))/\sqrt{2}$ must be a horizontal lift of $f|_U$. Finding all the hypersurfaces of $S^{n+1}(1)$ of which f is locally the Gauss map is hence equivalent to finding the local horizontal lifts of f for which the real part is an immersion.

It follows from [7] that for every simply connected open neighborhood U of p_0 in M^n there exists a horizontal lift $\hat{f}_0 : U \to V^{2n+1}$ of f. Moreover, any other horizontal lift of f on U can be written as $\hat{f}_t = e^{it}\hat{f}_0$ for some constant $t \in \mathbb{R}$. Remark that if we split $\hat{f}_t$ in a real and imaginary part as $\hat{f}_t = (a_t + ib_t)/\sqrt{2}$, then $a_t = (\cos t)a_0 - (\sin t)b_0$ and $b_t = (\sin t)a_0 + (\cos t)b_0$ for all t. If a_0 and a_t are immersions, they hence define parallel hypersurfaces of $S^{n+1}(1)$.

In order to investigate the derivative of a_t, we remark that

$$a_t = \sqrt{2}\,\mathrm{Re}\,\hat{f}_t = \frac{1}{\sqrt{2}}\left(\hat{f}_t + \overline{\hat{f}_t}\right), \quad b_t = \sqrt{2}\,\mathrm{Im}\,\hat{f}_t = -\frac{i}{\sqrt{2}}\left(\hat{f}_t - \overline{\hat{f}_t}\right). \tag{3.5}$$

Now choose A_t as in Remark 2.1 using the horizontal lift $\hat{f}_t$. If $\{e_1^{(t)}, \dots, e_n^{(t)}\}$ is a local orthonormal frame adapted to f in the sense of (2.1), then there are local functions $\theta_1^{(t)}, \dots, \theta_n^{(t)}$ such that $A_t(df)e_j^{(t)} = \cos(2\theta_j^{(t)})(df)e_j^{(t)} - \sin(2\theta_j^{(t)})J(df)e_j^{(t)}$ for $j = 1, \dots, n$. Taking the horizontal lift to the image of $\hat{f}_t$ on both sides of this equality yields

$$-\left(\overline{d\hat{f}_t}\right)e_j^{(t)} = \cos(2\theta_j^{(t)})(d\hat{f}_t)e_j^{(t)} - i\sin(2\theta_j^{(t)})(d\hat{f}_t)e_j^{(t)} = e^{-2i\theta_j^{(t)}}(d\hat{f}_t)e_j^{(t)} \tag{3.6}$$

for $j = 1, \dots, n$. Substituting $\hat{f}_t = e^{it}\hat{f}_0$ into (3.6) and then applying $d\pi$ gives $A_0(df)e_j^{(t)} = \cos(2(\theta_j^{(t)} - t))(df)e_j^{(t)} - \sin(2(\theta_j^{(t)} - t))J(df)e_j^{(t)}$, where A_0 is chosen as in Remark 2.1 using the horizontal lift $\hat{f}_0$. This implies that the frame $\{e_1^{(t)}, \dots, e_n^{(t)}\}$ does not depend on t –we will denote it by $\{e_1, \dots, e_n\}$ from now on– and that the corresponding angle functions of A_t and A_0 are related by

$$\theta_j^{(t)} = \theta_j^{(0)} + t. \tag{3.7}$$

From (3.5), (3.6) and (3.7) we obtain

$$(da_t)e_j = \frac{1}{\sqrt{2}}\left(1 - e^{-2i(\theta_j^{(0)}+t)}\right)(d\hat{f}_t)e_j \tag{3.8}$$

for $j = 1, \dots, n$. If we choose $t \in \mathbb{R}$ such that $\theta_j^{(0)}(p_0) + t$ is not an integer multiple of π for $j = 1, \dots, n$ and, if necessary, we shrink U to U_t such that none of the functions $\theta_j^{(0)} + t$ attains an integer multiple of π on U_t, then $a_t|_{U_t}$ is an immersion. There are hence infinitely many choices of $t \in \mathbb{R}$ for which a_t is an immersion in a neighborhood of p_0.

Now choose any $t \in \mathbb{R}$ such that $a_t : U_t \subseteq M^n \to S^{n+1}(1)$ is an immersion. In order to find the principal curvatures of a_t, we compute the derivative of the corresponding b_t. From (3.5), (3.6), (3.7) and (3.8) we find

$$(db_t)e_j = -\frac{i}{\sqrt{2}}\left(1 + e^{-2i(\theta_j^{(0)}+t)}\right)(d\hat{f}_t)e_j = -\cot(\theta_j^{(0)} + t)(da_t)e_j \tag{3.9}$$

This implies that $\{e_1, \dots, e_n\}$ is a local frame of principal directions for the hypersurface a_t and that the principal curvatures of a_t defined using the shape operator associated to b_t are given by $\lambda_j^{(t)} = \cot(\theta_j^{(0)} + t)$ for $j = 1, \dots, n$. A first issue is that the principal curvatures are only defined up to sign: if we change the orientation of the unit normal, the signs of the principal curvatures change. A second issue is that the local angle functions $\theta_j^{(0)}$ are only defined through the choice of the almost product structure A_0. If one chooses an $A \in \mathcal{A}$, which is at least defined along $f(U_t)$, then there exists a function $\varphi : U_t \to \mathbb{R}$ such that $A = \cos\varphi A_0 + \sin\varphi JA_0$ along $f(U_t)$ and it was shown in [2] that the local angle functions associated to A

are given by $\theta_j = \theta_j^{(0)} - \varphi/2$ for $j = 1, \ldots, n$. This implies that the difference of two local angle functions does not depend on the choice of A. Hence, using the formula for the cotangent of a difference, we can state that

$$\cot(\theta_j - \theta_k) = \cot((\theta_j^{(0)} + t) - (\theta_k^{(0)} + t)) = \frac{(\pm\lambda_j^{(t)})(\pm\lambda_k^{(t)}) + 1}{(\pm\lambda_j^{(t)}) - (\pm\lambda_k^{(t)})} = \pm\frac{\lambda_j^{(t)}\lambda_k^{(t)} + 1}{\lambda_j^{(t)} - \lambda_k^{(t)}}$$

for all $j, k = 1, \ldots, n$ in those points where $\lambda_j^{(t)} \neq \lambda_k^{(t)}$. In particular, the right hand side does not depend on t. In other words: it does not depend on the chosen horizontal lift of f, as long as the real part of this lift is an immersion, or, equivalently, it remains invariant when changing from a hypersurface of a sphere to a parallel hypersurface. This last fact can also be checked directly. □

References

[1] G. R. Jensen, *Homogeneous Einstein spaces of dimension four*, J. Differential Geometry **3** (1969), 309–349. MR261487

[2] H. Li, H. Ma, J. Van der Veken, L. Vrancken, and X. Wang, *Minimal Lagrangian submanifolds of the complex hyperquadric*, Sci. China Math. **63** (2020), no. 8, 1441–1462, DOI 10.1007/s11425-019-9551-2. MR4125728

[3] H. Ma and Y. Ohnita, *On Lagrangian submanifolds in complex hyperquadrics and isoparametric hypersurfaces in spheres*, Math. Z. **261** (2009), no. 4, 749–785, DOI 10.1007/s00209-008-0350-5. MR2480757

[4] H. Ma and Y. Ohnita, *Hamiltonian stability of the Gauss images of homogeneous isoparametric hypersurfaces. I*, J. Differential Geom. **97** (2014), no. 2, 275–348. MR3263508

[5] H. Ma and Y. Ohnita, *Hamiltonian stability of the Gauss images of homogeneous isoparametric hypersurfaces II*, Tohoku Math. J. (2) **67** (2015), no. 2, 195–246, DOI 10.2748/tmj/1435237041. MR3365370

[6] B. Palmer, *Hamiltonian minimality and Hamiltonian stability of Gauss maps*, Differential Geom. Appl. **7** (1997), no. 1, 51–58, DOI 10.1016/S0926-2245(96)00035-6. MR1441918

[7] H. Reckziegel, *Horizontal lifts of isometric immersions into the bundle space of a pseudo-Riemannian submersion*, Global differential geometry and global analysis 1984 (Berlin, 1984), Lecture Notes in Math., vol. 1156, Springer, Berlin, 1985, pp. 264–279, DOI 10.1007/BFb0075098. MR824074

[8] H. Reckziegel, *On the geometry of the complex quadric*, Geometry and topology of submanifolds, VIII (Brussels, 1995/Nordfjordeid, 1995), World Sci. Publ., River Edge, NJ, 1996, pp. 302–315. MR1434581

[9] A. Siffert, *A new structural approach to isoparametric hypersurfaces in spheres*, Ann. Global Anal. Geom. **52** (2017), no. 4, 425–456, DOI 10.1007/s10455-017-9563-3. MR3735906

[10] B. Smyth, *Differential geometry of complex hypersurfaces*, Ann. of Math. (2) **85** (1967), 246–266, DOI 10.2307/1970441. MR206881

KU Leuven, Department of Mathematics, Celestijnenlaan 200B - Box 2400, 3001 Leuven, Belgium

Email address: joeri.vanderveken@kuleuven.be

KU Leuven, Department of Mathematics, Celestijnenlaan 200B - Box 2400, 3001 Leuven, Belgium

Email address: anne.wijffels@kuleuven.be

Contemporary Mathematics
Volume **756**, 2020
https://doi.org/10.1090/conm/756/15215

Growth estimates for generalized harmonic forms on noncompact manifolds with geometric applications

Shihshu Walter Wei

ABSTRACT. We introduce Condition W (1.2) for a smooth differential form ω on a complete noncompact Riemannian manifold M. We prove that ω is a harmonic form on M if and only if ω is both closed and co-closed on M, where ω has 2-balanced growth either for $q = 2$, or for $1 < q(\neq 2) < 3$ with ω satisfying Condition W (1.2). In particular, every L^2 harmonic form, or every L^q harmonic form, $1 < q(\neq 2) < 3$ satisfying Condition W (1.2) is both closed and co-closed (cf. Theorem 1.1). This generalizes the work of A. Andreotti and E. Vesentini [**AV**] for every L^2 harmonic form ω. In extending ω in L^2 to L^q, for $q \neq 2$, Condition W (1.2) has to be imposed due to counter-examples of D. Alexandru Rugina([**AR**] p. 81, Remarque 3). We then study nonlinear partial differential inequalities for differential forms $\langle \omega, \Delta\omega \rangle \geq 0$, in which solutions ω can be viewed as generalized harmonic forms. We prove that under the same growth assumption on ω (as in Theorem 1.1, 1.2, or 1.3, the following six statements: (i) $\langle \omega, \Delta\omega \rangle \geq 0$, (ii) $\Delta\omega = 0$, (iii) $d\omega = d^\star\omega = 0$, (iv) $\langle \star\omega, \Delta \star \omega \rangle \geq 0$, (v) $\Delta \star \omega = 0$, and (vi) $d \star \omega = d^\star \star \omega = 0$ are equivalent (cf. Theorem 4.1). We also show the nonexistence of eigenform associated with positive eigenvalue for $\Delta\omega = \lambda\omega$, and the nonexistence of solution of $\langle \omega, \Delta\omega \rangle > 0$, under the same growth assumption on ω. As a geometric application, we observe that a both closed and co-closed form on M satisfies a conservation law (8.5) and can apply the theory we developed in [**DW**] and [**W3**]. Based on comparison theorem in Riemannian geometry under a curvature assumption, a conservation law, variational method and coarea formula, we solve constant Dirichelet problems for generalized harmonic 1-forms and harmonic maps on starlike domains in M (cf. Theorem 10.3 and Corollary 10.1, generalizing and extending the work of Karcher and Wood for harmonic maps on disc domains in $\mathbb{R}^n$ [**KW**], derive monotonicity formulas for 2-balanced solutions, and vanishing theorems for 2-moderate solutions of $\langle \omega, \Delta\omega \rangle \geq 0$ on M (cf. Theorem 8.2 and Theorem 9.3).

1. Introduction

The study of harmonic forms, or more generally p-harmonic geometry is an active research area that interacts with or enters into many branches of mathematics. Harmonic forms generalize harmonic functions in the study of partial differential equations and potential theory, and are privileged representatives in a de Rham

2000 *Mathematics Subject Classification.* Primary: 53C40.
Key words and phrases. Harmonic forms, 2-balanced growth.
Research was partially supported by NSF Award No DMS-0508661, OU Presidential International Travel Fellowship, and OU Faculty Enrichment Grant.

cohomology class picked out by the Hodge Laplacian. It is well-known that on a compact Riemannian manifold, a smooth differential form ω is harmonic if and only if it is closed and co-closed. That is,

$$\Delta\omega = 0 \quad \text{if and only if} \quad d\omega = 0 \quad \text{and} \quad d^\star\omega = 0. \tag{1.1}$$

holds, where d denotes the exterior differential operator, $d^\star$ is the codifferential operator, and $\Delta := -(dd^* + d^*d)$ denotes the Hodge Laplacian.

On *complete noncompact* Riemannian manifolds, although (1.1) holds for smooth ω with compact support, it does not hold in general. Simple examples include, in $\mathbb{R}^n$, a closed, non-co-closed, harmonic form $\omega_1 = x_1 dx_1$, a non-closed, co-closed, harmonic form $\omega_2 = x_n dx_1$, and a non-closed, non-co-closed, harmonic form $\omega_3 = (x_1 + x_n)dx_1$, or $\omega_4 = x_1 x_n dx_1 + x_n dx_n$. However, it is proved in [**AV**]

Theorem A (A. Andreotti and E. Vesentini [**AV**]). *On a complete noncompact Riemannian manifold M,* (1.1) *holds for every smooth L^2 differential form ω.*

It is interesting to explore any possible generalizations of Theorem A, in particular, to discuss whether or not (1.1) holds for L^q differential form ω, where $q \neq 2$. Some study of this generalization can be found in [[**Y1**], p.663, Proposition 1], its counter-examples are given by D. Alexandru Rugina on the hyperbolic space $H^m_{-1}, m \geq 3$, (cf. [[**AR**], p. 81, Remarque 3]) and relevant remarks of S. Pigola, A.G. Rigoli, and M. Setti are discussed in [[**PRS**], p.260, Remark B.8]"

In the first part of this paper, we introduce and add Condition W (1.2) to the above study in extending ω in L^2 to ω in $L^q, q \neq 2$ so that the counter-examples cannot prevail, the conclusion of the Proposition still holds, and works in a more general setting with geometric and analytic applications. From now on, unless specified otherwise, we let M be an n-dimensional complete noncompact Riemannian manifold, A^k be the space of smooth differential k-forms on M, and ω be a smooth differential k-form, $k \geq 0$ on M. Denote by $\langle \cdot, \cdot \rangle$, $|\cdot|$, and $\star : A^k \to A^{n-k}$, the *inner product*, the *norm* induced in fibers of various tensor bundles by the metric of M, and the *linear operator* which assigns to each k-form on M an $(n-k)$-form and which satisfies $\star\star = (-1)^{nk+k+1}$ respectively.

Definition 1.1. A differential form ω on M is said to satisfy **Condition W** if

$$\begin{aligned} &|\langle d(|\omega|^2) \wedge \omega, d\omega\rangle| \leq 2|\omega|^2 |d\omega|^2 \\ &\big|\langle d(|\star\omega|^2) \wedge \star\omega, d\star\omega\rangle\big| \leq 2|\star\omega|^2 |d\star\omega|^2. \end{aligned} \tag{1.2}$$

Simple examples of differential forms that satisfy Condition W (1.2) include smooth 0-forms (or smooth functions) and differential n-forms on M. In general, there is a *hidden* duality involved. Namely, $\omega \in A^k$ satisfies Condition W (1.2) if and only if $\star\omega \in A^{n-k}$ satisfies Condition W (1.2) (cf. Proposition 2.5).

We then extend the differential form in L^2 space in Theorem A in several ways. To this end, recall in extending functions in L^2 space (resp. in L^q space), we introduce and study the notion of function growth: "*p-finite, p-mild, p-obtuse, p-moderate,* and *p-small*" growth $\big($ for $q = 2$ (resp. for the same value of q)$\big)$, and their counter-parts "*p-infinite, p-severe, p-acute, p-immoderate,* and *p-large*" growth [**WLW**] (cf. Definition 2.2). We then introduce the notion of *p-balanced* and *p-imbalanced* growth for functions and differential forms on complete noncompact

Riemannian manifold M in [**W1**]. Namely, a function or a differential form f has *p-balanced growth* (or, simply, *is p-balanced*) if f has one of the following: *p-finite, p-mild, p-obtuse, p-moderate*, or *p-small* growth on M, and has *p-imbalanced growth* (or, simply, is *p-imbalanced*) otherwise (cf. Definition 2.3).

Furthermore, extending our techniques for function growth (cf. [**CW1**, **CW2**]) to differential form growth, we use direct simple new methods (cf. Proofs of Theorems 3.4 and 3.5) and obtain

THEOREM 1.1. *If a smooth differential k-form ω on M has* 2*-balanced growth for*

$$\begin{cases} q = 2 & \text{or} \\ 1 < q(\neq 2) < 3 & \textit{with } \omega \textit{ satisfying } \text{Condition W (1.2)}, \end{cases} \tag{1.3}$$

then (1.1) *holds.*

When ω is in L^2, we recapture Theorem A of Andreotti and Vesentini, as an L^2 form has 2*-finite*, 2*-mild*, 2*-obtuse*, 2*-moderate*, and 2*-small* growth for $q = 2$, by Proposition 2.2, and hence has 2*-balanced growth* for $q = 2$, by Definition 2.3. Theorem 1.1 also extends ω in L^2 to ω in $L^q, 1 < q(\neq 2) < 3$, and to 2-balanced growth for $1 < q(\neq 2) < 3$. But this extension, in sharp contrast to the 2-balanced growth for $q = 2$ extension, requires an additional assumption that ω satisfies Condition W (1.2). Or there would be counter-examples ([**AR**] p. 81, Remarque 3 in reference to [**Y1**]). For clarity, it will be discussed in Remarks 3.2 and 3.3, that why, where, and how Condition W (1.2) is used for the case $q \neq 2$, and is not needed for the case $q = 2$ in the unified proof of Theorems 1.1 and 1.2. Furthermore, examples and counter-examples of differential forms that satisfy Condition W (1.2) are also provided (cf. Propositions 2.3 and 2.4, and Remark 2.1). As an immediate consequence of Theorem 1.1, we have

COROLLARY 1.1. (*i*) ([**AV**]) *Let ω be an L^2 differential k-form on M. Then* (1.1) *holds.* (ii) *Let ω be an $L^q, 1 < q(\neq 2) < 3$ differential k-form on M satisfying* Condition W (1.2). *Then* (1.1) *holds.*
(iii) *Let ω be an $L^q, 1 < q < 3$ differential* 0*-form on M. Then ω is a harmonic function if and only if ω is constant on M.*

In view of Propositions 2.3 and 2.4, we have

COROLLARY 1.2. *Let ω be a simple k-form or a differential k-form satisfying Kato's type inequality* (2.7) *on M. If ω has* 2*-balanced growth for $1 < q < 3$, then ω is harmonic if and only if it is both closed and co-closed on M. In particular, for every L^q, $1 < q < 3$ simple k-form or a differential k-form ω satisfying* (2.7) *on M,* (1.1) *holds.*

The above Kato's type inequality (2.7) in Corollary 1.2 has to be assumed, as Kato's type inequality (2.7) holds for smooth functions, but does not hold for differential forms in general, cf. Remark 2.2 for a counter-example of Kato's type inequality.

In the second part of this paper, we study nonlinear partial differential inequalities for differential forms

$$\langle \omega, \Delta\omega \rangle \geq 0 \tag{1.4}$$

and explore related eigenvalues and eigenforms. Recall given a differential operator Δ on the space of differential forms, an eigenform is a differential form ω such that

$$\Delta\omega = \lambda\omega \tag{1.5}$$

for some constant real number λ.

Apparently, solutions of (1.4) can be viewed as generalized harmonic forms as they include harmonic functions, harmonic forms, eigenfunctions and eigenforms with positive eigenvalues, nonnegative subharmonic functions, nonpositive superharmonic functions, etc. The following results manifest their interconnectedness, and our technique is sufficient general to provide a unified proof of Theorems 1.1 and 1.2.

THEOREM 1.2. *Suppose $\omega \in A^k$ has 2-balanced growth, for $q = 2$, or for $1 < q(\neq 2) < 3$ with ω satisfying* Condition W (1.2) (cf. (1.3)). *Then ω is a solution of $\langle \omega, \Delta\omega\rangle \geq 0$ on M if and only if ω is closed and co-closed. Or equivalently,*

$$\begin{aligned} \langle \omega, \Delta\omega\rangle \geq 0 \qquad &\text{if and only} \qquad \Delta\omega = 0 \\ &\text{if and only if} \qquad d\omega = d^\star\omega = 0\,. \end{aligned} \tag{1.6}$$

COROLLARY 1.3. i) *Let ω be an L^2 differential k-form on M. Then* (1.6) *holds.*

ii) *Let ω be an $L^q, 1 < q(\neq 2) < 3$ differential k-form on M satisfying* Condition W (1.2). *Then* (1.6) *holds.*

iii) *Let ω be an $L^q, 1 < q < 3$ differential 0-form on M. Then $\omega\Delta\omega \geq 0$(e.g. ω is a nonnegative subharmonic function) if and only if ω is constant on M.*

Analogously,

COROLLARY 1.4. *Let ω be a simple k-form or a differential k-form satisfying Kato's type inequality* (2.7) *on M. If ω has 2-balanced growth for $1 < q < 3$, then* (1.6) *holds. In particular, for every L^q, $1 < q < 3$ simple k-form or a differential k-form ω satisfying* (2.7) *on M,* (1.6) *holds.*

The duality in Condition W (1.2) leads to Duality Theorem 1.3 for Theorem 1.2.

THEOREM 1.3 (Duality Theorem). *If $\omega \in A^k$ has 2-balanced growth on M, for $q = 2$, or for $1 < q(\neq 2) < 3$ with ω satisfying* Condition W (1.2) (cf. (1.3)), *then so is $\star\omega \in A^{n-k}$, and $\star\omega$ is a solution of $\langle \star\omega, \Delta\star\omega\rangle \geq 0$ on M if and only if $\star\omega$ is closed and co-closed. Or equivalently,*

$$\begin{aligned} \langle \star\omega, \Delta\star\omega\rangle \geq 0 \qquad &\text{if and only if} \qquad \Delta\star\omega = 0 \\ &\text{if and only if} \qquad d\star\omega = d^\star\star\omega = 0\,. \end{aligned} \tag{1.7}$$

Striking the delicate balance between Dual Theorems 1.2 and 1.3, we obtain **Theorem 4.1.** (Unity Theorem) *Under the same growth assumption on ω (as in Theorem 1.1, or 1.2, or 1.3), the following six statements are equivalent.*

(i) $\langle \omega, \Delta\omega\rangle \geq 0$.
(ii) $\Delta\omega = 0$.
(iii) $d\omega = d^\star\omega = 0$.
(iv) $\langle \star\omega, \Delta\star\omega\rangle \geq 0$.

(v) $\Delta \star \omega = 0$.
(vi) $d \star \omega = d^\star \star \omega = 0$.

Theorem 5.1. (Nonexistence of eigenforms associated with positive eigenvalues) *Under the assumption of Theorem 1.2, there does not exist an eigenform ω satisfying (1.5) $\Delta\omega = \lambda\omega$ associated with any eigenvalue $\lambda > 0$.*

Theorem 5.2. (Nonexistence of solution of $\langle\omega, \Delta\omega\rangle > 0$) *Under the assumption of Theorem 1.2, there does not exist a solution of $\langle\omega, \Delta\omega\rangle > 0$.*

Theorem 5.3. (Vanishing Theorem) *Let ω be a solution of $\langle\omega, \Delta\omega\rangle > 0$ and satisfy Kato's type inequality (2.7) on M. Suppose M has the volume growth*

$$\liminf_{r\to\infty} \frac{1}{r^2} \operatorname{Vol}(B(x_0; r)) = \infty \text{ and } \int_a^\infty \Big(\operatorname{Vol}(\partial B(x_0; r))\Big)^{-1} dr < \infty, \tag{1.8}$$

for every $x_0 \in M$ and every $a > 0$. If ω has 2-balanced growth for $1 < q < 3$, then $\omega \equiv 0$ on M.

Just as a warping function (cf. [**CW1, CW2**]), or a function u satisfying $u\Delta u \ge 0$ has a dichotomy between constancy ($u \equiv$ constant) and "infinity" (e.g., $\infty = \liminf_{r\to\infty} \frac{1}{r^2}\int_{B(x_0;r)} |u|^q\, dv$ or more generally, u is 2-imbalanced, for $q > 1$), so does a differential form ω satisfy $\langle\omega, \Delta\omega\rangle \ge 0$ on M have the following dichotomy.

Theorem 6.1. *Let $\omega \in A^k$ be a solution of $\langle\omega, \Delta\omega\rangle \ge 0$, satisfying* Condition W (1.2). *Then either (i) ω is both closed and co-closed, or (ii) ω has 2-imbalanced growth for $1 < q < 3$.*

When ω is a function or 0-form, ω is automatically co-closed and satisfies Condition W (1.2). That ω is closed implies ω is constant. We recapture the dichotomy of function u satisfying $u\Delta u \ge 0$, between constancy and "infinity".

The following theorem is a dual version of the above dichotomy theorem and yields information on the growth of a nontrivial solution of (1.4). (Recall A^k denotes the space of smooth differential k-forms on M, as introduced prior to Definition 1.1.)

Theorem 7.1. *Let $\omega \in A^k$ be a solution of $\langle\omega, \Delta\omega\rangle \ge 0$ with either $d\omega \ne 0$ or $d^*\omega \ne 0$ on M. Then ω has 2-imbalanced growth for $q = 2$, and if ω satisfies* Condition W (1.2), *then ω has 2-imbalanced growth for $1 < q (\ne 2) < 3$.*

In the last part of this paper, we discuss applications of these results. Let $A^k(\xi) = \Gamma(\wedge^k T^*M \otimes E)$ denote the space of smooth differential k-forms on M with values in the Riemannian vector bundle $\xi : E \to M$ over M. Note that if E is the trivial bundle $M \times \mathbb{R}$ equipped with the canonical metric, then A^k is isometric to $A^k(\xi)$, $d = d^\nabla$ (as in (8.1)), and $d^* = \delta^\nabla$ (as in (8.2)). We observe that once a differential k-form $\omega \in A^k$ is found to be both closed and co-closed on M, then ω satisfies a conservation law (8.5) and is ready to apply the theory we developed in [**DW**] and [**W3**]. Based on a variational method, conservation law, comparison theorems in Riemannian geometry under a curvature assumption (8.6) and coarea formula, we solve constant Dirichelet problems for generalized harmonic 1-forms and harmonic maps on starlike domains in M (cf. Theorem 10.3 and Corollary 10.1, generalizing and extending the work of Karcher and Wood for harmonic maps on disc domains in $\mathbb{R}^n$ [**KW**]) and derive monotonicity formulas for ω (cf. Theorems 8.2). We further observed that when ω has 2-moderate growth $\big((2.4), p = 2\big)$ in which the auxiliary function ψ is monotone decreasing or under a more general condition (9.3), we obtain $o(\rho^\lambda)$ growth estimate on the L^2 norm of ω over the geodesic ball $B(\rho)$, as $\rho \to \infty$ (cf. Theorems 9.1). In contrast to a "microscopic" viewpoint in geometric measure theory, this growth estimate provides a "macroscopic" viewpoint of monotonicity formula, and yields a vanishing theorem for a solution ω of $\langle \omega, \Delta\omega\rangle \geq 0$ on M (cf. Theorems 9.3).

The author wishes to thank the referee for his comments and suggestions which helped the author prepared for the final version of this paper.

2. Preliminary

We denote the volume element of n-manifold M by dv, the geodesic ball of radius r centered at x_0 in M by $B(x_0;r)$ or $B(r)$, and its boundary by $\partial B(x_0;r)$ or $\partial B(r)$. Let $d : A^k \to A^{k+1}$ be the exterior differential operator and d^* be the codifferential operator $d^* : A^k \to A^{k-1}$ given by

$$d^* = (-1)^{nk+n+1} \star d \star$$

In particular, if $\omega \in A^1$, d^* is defined by $d^*\omega = -\operatorname{trace} \nabla\omega = -\operatorname{div} \omega$. The Hodge Laplacian Δ is defined on the differential forms by

$$\Delta = -(dd^* + d^*d) : A^k \to A^k$$

Thus, by our convention, on the space of smooth real-valued functions f on M, the Hodge Laplacian agrees with the connection Laplacian, or the Laplace-Beltrami operator; that is,

$$\Delta f = -d^*df = \operatorname{trace} \nabla df = \operatorname{div}(\nabla f).$$

DEFINITION 2.1. A differential k-form ω is said to be *harmonic* if $\Delta\omega = 0$, *closed* if $d\omega = 0$, and *co-closed* if $d^*\omega = 0$.

It follows from the Stokes' Theorem that on a compact Riemannian manifold, (1.1) holds. We recall the following definitions from [**WLW**]:

DEFINITION 2.2. A function or a differential form f has *p-finite growth* (or, simply, *is p-finite*) if there exists $x_0 \in M$ such that

$$\liminf_{r\to\infty} \frac{1}{r^p} \int_{B(x_0;r)} |f|^q \, dv < \infty \tag{2.1}$$

and has *p-infinite growth* (or, simply, *is p-infinite*) otherwise.

A function or a differential form f has *p-mild growth* (or, simply, *is p-mild*) if there exist $x_0 \in M$, and a strictly increasing sequence of $\{r_j\}_0^\infty$ going to infinity, such that for every $l_0 > 0$, we have

$$\sum_{j=\ell_0}^{\infty} \left(\frac{(r_{j+1}-r_j)^p}{\int_{B(x_0;r_{j+1})\setminus B(x_0;r_j)} |f|^q \, dv} \right)^{\frac{1}{p-1}} = \infty, \tag{2.2}$$

and has *p-severe growth* (or, simply, *is p-severe*) otherwise.

A function or a differential form f has *p-obtuse growth* (or, simply, *is p-obtuse*) if there exists $x_0 \in M$ such that for every $a > 0$, we have

$$\int_a^\infty \left(\frac{1}{\int_{\partial B(x_0;r)} |f|^q ds} \right)^{\frac{1}{p-1}} dr = \infty, \tag{2.3}$$

and has *p-acute growth* (or, simply, *is p-acute*) otherwise.

A function or a differential form f has *p-moderate growth* (or, simply, *is p-moderate*) if there exist $x_0 \in M$, and $\psi(r) \in \mathcal{F}$, such that

$$\limsup_{r\to\infty} \frac{1}{r^p \psi^{p-1}(r)} \int_{B(x_0;r)} |f|^q \, dv < \infty, \tag{2.4}$$

and has *p-immoderate growth* (or, simply, *is p-immoderate*) otherwise, where

$$\mathcal{F} = \{\psi : [a,\infty) \longrightarrow (0,\infty) | \int_a^\infty \frac{dr}{r\psi(r)} = \infty \quad for \quad some \quad a \geq 0\}. \tag{2.5}$$

(Notice that the functions in $\mathcal{F}$ are not necessarily monotone.)

A function or a differential form f has *p-small growth* (or, simply, *is p-small*) if there exists $x_0 \in M$, such that for every for every $a > 0$, we have

$$\int_a^\infty \left(\frac{r}{\int_{B(x_0;r)} |f|^q \, dv}\right)^{\frac{1}{p-1}} dr = \infty, \tag{2.6}$$

and has *p-large growth* (or, simply, *is p-large*) otherwise.

DEFINITION 2.3. A function f has *p-balanced growth* (or, simply, *is p-balanced*) if f has one of the following: *p-finite, p-mild, p-obtuse, p-moderate,* or *p-small* growth, and has *p-imbalanced growth* (or, simply, is *p-imbalanced*) otherwise.

The above definitions of "p-balanced, p-finite, p-mild, p-obtuse, p-moderate, p-small" and their counter-parts "p-imbalanced, p-infinite, p-severe, p-acute, p-immoderate, p-large" growth depend on q, and q will be specified in the context in which the definition is used.

PROPOSITION 2.1. (cf.[**W 1**, **W 4**]) *For a given q, f is*

$$p-moderate\,(2.4) \quad \Leftrightarrow \quad p-small\,(2.5) \quad \Rightarrow \quad p-mild\,(2.2) \quad \Rightarrow \quad p-obtuse\,(2.3)$$

or equivalently,

$$p-acute \quad \Rightarrow \quad p-severe \quad \Rightarrow \quad p-large \quad \Leftrightarrow \quad p-immoderate.$$

Hence, for a given q, f is

$$\begin{array}{lllllll} p-\text{balanced} & \Rightarrow & \text{either} & p-\text{finite}\,(2.1) & \text{or} & p-\text{obtuse}\,(2.3) \\ p-\text{imbalanced} & \Rightarrow & \text{both} & p-\text{infinite} & \text{and} & p-\text{immoderate}. \end{array}$$

If in addition, $\int_{B(x_0;r)} |f|^q dv$ is convex in r, then f is p-mild, p-obtuse, p-moderate, *and* p-small (resp. p-severe, p-acute, p-immoderate, *and* p-large) *are all equivalent.*

PROOF. Please see [**W4**], Theorem 5.3. □

PROPOSITION 2.2. *Every L^q differential form ω has* p-balanced *growth, and in fact, has* p-finite, p-mild, p-obtuse, p-moderate, *and* p-small *growth with the same value of q*

PROOF. It follows from Definition 2.2 that an L^q differential form ω is both *p-finite* and *p-moderate.* Now the assertion follows from Definition 2.3 and Proposition 2.1. □

Recall a differential k-form ω on M is said to be *simple* if there exists a smooth function f on M, and $1 \leq i_1 < \cdots < i_k \leq n$ such that $\omega = f dx_{i_1} \wedge \cdots \wedge dx_{i_k}$, where $(x_1, \cdots, x_n)$ is local coordinate on M.

PROPOSITION 2.3. *Every simple k-form satisfies* Condition W (1.2).

PROOF. Let $\omega = f dx_{i_1} \wedge \cdots \wedge dx_{i_k}$. Then $|\omega|^2 = f^2$, $d(|\omega|^2) = d(|\star\omega|^2) = 2f df$, $d\omega = df \wedge dx_{i_1} \wedge dx_{i_2} \cdots \wedge dx_{i_k}$, $d \star \omega = df \wedge dx_{i_{k+1}} \wedge dx_{i_{k+2}} \cdots \wedge dx_{i_n}$, where

$dx_{i_1} \wedge dx_{i_2} \cdots \wedge dx_{i_k} \wedge dx_{i_{k+1}} \wedge dx_{i_{k+2}} \cdots \wedge dx_{i_n}$ lies in the component of $A^n \backslash \{0\}$ determined by the orientation of M. Hence,

$$\begin{aligned} \langle d(|\omega|^2) \wedge \omega, d\omega \rangle &= \langle 2f df \wedge f dx_{i_1} \wedge \cdots \wedge dx_{i_k}, d\omega \rangle \\ &= 2f^2 \langle d \star \omega, d \star \omega \rangle \\ &= 2|\omega|^2 |d\omega|^2 \end{aligned}$$

and

$$\begin{aligned} \langle d(|\star \omega|^2) \wedge \star\omega, d \star \omega \rangle &= \langle 2f df \wedge f dx_{i_{k+1}} \wedge dx_{i_{k+2}} \cdots \wedge dx_{i_n}, d \star \omega \rangle \\ &= 2f^2 \langle d \star \omega, d \star \omega \rangle \\ &= 2|\star \omega|^2 |d \star \omega|^2 \end{aligned}$$

□

A differential k-form ω on M is said to satisfy Kato's type inequality if

$$|d|\omega|| \leq |d\omega| \quad \text{on} \quad M. \tag{2.7}$$

PROPOSITION 2.4. *If a differential k-form ω satisfies Kato's type inequality* (2.7), *then ω satisfies* Condition W (1.2).

PROOF. This follows from a direct computation. □

On the other hand, Condition W (1.2) does not hold in general:

REMARK 2.1. (Counter-Example of Condition W (1.2)) Let the differential 1-form $\omega = x_1 x_n dx_1 + x_n dx_n$ in $\mathbb{R}^n$. Then $|\omega|^2 = (x_1^2+1)x_n^2$, $d(|\omega|^2) = 2x_1 x_n^2 dx_1 + 2(x_1^2+1)x_n dx_n$, and $d\omega = -x_1 dx_1 \wedge dx_n$. Consequently,

$$\begin{aligned} &\langle d(|\omega|^2) \wedge \omega, d\omega \rangle \\ &= \langle (2x_1 x_n^3 - 2(x_1^2+1)x_1 x_n^2) dx_1 \wedge dx_n, d\omega \rangle \\ &= \langle \big(-2x_n^3 + 2(x_1^2+1)x_n^2 \big) d\omega, d\omega \rangle \\ &= -2x_n^3 |d\omega|^2 + 2|\omega|^2 |d\omega|^2 \\ &> 2|\omega|^2 |d\omega|^2 \quad \text{if} \quad x_n < 0 \quad \text{and} \quad x_1 \neq 0. \end{aligned}$$

REMARK 2.2. (Counter-Example of Kato's type inequality) Let $\omega = x_1 x_n dx_1 + x_n dx_n$ in $\mathbb{R}^n$. Then

$$|d|\omega|| = \sqrt{x_1^2 + 1 + \frac{x_1^2 x_2^2}{x_1^2+1}} > |x_1| = |d\omega|.$$

PROPOSITION 2.5. (*Duality for* Condition W (1.2)) $\omega \in A^k$ *satisfies* Condition W (1.2) *if and only if* $\star\omega \in A^{n-k}$ *satisfies* Condition W (1.2).

PROOF. This follows at once from Definition 1.1 of Condition W (1.2), and the fact that $\star\star = (-1)^{nk+k+1}$. □

In the next section, we are interested in differential forms of p-balanced growth when $p = 2$.

3. 2-balanced harmonic forms and solutions of $\langle \omega, \Delta\omega \rangle \geq 0$

We will use the following result:

LEMMA 3.1. *Suppose ω is a differential k-form on an n-dimensional manifold M. Then, for any $f \in C^\infty(M)$, we have*

$$d^*(f\omega) = f d^*(\omega) + (-1)^{nk+n+1} \star (df \wedge \star\omega) \tag{3.1}$$

PROOF. Since the operator $\star$ is linear with respect to multiplication by functions and $d^* = (-1)^{nk+n+1} \star d \star$,

$$\begin{aligned} d^*(f\omega) &= (-1)^{nk+n+1} \star d \star (f\omega) \\ &= (-1)^{nk+n+1} \star d(f \star \omega) \\ &= (-1)^{nk+n+1} \star (fd \star \omega + df \wedge \star\omega) \\ &= f(-1)^{nk+n+1} \star d \star \omega + (-1)^{nk+n+1} \star (df \wedge \star\omega) \\ &= fd^*\omega + (-1)^{nk+n+1} \star (df \wedge \star\omega) \end{aligned}$$

□

REMARK 3.1. Lemma 3.1 shows that the codifferential operator d^* is $\mathbb{R}$-linear, but neither linear with respect to multiplication by functions, nor satisfies the product rule $\Big($ i.e., $d^*(f\omega) \neq (d^*f)\omega + fd^*\omega\Big)$, unless $f \equiv$ constant on M.

THEOREM 3.1. *Suppose $\omega \in A^k$ has 2-finite growth $\big((2.1), p = 2\big)$, for $q = 2$, or for $1 < q(\neq 2) < 3$ with* Condition W (1.2) $\big($cf. (1.3)$\big)$. *Then ω is a solution of $\langle \omega, \Delta\omega\rangle \geq 0$ (resp. ω is harmonic) on M if and only if ω is closed and co-closed. Or equivalently,* (1.6) $\big($resp. (1.1)$\big)$ *holds.*

PROOF. ($\Leftarrow$) It is obvious. ($\Rightarrow$) Choose a smooth cut-off function $\psi(x)$ as in [**W2**, (3.1)], i.e. for any $x_0 \in M$ and any pair of positive numbers s, t with $s < t$, a rotationally symmetric Lipschitz continuous nonnegative function $\psi(x) = \psi(x; s, t)$ satisfies $\psi \equiv 1$ on $B(s)$, $\psi \equiv 0$ off $B(t)$, and $|\nabla\psi| \leq \frac{C_1}{t-s}$, *a.e. on* M, where $C_1 > 0$ is a constant (independent of x_0, s, t). Let $1 < q < \infty$. to be determined later. By (1.4) or the harmonicity of ω, $\Delta = -(d^*d + dd^*)$, and the adjoint relationship of d and d^* on differential forms with compact support, we have for any constant $\epsilon > 0$,

(3.2)

$$\begin{aligned} 0 &\leq \int_{B(t)} \langle \psi^2(|\omega|^2+\epsilon)^{\frac{q}{2}-1}\omega, \Delta\omega\rangle\, dv \\ &= \int_{B(t)} -\langle \psi^2(|\omega|^2+\epsilon)^{\frac{q}{2}-1}\omega, d^*d\omega\rangle - \langle \psi^2(|\omega|^2+\epsilon)^{\frac{q}{2}-1}\omega, dd^*\omega\rangle\, dv \\ &= \int_{B(t)} -\langle d\big(\psi^2(|\omega|^2+\epsilon)^{\frac{q}{2}-1}\omega\big), d\omega\rangle\, dv - \int_{B(t)} \langle d^*\big(\psi^2(|\omega|^2+\epsilon)^{\frac{q}{2}-1}\omega\big), d^*\omega\rangle\, dv \end{aligned}$$

Applying Lemma 3.1 in which $f = \psi^2(|\omega|^2+\epsilon)^{\frac{q}{2}-1}$ to (3.2), we have

$$\begin{aligned} 0 \leq & -\int_{B(t)} \langle 2\psi(|\omega|^2+\epsilon)^{\frac{q}{2}-1} d\psi \wedge \omega + \psi^2 d\big((|\omega|^2+\epsilon)^{\frac{q}{2}-1}\big) \wedge \omega \\ & \qquad + \psi^2(|\omega|^2+\epsilon)^{\frac{q}{2}-1} d\omega, d\omega\rangle\, dv \\ & - \int_{B(t)} \psi^2(|\omega|^2+\epsilon)^{\frac{q}{2}-1}|d^*\omega|^2\, dv \\ & - \int_{B(t)} \langle (-1)^{nk+n+1} \star \Big(d\big(\psi^2(|\omega|^2+\epsilon)^{\frac{q}{2}-1}\big) \wedge \star\omega\Big), d^*\omega\rangle\, dv \end{aligned} \tag{3.3}$$

The first integral in (3.3), via a generalized Hadmard Theorem $|d\psi \wedge \omega| \leq |d\psi||\omega|$, Condition W (1.2) (applying to $\langle d(|\omega|^2+\epsilon) \wedge \omega, d\omega\rangle$) and Cauchy-Schwarz inequality $\Big($applying to $\int_{B(t)\setminus B(s)} \psi(|\omega|^2+\epsilon)^{\frac{q-1}{2}}|d\psi||d\omega|\, dv = \int_{B(t)\setminus B(s)} \big(|d\psi|(|\omega|^2+$

$\epsilon)^{\frac{q}{4}}) \cdot (\psi(|\omega|^2+\epsilon)^{\frac{q-2}{4}}|d\omega|)\, dv\Big)$, becomes

(3.4)

$$\begin{aligned}
&-\int_{B(t)}\langle 2\psi(|\omega|^2+\epsilon)^{\frac{q}{2}-1}d\psi\wedge\omega+\psi^2 d\big((|\omega|^2+\epsilon)^{\frac{q}{2}-1}\big)\wedge\omega \\
&\qquad\qquad +\psi^2(|\omega|^2+\epsilon)^{\frac{q}{2}-1}d\omega, d\omega\rangle\, dv \\
&\le \int_{B(t)\setminus B(s)} 2\psi(|\omega|^2+\epsilon)^{\frac{q-1}{2}}|d\psi||d\omega|\, dv \\
&\quad+\int_{B(t)}(\tfrac{q}{2}-1)\psi^2(|\omega|^2+\epsilon)^{\frac{q}{2}-2}\langle d|\omega|^2\wedge\omega, d\omega\rangle\, dv \\
&\quad-\int_{B(t)}\psi^2(|\omega|^2+\epsilon)^{\frac{q}{2}-1}|d\omega|^2\, dv \\
&\le 2\big(\int_{B(t)\setminus B(s)}|d\psi|^2(|\omega|^2+\epsilon)^{\frac{q}{2}}\, dv\big)^{\frac{1}{2}}\big(\int_{B(t)\setminus B(s)}\psi^2(|\omega|^2+\epsilon)^{\frac{q}{2}-1}|d\omega|^2\, dv\big)^{\frac{1}{2}} \\
&\quad+(|q-2|-1)\int_{B(t)}\psi^2(|\omega|^2+\epsilon)^{\frac{q}{2}-1}|d\omega|^2\, dv
\end{aligned}$$

On the other hand, by Condition W (1.2) we have

$$\begin{aligned}
|\langle d(|\omega|^2+\epsilon)\wedge\star\omega,\star d^*\omega\rangle| &= |\langle d(|\star\omega|^2)\wedge\star\omega, d\star\omega\rangle| \\
&\le 2|\star\omega|^2|d\star\omega|^2 \\
&= 2|\omega|^2||\star d^*\omega|^2 \\
&= 2|\omega|^2|d^*\omega|^2
\end{aligned} \tag{3.5}$$

The third integral in (3.3), via (3.5) and Cauchy-Schwarz inequality$\Big($applying to $\int_{B(t)\setminus B(s)}\psi(|\omega|^2+\epsilon)^{\frac{q-1}{2}}|d\psi||d^*\omega|\, dv = \int_{B(t)\setminus B(s)}\big(|d\psi|(|\omega|^2+\epsilon)^{\frac{q}{4}}\big)\cdot\big(\psi(|\omega|^2+\epsilon)^{\frac{q-2}{4}}|d^*\omega|\big)\, dv\Big)$, becomes

(3.6)

$$\begin{aligned}
&-\int_{B(t)}\langle(-1)^{nk+n+1}\star\Big(d(\psi^2(|\omega|^2+\epsilon)^{\frac{q}{2}-1})\wedge\star\omega\Big), d^*\omega\rangle\, dv \\
&\le |\int_{B(t)}\langle d(\psi^2(|\omega|^2+\epsilon)^{\frac{q}{2}-1})\wedge\star\omega,\star d^*\omega\rangle|\, dv \\
&\le \int_{B(t)\setminus B(s)}|\langle 2\psi(|\omega|^2+\epsilon)^{\frac{q}{2}-1}d\psi\wedge\star\omega,\star d^*\omega\rangle|\, dv \\
&\quad+\int_{B(t)}|\tfrac{q}{2}-1|\psi^2(|\omega|^2+\epsilon)^{\frac{q}{2}-2}|\langle d(|\omega|^2+\epsilon)\wedge\star\omega,\star d^*\omega\rangle|\, dv \\
&\le 2\int_{B(t)\setminus B(s)}\psi(|\omega|^2+\epsilon)^{\frac{q-1}{2}}|d\psi||d^*\omega|\, dv+\int_{B(t)}|q-2|\psi^2(|\omega|^2+\epsilon)^{\frac{q}{2}-1}|d^*\omega|^2\, dv \\
&\le |q-2|\int_{B(t)}\psi^2(|\omega|^2+\epsilon)^{\frac{q}{2}-1}|d^*\omega|^2\, dv \\
&\quad+2\big(\int_{B(t)\setminus B(s)}|d\psi|^2(|\omega|^2+\epsilon)^{\frac{q}{2}}\, dv\big)^{\frac{1}{2}}\big(\int_{B(t)\setminus B(s)}\psi^2(|\omega|^2+\epsilon)^{\frac{q}{2}-1}|d^*\omega|^2\, dv\big)^{\frac{1}{2}}
\end{aligned}$$

Substituting (3.4) and (3.6) into (3.3), we obtain the following inequality:

$$\begin{aligned}
&(1-|q-2|)\int_{B(t)}\psi^2(|\omega|^2+\epsilon)^{\frac{q}{2}-1}\big(|d\omega|^2+|d^*\omega|^2\big)\, dv \\
&\le 2\Big(\int_{B(t)\setminus B(s)}|d\psi|^2(|\omega|^2+\epsilon)^{\frac{q}{2}}\, dv\Big)^{\frac{1}{2}} \\
&\qquad\cdot\Big(\big(\int_{B(t)\setminus B(s)}\psi^2(|\omega|^2+\epsilon)^{\frac{q}{2}-1}|d\omega|^2\, dv\big)^{\frac{1}{2}} \\
&\qquad\qquad+\big(\int_{B(t)\setminus B(s)}\psi^2(|\omega|^2+\epsilon)^{\frac{q}{2}-1}|d^*\omega|^2\, dv\big)^{\frac{1}{2}}\Big) \\
&\le 2\sqrt{2}\Big(\int_{B(t)\setminus B(s)}|d\psi|^2(|\omega|^2+\epsilon)^{\frac{q}{2}}\, dv\Big)^{\frac{1}{2}} \\
&\qquad\cdot\Big(\int_{B(t)\setminus B(s)}\psi^2(|\omega|^2+\epsilon)^{\frac{q}{2}-1}(|d\omega|^2+|d^*\omega|^2)\, dv\Big)^{\frac{1}{2}}
\end{aligned} \tag{3.7}$$

where in the last step we have applied the inequality

$$\mathcal{A}^{\frac{1}{2}} + \mathcal{B}^{\frac{1}{2}} \leq \sqrt{2}(\mathcal{A} + \mathcal{B})^{\frac{1}{2}}$$

in which

$$\mathcal{A} = \int_{B(t)\setminus B(s)} \psi^2(|\omega|^2 + \epsilon)^{\frac{q}{2}-1}|d\omega|^2\, dv \quad \text{and}$$

$$\mathcal{B} = \int_{B(t)\setminus B(s)} \psi^2(|\omega|^2 + \epsilon)^{\frac{q}{2}-1}|d^*\omega|^2\, dv\,.$$

Choosing

$$1 < q < 3, \quad \text{i.e.} \quad 1 - |q-2| > 0\,, \tag{3.8}$$

squaring both sides of (3.7), and using $|\nabla\psi| \leq \frac{C_1}{t-s}$, *a.e. on* M, we have

$$\begin{aligned}&\Big(\int_{B(t)} \psi^2(|\omega|^2 + \epsilon)^{\frac{q}{2}-1}(|d\omega|^2 + |d^*\omega|^2)\, dv\Big)^2 \\ &\leq \left(\Big(\frac{2\sqrt{2}C_1}{1-|q-2|}\Big)^2\Big(\frac{1}{(t-s)^2}\int_{B(t)\setminus B(s)}(|\omega|^2+\epsilon)^{\frac{q}{2}}\, dv\right) \\ &\cdot\left(\int_{B(t)\setminus B(s)} \psi^2(|\omega|^2 + \epsilon)^{\frac{q}{2}-1}(|d\omega|^2 + |d^*\omega|^2)\, dv\right)\end{aligned} \tag{3.9}$$

Let $\{r_j\}$ be a strictly increasing sequence of positive real numbers going to infinity and define:

$$\begin{aligned}&A_j(\epsilon) = \frac{1}{r_j^2}\int_{B(r_j)} (|\omega|^2+\epsilon)^{\frac{q}{2}}\, dv \\ &\varphi_j(x) = \psi(x; r_j, r_{j+1}) \\ &Q_{j+1}(\epsilon) = \int_{B(r_{j+1})} \varphi_j^2(|\omega|^2 + \epsilon)^{\frac{q}{2}-1}(|d\omega|^2 + |d^*\omega|^2)\, dv \\ &C = \Big(\frac{2\sqrt{2}C_1}{1-|q-2|}\Big)^2\end{aligned} \tag{3.10}$$

Using the above notations, substituting $s = r_j$, $t = r_{j+1}$ and $\psi = \varphi_j$ into (3.9), and using the observation,

$\varphi_{j-1} \leq \varphi_j \Rightarrow Q_j(\epsilon) \leq \int_{B(r_j)} \varphi_j^2(|\omega|^2 + \epsilon)^{\frac{q}{2}-1}(|d\omega|^2 + |d^*\omega|^2)\, dv$, we have

$$Q_{j+1}^2(\epsilon) \leq C\left(\frac{r_{j+1}^2 A_{j+1}(\epsilon) - r_j^2 A_j(\epsilon)}{(r_{j+1}-r_j)^2}\right)\Big(Q_{j+1}(\epsilon) - Q_j(\epsilon)\Big) \tag{3.11}$$

Choosing $\{r_j\}$ such that $r_{j+1} \geq 2r_j$, we have

$$\frac{r_{j+1}^2 A_{j+1}(\epsilon) - r_j^2 A_j(\epsilon)}{(r_{j+1}-r_j)^2} \leq 4A_{j+1}(\epsilon) \tag{3.12}$$

and so

$$Q_{j+1}^2(\epsilon) \leq 4CA_{j+1}(\epsilon)\Big(Q_{j+1}(\epsilon) - Q_j(\epsilon)\Big) \tag{3.13}$$

Since $Q_{j+1}(\epsilon) \neq \infty$, it follows from (3.13) that

$$0 \leq Q_{j+1}(\epsilon) \leq 4CA_{j+1}(\epsilon) \tag{3.14}$$

For $2 \leq q < 3$ (resp. $1 < q < 2$), $\{Q_j(\epsilon)\}$ is an increasing (resp. deacreasing) sequence in ϵ, and for $1 < q < 3$, $\{A_j(\epsilon)\}$ is an increasing sequence in ϵ. We define

$Q_j := Q_j(0)$ and $A_j := A_j(0)$. Then by the Monotone Convergence Theorem, $\lim_{\epsilon\to 0} A_j(\epsilon) = A_j$, $\lim_{\epsilon\to 0} Q_j(\epsilon) = Q_j$ and as $\epsilon \to 0$ in (3.14) and (3.11), we have

$$0 \le Q_{j+1} \le 4CA_{j+1} \tag{3.15}$$

and

$$Q_{j+1}^2 \le C\left(\frac{r_{j+1}^2 A_{j+1} - r_j^2 A_j}{(r_{j+1}-r_j)^2}\right)\left(Q_{j+1} - Q_j\right) \tag{3.16}$$

respectively.

We claim that $Q_{j+1} \to 0$ as $j \to \infty$. To complete the proof, assume that ω has 2-finite growth (i.e., $\liminf_{r\to\infty} \frac{1}{r^2}\int_{B(x_0;r)} |\omega|^q \, dv = \liminf_{r\to\infty} A_r < \infty$) for $1 < q < 3$, and ω is either $d\omega \ne 0$, or $d^*\omega \ne 0$. Thus, there exist a constant $K > 0$, and a sequence $\{r_j\}$ with $r_{j+1} \ge 2r_j$, such that $A_{j+1} \le K$. As $\epsilon \to 0$ in (3.13), summing over j, we have via (3.15) for $\forall N > 1$,

$$\sum_{j=1}^{N} Q_{j+1}^2 \le 4CK(Q_{N+1} - Q_1) \le 16C^2K^2$$

Therefore, the convergent infinite series $\sum_{j=1}^\infty Q_{j+1}^2$ implies that $Q_{j+1} \to 0$ as $j \to \infty$, i.e., $\int_M |\omega|^{q-2}(|d\omega|^2 + |d^*\omega|^2)\, dv = 0$. This would lead to $d\omega = d^*\omega \equiv 0$ on M, a contradiction. Indeed, if $d\omega$ (resp. $d^*\omega$) $\ne 0$ at a point, then $d\omega$ (resp. $d^*\omega$) $\ne 0$ in an open set U containing that point by the continuity. For $q > 2$, this would force $\omega = 0$ and lead to a contradiction that $d\omega$ (resp. $d^*\omega$) $= 0$ in U. For $q = 2$, $\int_M |\omega|^{q-2}(|d\omega|^2 + |d^*\omega|^2)\, dv \ge \int_U (|d\omega|^2 + |d^*\omega|^2)\, dv > 0$, a contradiction. For $1 < q < 2$, either $\omega \ne 0$ at a point in U, and hence in an open set $V \subset U$, or $\omega \equiv 0$ in U would lead to $\int_M |\omega|^{q-2}(|d\omega|^2 + |d^*\omega|^2)\, dv \ge \int_V |\omega|^{q-2}(|d\omega|^2 + |d^*\omega|^2)\, dv > 0$, a contradiction. This completes the proof. □

REMARK 3.2. The case ω is a harmonic form, the first integral and second integral in (3.3) are treated in [**Y1**, p.664, (2.25), (2.26)]. Since the codifferential d^* is not linear with respect to multiplication by functions (cf. Remark 3.1), based on Lemma 3.1, we include the third integral $\big($in (3.3)$\big)$ $\int_{B(t)} \langle (-1)^{nk+n+1} \star \Big(d(\psi^2(|\omega|^2 + \epsilon)^{\frac{q}{2}-1}) \wedge \star\omega \Big), d^*\omega\rangle \, dv$ and make estimates on this third integral in (3.6). In contrast to the case $q = 2$, the term $\int_{B(t)} |\frac{q}{2} - 1|\psi^2(|\omega|^2 + \epsilon)^{\frac{q}{2}-2}|\langle d(|\omega|^2 + \epsilon) \wedge \star\omega, \star d^*\omega\rangle|\, dv$ in (3.6) does not vanish for $q \ne 2$ and we use Condition W (1.2) to evaluate the factor $|\langle d(|\omega|^2 + \epsilon) \wedge \star\omega, \star d^*\omega\rangle|$ (cf. (3.5)) in the integrand of this term to obtain the estimates in (3.6). Similarly, in computing the first integral in (3.3), $-\int_{B(t)} \langle 2\psi(|\omega|^2 + \epsilon)^{\frac{q}{2}-1} d\psi \wedge \omega + \psi^2 d(|\omega|^2 + \epsilon)^{\frac{q}{2}-1} \wedge \omega + \psi^2(|\omega|^2 + \epsilon)^{\frac{q}{2}-1} d\omega, d\omega\rangle\, dv$, the term $\int_{B(t)} (\frac{q}{2} - 1)\psi^2(|\omega|^2 + \epsilon)^{\frac{q}{2}-2}\langle d|\omega|^2 \wedge \omega, d\omega\rangle\, dv$ in (3.4) does not vanish for $q \ne 2$ and we use Condition W (1.2) to evaluate the factor $\langle d|\omega|^2 \wedge \omega, d\omega\rangle$ to obtain the estimates in (3.4).

REMARK 3.3. When $q = 2$, the above term in the first integral, $\int_{B(t)} |\frac{q}{2} - 1|\psi^2(|\omega|^2 + \epsilon)^{\frac{q}{2}-2}|\langle d(|\omega|^2 + \epsilon) \wedge \star\omega, \star d^*\omega\rangle|\, dv = 0$, and the term in the third integral, $\int_{B(t)} (\frac{q}{2} - 1)\psi^2(|\omega|^2 + \epsilon)^{\frac{q}{2}-2}\langle d|\omega|^2 \wedge \omega, d\omega\rangle\, dv = 0$. Thus, no Condition W (1.2) is needed to make estimates. This explains why if ω is in L^2 or is 2-balanced for $q=2$, Condition W (1.2) is not involved at all.

THEOREM 3.2. *Suppose $\omega \in A^k$ has 2-mild growth $\big((2.2), p=2\big)$, for $q=2$, or for $1<q(\neq 2)<3$ with* Condition W (1.2) $\big($cf. (1.3)$\big)$. *Then ω is a solution of* $\langle \omega, \Delta\omega\rangle \geq 0$ (resp. *ω is harmonic) on M if and only if ω is closed and co-closed. Or equivalently,* (1.6) $\big($resp. (1.1)$\big)$ *holds.*

PROOF. ($\Leftarrow$) It is evident. ($\Rightarrow$) If either $d\omega \neq 0$, or $d^*\omega \neq 0$, then via (3.10), for every strictly increasing sequence $\{r_j\}_1^\infty$ going to infinity, there exists an integer $l_0>0$, with $r_{l_0}\geq a$, (where a is a fixed positive number independent of the sequence) such that

$$
\begin{aligned}
&Q_{j+1} \geq Q_{l_0} \geq Q_a := \textstyle\int_{B(a)} |\omega|^{q-2}(|d\omega|^2+|d^*\omega|^2)\,dv > 0,\\
&Q_{j+1}-Q_j>0,\\
&A_j>0, \quad \text{and} \quad r_{j+1}^2A_{j+1}-r_j^2A_j>0 \quad \text{whenever} \quad j\geq l_0 .
\end{aligned}
\tag{3.17}
$$

Let n_j be the largest nonnegative integer such that $(Q_{j+1}-n_j-1<)\,Q_j\leq Q_{j+1}-n_j$. Multiplying (3.16) by $\frac{(r_{j+1}-r_j)^2}{Q_{j+1}^2(r_{j+1}^2A_{j+1}-r_j^2A_j)}$, and summing over j from $j=l_0$ to l, we have

(3.18)
$$
\begin{aligned}
&\textstyle\sum_{j=l_0}^{l} \frac{(r_{j+1}-r_j)^2}{r_{j+1}^2A_{j+1}-r_j^2A_j}\\
\leq\ & C\textstyle\sum_{j=l_0}^{l} \frac{Q_{j+1}-Q_j}{Q_{j+1}^2}\\
\leq\ & C\textstyle\sum_{j=l_0}^{l} \frac{\big(Q_{j+1}-(Q_{j+1}-1)\big)+\cdots+\big((Q_{j+1}-n_j+1)-(Q_{j+1}-n_j)\big)+\big((Q_{j+1}-n_j)-Q_j\big)}{Q_{j+1}^2}\\
\leq\ & C\textstyle\sum_{j=l_0}^{l} \frac{1}{Q_{j+1}^2}+\cdots+\frac{1}{(Q_{j+1}-n_j+1)^2}+\frac{(Q_{j+1}-n_j)-Q_j}{(Q_{j+1}-n_j)^2}\\
\leq\ & C\textstyle\sum_{j=l_0}^{l} \int_{Q_j}^{Q_{j+1}} \frac{1}{r^2}\,dr\\
<\ & C\textstyle\int_{Q_{l_0}}^{\infty} \frac{1}{r^2}\,dr\\
=\ & C\frac{1}{Q_{l_0}}\\
<\ & \frac{C}{Q_a}
\end{aligned}
$$

Letting $l\to\infty$, we obtain

$$
\sum_{j=\ell_0}^{\infty}\left(\frac{(r_{j+1}-r_j)^2}{\int_{B(x_0;r_{j+1})\backslash B(x_0;r_j)}|\omega|^q\,dv}\right)=\sum_{j=l_0}^{\infty}\frac{(r_{j+1}-r_j)^2}{r_{j+1}^2A_{j+1}-r_j^2A_j}\leq\frac{C}{Q_a}<\infty
$$

which contradicts the assumption that ω has 2-mild growth for $1<q<3$. Consequently, $d\omega=d^*\omega=0$ on M. □

THEOREM 3.3. *Suppose $\omega\in A^k$ has 2–obtuse growth $\big((2.3), p=2\big)$, for $q=2$, or for $1<q(\neq 2)<3$ with* Condition W (1.2) $\big($cf. (1.3)$\big)$. *Then ω is a solution of* $\langle\omega,\Delta\omega\rangle\geq 0$ (resp. *ω is harmonic) on M if and only if ω is closed and co-closed. Or equivalently,* (1.6) $\big($resp. (1.1)$\big)$ *holds.*

PROOF. ($\Leftarrow$) It is apparent. ($\Rightarrow$) If either $d\omega\neq 0$, or $d^*\omega\neq 0$, then based on (3.10) and (3.16), we have

$$
\frac{1}{\frac{r_{j+1}^2A_{j+1}-r_j^2A_j}{r_{j+1}-r_j}}\leq C\frac{\frac{Q_{j+1}-Q_j}{r_{j+1}-r_j}}{Q_{j+1}^2}
$$

Since $\{r_j\}$ is arbitrary in the above inequality, we can set a variable $r = r_j$, and let $r_{j+1} \to r = r_j$, and obtain

$$\frac{1}{\frac{d}{dr}(r^2 A_r)} \le C \frac{\frac{d}{dr} Q_r}{Q_r^2}$$

Integrating the above inequality over the interval $[a, t]$ yields:

$$\int_a^t \Big(\frac{1}{\frac{d}{dr}(r^2 A_r)}\Big) dr \le \frac{C}{Q_a} < \infty$$

Letting $t \to \infty$, by the Coarea formula, we get:

$$\begin{aligned} \int_a^\infty \Big(\frac{1}{\int_{\partial B(x_0;r)} |\omega|^q ds}\Big)\, dr &= \int_a^\infty \Big(\frac{1}{\frac{d}{dr}\int_{B(x_0;r)} |\omega|^q\, dv}\Big)\, dr \\ &= \int_a^\infty \Big(\frac{1}{\frac{d}{dr}(r^2 A_r)}\Big)\, dr \\ &< \infty \end{aligned}$$

which contradicts the assumption that ω has 2-obtuse growth, for $1 < q < 3$. Consequently, $d\omega = d^*\omega = 0$ on M. □

THEOREM 3.4. *Suppose $\omega \in A^k$ has 2-moderate growth $\big((2.4), p = 2\big)$, for $q = 2$, or for $1 < q (\neq 2) < 3$ with* Condition W (1.2) $\big($cf. (1.3)$\big)$. *Then ω is a solution of $\langle \omega, \Delta\omega\rangle \ge 0$* (resp. *$\omega$ is harmonic) on M if and only if ω is closed and co-closed. Or equivalently,* (1.6) $\big($resp. (1.1)$\big)$ *holds.*

PROOF. ($\Leftarrow$) It is clear. ($\Rightarrow$) Since ω has 2-moderate growth, for $1 < q < 3$, we may assume, by the definition of the limit superior of functions, there exists a $\psi \in \mathcal{F}\big((2.5), p = 2\big)$, and a constant $K' > 0$ such that $\frac{1}{\psi(r)} A_r \le K'$ for $r > \ell_0$. This implies that

$$\frac{1}{rA_r} \ge \frac{1}{K'} \frac{1}{r\psi(r)} \quad \text{for} \quad r > \ell_0\,. \tag{3.19}$$

If either $d\omega \neq 0$, or $d^*\omega \neq 0$, then in view of (3.17) and (3.19) for a strictly increasing sequence $\{r_j\}$ with $r_{j+1} = 2r_j$, and for any $\psi(r) > 0$, we obtain

$$\begin{aligned} \textstyle\sum_{j=l_0}^l \frac{(r_{j+1}-r_j)^2}{r_{j+1}^2 A_{j+1} - r_j^2 A_j} &\ge \textstyle\sum_{j=l_0}^l \frac{r_{j+1}-r_j}{r_{j+1}^2 A_{j+1}} \cdot (r_{j+1} - r_j) \\ &= \textstyle\sum_{j=l_0}^l \frac{\frac{1}{2}}{r_{j+1} A_{j+1}} \cdot \frac{1}{2}(r_{j+2} - r_{j+1}) \\ &\ge \textstyle\frac{1}{4}\sum_{j=l_0}^l \int_{r_{j+1}}^{r_{j+2}} \frac{1}{rA_r}\, dr \\ &\ge \textstyle\frac{1}{4K'} \int_{r_{l_0+1}}^{r_{l+2}} \frac{1}{r\psi(r)} dr \end{aligned} \tag{3.20}$$

where we have applied the Mean Value Theorem for integrals to $\int_{r_{j+1}}^{r_{j+2}} \frac{1}{rA_r} dr$ in the third step. Combining (3.20) and (3.18), and letting $l \to \infty$, we would have

$$\int_{r_{l_0+1}}^\infty \frac{1}{r\psi(r)} dr < \infty$$

contradicting $\psi \in \mathcal{F}$. Consequently, $d\omega = d^*\omega = 0$ on M. □

THEOREM 3.5. *Suppose $\omega \in A^k$ has 2-small growth $\big((2.6), p = 2\big)$, for $q = 2$, or for $1 < q (\neq 2) < 3$ with* Condition W (1.2) $\big($cf. (1.3)$\big)$. *Then ω is a solution of $\langle \omega, \Delta\omega\rangle \ge 0$* (resp. *$\omega$ is harmonic) on M if and only if ω is closed and co-closed. Or equivalently,* (1.6) $\big($resp. (1.1)$\big)$ *holds.*

PROOF. ($\Leftarrow$) It is transparent. ($\Rightarrow$) If either $d\omega \neq 0$, or $d^*\omega \neq 0$, then in view of (3.20) and (3.18) for a strictly increasing sequence $\{r_j\}$ with $r_{j+1} = 2r_j$, and for any $\psi(r) > 0$, we would have

$$\int_{r_{l_0+1}}^{r_{l+2}} \frac{1}{rA_r}\, dr \leq 4 \sum_{j=l_0}^{l} \frac{(r_{j+1} - r_j)^2}{r_{j+1}^2 A_{j+1} - r_j^2 A_j} \leq \frac{4C}{Q_a} \tag{3.21}$$

Letting $l \to \infty$ would lead to

$$\int_{r_{l_0+1}}^{\infty} \left(\frac{r}{\int_{B(x_0;r)} |\omega|^q\, dv} \right) dr = \int_{r_{l_0+1}}^{\infty} \frac{1}{rA_r}\, dr < \infty ,$$

contradicting the assumption that ω has 2-small growth, for $1 < q < 3$. Consequently, ω is closed and co-closed. □

4. Duality, unity and proof of Theorems 1.1 and 1.2

We begin with

PROOF OF THEOREMS 1.1 AND 1.2. This follows at once from Theorems 3.1-3.5, Definition 2.3, and Definition 2.1. □

PROOF OF COROLLARY 1.3. i), ii) From Proposition 2.2, an L^2 and an $L^q, 1 < q(\neq 2) < 3$ form are 2-balanced for $q = 2$ and for $1 < q(\neq 2) < 3$ respectively. Now Theorem 1.2 completes the proof. iii) A 0-form is a function, hence is co-closed, and satisfies Condition W (1.2). A closed 0-form is a constant. Hence, it follows at once from Theorem 1.2. □

PROOF OF COROLLARY 1.4. This follows from Propositions 2.3 and 2.4, and Theorem 1.2 □

PROOF OF COROLLARY 1.1. This is an immediately consequence of Corollary 1.3. □

PROOF OF COROLLARY 1.2. This follows at once from Corollary 1.4. □

PROOF OF DUAL THEOREM 1.3. In view of Proposition 2.5 and the fact $|\omega|^2 = |\star\, \omega|^2$, $\star\omega$ also satisfies Condition W (1.2), and has the same 2-balanced growth, for q in the same range as ω. Hence $\star\omega$ satisfies the assumption of Theorem 1.2. Now the desired (1.7) follows from applying Theorem 1.2 to $\star\omega$. □

THEOREM 4.1 (Unity Theorem). *Under the same growth assumption on* ω (*as in Theorem* 1.1, 1.2, *or* 1.3), *the following six statements are equivalent:*
i $\langle \omega, \Delta\omega\rangle \geq 0$, ii $\Delta\omega = 0$, iii $d\omega = d^\star\omega = 0$, iv $\langle \star\omega, \Delta \star \omega\rangle \geq 0$, v $\Delta \star \omega = 0$, *and* vi $d \star \omega = d^\star \star \omega = 0$.

PROOF. From Theorem 1.2, (i) $\Leftrightarrow$ (ii) $\Leftrightarrow$ (iii). By Duality Theorem 1.3, (iv) $\Leftrightarrow$ (v) $\Leftrightarrow$ (vi). On the other hand, (ii) $\Leftrightarrow$ (v). Consequently, (i) through (vi) are all equivalent. □

5. Eigenforms, solution of $\langle\omega,\Delta\omega\rangle > 0$, and vanishing theorems

THEOREM 5.1 (Nonexistence of eigenforms associated with positive eigenvalues). *Under the assumption of Theorem 1.2, there does not exist an eigenform ω satisfying (1.5) associated with eigenvalue $\lambda > 0$.*

PROOF. Suppose contrary, there would exist an eigenform ω satisfying

$$\langle\omega,\Delta\omega\rangle = \lambda|\omega|^2 > 0 .$$

This would imply by Theorem 1.2, $d\omega = d^*\omega = 0$, and hence $\langle\omega,\Delta\omega\rangle = 0$, a contradiction. □

THEOREM 5.2 (Nonexistence of solution of $\langle\omega,\Delta\omega\rangle > 0$). *Under the assumption of Theorem 1.2, there does not exist a solution of $\langle\omega,\Delta\omega\rangle > 0$.*

PROOF. Suppose the contrary, there would exist an ω satisfying $\langle\omega,\Delta\omega\rangle > 0$, Then from Theorem 1.2 , $d\omega = d^*\omega = 0$, and hence $\langle\omega,\Delta\omega\rangle = 0$, a contradiction. □

THEOREM 5.3 (Vanishing Theorem). *Let ω be a solution of (1.4) satisfying Kato's type inequality (2.7) on M. Suppose M has the volume growth (1.8) for every $x_0 \in M$ and every $a > 0$. If ω has 2-balanced growth for $1 < q < 3$, then $\omega \equiv 0$ on M.*

PROOF. By Proposition 2.4, if ω satisfies Kato's type inequality, then ω satisfies Condition W (1.2). It follows from Corollary 1.4 that ω is closed and hence by (2.7), $|\omega| \equiv \text{Const}$. Proposition 2.1 implies that if ω is 2-balanced, then ω is either 2-finite $\big((2.1), p = 2\big)$ or 2-obtuse $\big((2.3), p = 2\big)$ for the same value of q. Suppose $|\omega|$ were a nonzero constant, then via (2.1) and (2.3), $p = 2$ there would exist $x_0 \in M$ and for every $a > 0$,

$$\liminf_{r\to\infty} \frac{1}{r^2}\operatorname{Vol}(B(x_0;r)) < \infty \quad \text{or} \quad \int_a^\infty \Big(\operatorname{Vol}(\partial B(x_0;r))\Big)^{-1} dr = \infty .$$

This contradicts (1.8). □

6. On a dichotomy of solutions of $\langle\omega,\Delta\omega\rangle \geq 0$

Just as a warping function (cf.[**CW1, CW2**]), or nonnegative subharmonic function has a dichotomy, so does differential forms satisfying (1.4) :

THEOREM 6.1. *Let ω be a solution of (1.4) and satisfy* Condition W (1.2), *Then either (i) ω is both closed and co-closed, or (ii) ω has 2-imbalanced growth for $1 < q < 3$.*

PROOF. Suppose contrary, $(\sim i)$ ω were not both closed and co-closed, and $(\sim ii)$ ω had 2-balanced growth for $1 < q < 3$. then $(\sim ii)$, via Theorem 1.2 and the hypothesis, ω satisfies Condition W (1.2) would imply ω were both closed and co-closed. This contradicts $(\sim i)$. □

By a similar argument, we can prove

THEOREM 6.2. *Let ω be a solution of (1.4) and have 2-balanced growth for $1 < q < 3$. Then either (i) ω is both closed and co-closed, or (ii) ω does not satisfy* Condition W (1.2),

The following Theorem is equivalent to Theorem 6.1 and yield information on a nontrivial solution of (1.4)

THEOREM 6.3. *Let ω be either a non-closed or non-co-closed solution of* (1.4) *and satisfy* Condition W (1.2), *Then ω has 2-imbalanced growth for $1 < q < 3$.*

7. The growth of solutions of $\langle \omega, \Delta\omega\rangle \geq 0$

Applying Theorem 1.2, we obtain immediately

THEOREM 7.1. *Let $\omega \in A^k$ be a non-closed or non-co-closed solution of* (1.4) *on M. Then ω has 2-imbalanced growth for $q = 2$, and if ω satisfies* Condition W (1.2), *then ω has 2-imbalanced growth for $1 < q(\neq 2) < 3$.*

THEOREM 7.2. *Let $\omega \in A^k$ be a non-closed or non-co-closed solution of* (1.4). *Then ω has 2-imbalanced growth for $q = 2$, and if ω has 2-balanced growth for $1 < q(\neq 2) < 3$, then ω does not satisfy* Condition W (1.2).

PROOF OF THEOREMS 7.1 AND 7.2. Suppose contrary, i.e., ω had 2-balanced growth for $q = 2$, or if ω satisfies Condition W (1.2), (resp. if ω has 2-balanced growth for $1 < q(\neq 2) < 3$,) then ω would have 2-balanced growth for $1 < q(\neq 2) < 3$ (resp. ω would satisfy Condition W (1.2)). By Theorem 1.2, this would imply $d\omega = 0$ and $d^*\omega = 0$, a contradiction. □

8. Monotonicity formulas for 2-balanced solutions of $\langle \omega, \Delta\omega\rangle \geq 0$

We recall in [**DW, W3**] the *exterior differential operator* $d^\nabla : A^k(\xi) \to A^{k+1}(\xi)$ relative to the connection ∇^E is given by

$$(d^\nabla\sigma)(X_1, ..., X_{k+1}) = \sum_{i=1}^{k+1}(-1)^{i+1}(\nabla_{X_i}\sigma)(X_1, ..., \widehat{X}_i, ..., X_{k+1}) \tag{8.1}$$

where the symbols covered by the circumflex $\widehat{}$ are omitted. Relative to the Riemannian structures of E and TM, the *codifferential operator* $\delta^\nabla : A^k(\xi) \to A^{k-1}(\xi)$ is characterized as

$$\delta^\nabla = (-1)^{nk+n+1} \star d^\nabla \star \tag{8.2}$$

Then δ^∇ is the adjoint of d^∇ via the formula

$$\int_M \langle d^\nabla\sigma, \rho\rangle \, dv_g = \int_M \langle \sigma, \delta^\nabla\rho\rangle \, dv_g \tag{8.3}$$

where $\sigma \in A^{k-1}(\xi), \rho \in A^p(\xi)$, one of which has compact support, and dv_g is the volume element associated with the metric g on TM.

And

$$(\delta^\nabla\omega)(X_1, ..., X_{k-1}) = -\sum_{i=1}^{k}(\nabla_{e_i}\omega)(e_i, X_1, ..., X_{k-1}) \tag{8.4}$$

where $\omega \in A^k(\xi)$ and $\{e_1, \cdots e_n\}$ is a local orthonormal frame field on (M, g).

DEFINITION 8.1. $\omega \in A^k(\xi)$ $(k \geq 1)$ is said to satisfy a *conservation law* if for any vector field X on M,

$$\langle i_X d^\nabla\omega, \omega\rangle + \langle \delta^\nabla\omega, i_X\omega\rangle = 0. \tag{8.5}$$

where i_X is the interior multiplication by X.

THEOREM 8.1. *Let M be an $n-$dimensional complete Riemannian manifold with a pole x_0. Let $\xi : E \to M$ be a Riemannian vector bundle on M and $\omega \in A^k(\xi)$. Assume that the radial curvature $K(r)$ of M satisfies one of the following five conditions:*

(8.6)

$(i) - \alpha^2 \leq K(r) \leq -\beta^2 \quad \text{with} \quad \alpha > 0, \beta > 0 \quad \text{and} \quad (n-1)\beta - 2k\alpha \geq 0;$

$(ii)\, K(r) = 0 \quad \text{with} \quad n - 2k > 0;$

$(iii) - \dfrac{A}{(1+r^2)^{1+\epsilon}} \leq K(r) \leq \dfrac{B}{(1+r^2)^{1+\epsilon}} \quad \text{with} \quad \epsilon > 0\,, A \geq 0\,, 0 < B < 2\epsilon \quad and$

$\quad n - (n-1)\dfrac{B}{2\epsilon} - 2ke^{\frac{A}{2\epsilon}} > 0;$

$(iv) - \dfrac{A(A-1)}{r^2} \leq K(r) \leq -\dfrac{A_1(A_1-1)}{r^2} \quad \text{with}$

$\quad 1 \leq A_1 \leq A \quad and \quad 1 + (n-1)A_1 - 2kA > 0;$

$(v) - \dfrac{A}{r^2} \leq K(r) \leq -\dfrac{A_1}{r^2} \quad \text{with}$

$\quad 0 \leq A_1 \leq A \quad and \quad 1 + (n-1)\dfrac{1+\sqrt{1+4A_1}}{2} - k(1+\sqrt{1+4A}) > 0.$

If ω satisfies a conservation law (8.5), *then the following monotonicity formula holds.*

$$\frac{1}{\rho_1^\lambda}\int_{B_{\rho_1}(x_0)} \frac{|\omega|^2}{2}\, dv \leq \frac{1}{\rho_2^\lambda}\int_{B_{\rho_2}(x_0)} \frac{|\omega|^2}{2}\, dv \tag{8.7}$$

for any $0 < \rho_1 \leq \rho_2$, where

$$\lambda = \begin{cases} n - 2k\frac{\alpha}{\beta} & \text{if } K(r) \text{ satisfies (i)} \\ n - 2k & \text{if } K(r) \text{ satisfies (ii)} \\ n - (n-1)\frac{B}{2\epsilon} - 2ke^{\frac{A}{2\epsilon}} & \text{if } K(r) \text{ satisfies (iii)} \\ 1 + (n-1)A_1 - 2kA & \text{if } K(r) \text{ satisfies } (iv) \\ 1 + (n-1)\frac{1+\sqrt{1+4A_1}}{2} - k(1+\sqrt{1+4A}) & \text{if } K(r) \text{ satisfies } (v). \end{cases} \tag{8.8}$$

PROOF. This follows from [**DW**, p.343, Theorem 4.1] and [**W3**, p.203, Theorem 6.4], in which $F(\frac{|\omega|^2}{2}) = \frac{|\omega|^2}{2}$ and the F-degree $d_F = \sup_{t \geq 0} \frac{tF'(t)}{F(t)} = 1$. □

THEOREM 8.2. *Suppose the radial curvature $K(r)$ of M satisfies* (8.6). *If $\omega \in A^k$ is a 2-balanced solution of $\langle \omega, \Delta\omega \rangle \geq 0$ or a harmonic k-form, for $q = 2$, or for $1 < q (\neq 2) < 3$ with ω satisfying* Condition W (1.2) (cf. (1.3)), *then monotonicity formulas* (8.7) *holds for any $0 < \rho_1 \leq \rho_2$, where λ is as in* (8.8).

PROOF. As A^k is isometric to $A^k(\xi)$, when $\xi : E = M \times \mathbb{R} \to M$ is the trivial bundle equipped with the canonical metric, via Theorem 1.2, ω is both closed ($d^\nabla \omega = d\omega = 0$) and co-closed ($\delta^\nabla \omega = d^*\omega = 0$) on M. Thus ω satisfies a conservation law (8.5), The assertion follows from Theorem 8.1. □

9. Vanishing theorems of 2-moderate solutions (for $q = 2$) of $\langle \omega, \Delta\omega \rangle \geq 0$

LEMMA 9.1. *Let $\psi(r) > 0$ be a continuous function such that*

$$\int_{\rho_0}^{\infty} \frac{dr}{r\psi(r)} = +\infty \tag{9.1}$$

for some $\rho_0 > 0$. *Then*

(i) $\lim_{r\to\infty} \frac{\psi(r)}{r^\lambda} \neq \infty$, *for any* $\lambda > 0$.

(ii) *If* $\lim_{r\to\infty} \frac{\psi(r)}{r^\lambda}$ *exists for some* $\lambda > 0$, *then*

$$\lim_{r\to\infty} \frac{\psi(r)}{r^\lambda} = 0, \tag{9.2}$$

In particular, if $\psi(r) > 0$ *is a continuous, monotone decreasing function, then* (9.2) *holds.*

PROOF. Suppose on the contrary, i.e. $\lim_{r\to\infty} \frac{\psi(r)}{r^\lambda} = c < \infty$, where $c \neq 0$ (resp. $\lim_{r\to\infty} \frac{\psi(r)}{r^\lambda} = \infty$). Then there would exist $\rho_1 > 0$ such that if $r \geq \rho_1$, $\psi(r) > \frac{c}{2} r^\lambda$ (resp. $\psi(r) > \kappa r^\lambda$, where $\kappa > 0$ is a constant.) This would lead to

$$\int_{\rho_1}^\infty \frac{dr}{r\psi(r)} \leq \frac{2}{c} \int_{\rho_1}^\infty \frac{dr}{r^{1+\lambda}} \left(\text{resp.} \quad \kappa \int_{\rho_1}^\infty \frac{dr}{r^{1+\lambda}} \right) < \infty,$$

contradicting (9.1), by the continuity of $\psi(r)$ if $\rho_0 < \rho_1$. This proves (i) and (ii). Now we prove the last statement. Since $\psi(r) > 0$ is a continuous, monotone decreasing function, $\frac{\psi(\rho)}{\rho^\lambda}$, $\lambda > 0$ is monotone decreasing. Hence, the monotonicity implies that $\lim_{\rho\to\infty} \frac{\psi(\rho)}{\rho^\lambda}$ exists for $\lambda > 0$. By (ii), (9.2) follows. □

THEOREM 9.1. *Let* ω *have* 2*-moderate growth* (2.4) *for* $q = 2$ *with*

$$\lim_{\rho\to\infty} \frac{\psi(\rho)}{\rho^{\lambda-2}} < \infty \tag{9.3}$$

for some $\lambda > 2$ *and for some* ψ *as in* (2.5). *Then*

$$\int_{B_\rho(x_0)} \frac{|\omega|^2}{2}\, dv = o(\rho^\lambda) \quad \text{as} \quad \rho \to \infty \tag{9.4}$$

In particular, if ω *has* 2*-moderate growth* (2.4) *for* $q = 2$ *with some continuous, monotone decreasing function* $\psi(\rho) > 0$ *as in* (2.5), *then* (9.4) *holds*

PROOF. In view of Lemma 9.1(ii), the assumption (9.3) implies that $\limsup_{\rho\to\infty} \frac{\psi(\rho)}{\rho^{\lambda-2}} = \lim_{\rho\to\infty} \frac{\psi(\rho)}{\rho^{\lambda-2}} = 0$, $\lambda > 2$. It follows that

$$\begin{aligned} \limsup_{\rho\to\infty} \frac{1}{\rho^\lambda} \int_{B_\rho(x_0)} \frac{|\omega|^2}{2}\, dv &\leq \limsup_{\rho\to\infty} \frac{\psi(\rho)}{\rho^{\lambda-2}} \left(\frac{1}{\rho^2\psi(\rho)} \int_{B(x_0;\rho)} \frac{|\omega|^2}{2}\, dv \right) \\ &\leq \left(\limsup_{\rho\to\infty} \frac{\psi(\rho)}{\rho^{\lambda-2}} \right) \left(\limsup_{\rho\to\infty} \frac{1}{\rho^2\psi(\rho)} \int_{B(x_0;\rho)} \frac{|\omega|^2}{2}\, dv \right) \\ &= 0. \end{aligned} \tag{9.5}$$

On the other hand, we have via (9.5),

$$0 \leq \liminf_{\rho\to\infty} \frac{1}{\rho^\lambda} \int_{B_\rho(x_0)} \frac{|\omega|^2}{2}\, dv \leq \limsup_{\rho\to\infty} \frac{1}{\rho^\lambda} \int_{B_\rho(x_0)} \frac{|\omega|^2}{2}\, dv \leq 0 \tag{9.6}$$

This gives the desired (9.4). □

THEOREM 9.2. *Suppose the radial curvature $K(r)$ of M satisfies* (8.6). *If $\omega \in A^k(\xi)$ satisfies a conservation law* (8.5) *and*

$$\int_{B_\rho(x_0)} \frac{|\omega|^2}{2}\, dv = o(\rho^\lambda) \quad \text{as } \rho \to \infty \tag{9.4}$$

where λ is given by (8.8), *then $\omega \equiv 0$ on M. In particular, if ω is L^2 on M and satisfies a conservation law* (8.5), *then $\omega \equiv 0$.*

PROOF. From Theorem 8.1, the monotonicity formula (8.7) holds for any $0 < \rho_1 \le \rho_2$, and λ as in (8.8). Combining (8.7) and (9.4) completes the proof. □

THEOREM 9.3. *Let manifold M and the radial curvature $K(r)$ of M be as in* (8.6). *Let $\omega \in A^k$ be a solution of* (1.4) $\langle \omega, \Delta\omega\rangle \ge 0$, *or harmonic k-form, where*

$$k < \begin{cases} \frac{n-2}{2}\frac{\beta}{\alpha} & \text{if } K(r) \text{ satisfies (i)} \\ \frac{n-2}{2} & \text{if } K(r) \text{ satisfies (ii)} \\ (\frac{n-2}{2} - (n-1)\frac{B}{4\epsilon})e^{-\frac{A}{2\epsilon}} & \text{if } K(r) \text{ satisfies (iii)} \\ \frac{n-1}{2}\frac{A_1}{A} - \frac{1}{2A} & \text{if } K_r \text{ satisfies } (iv) \\ \frac{n-1}{2}\frac{1+\sqrt{1+4A_1}}{1+\sqrt{1+4A}} - \frac{1}{1+\sqrt{1+4A}} & \text{if } K_r \text{ satisfies } (v) \end{cases} \tag{9.7}$$

Suppose ω has 2-moderate growth (2.4) *for $q = 2$ with ψ satisfying* (9.3) *for some ψ in* (2.5) *and λ as in* (8.8). *Then* (9.4) *holds and $\omega \equiv 0$ on M. In particular, every L^2 solution of* (1.4) $\langle \omega, \Delta\omega\rangle \ge 0$ *vanishes.*

PROOF. *First Proof:* From Theorem 1.2, or Theorem 3.4, ω is closed and co-closed. Hence, ω satisfies a conservation law (8.5). If k satisfies (9.7), then λ in (8.8) satisfies $\lambda > 2$. By Theorem 9.1, (9.4) holds, where λ is given by (8.8). Applying Theorem 9.2, we conclude $\omega \equiv 0$ on M.
Second Proof: By Definition 2.3, ω has 2-balanced growth for $q = 2$. Applying Theorem 8.2, we have monotonicity formula (8.7) for any $0 < \rho_1 \le \rho_2$, where λ is as in (8.8). If k satisfies (9.7), then λ in (8.8) satisfies $\lambda > 2$. By Theorem 9.1, (9.4) holds. Combining (9.4) and (8.7), we conclude $\omega \equiv 0$ on M. □

COROLLARY 9.1. *Let $M, K(r), \lambda, k$ as in Theorem 9.3. Let $\omega \in A^k$ be a solution of* (1.4) $\langle \omega, \Delta\omega\rangle \ge 0$, *or harmonic k-form on M. Suppose ω has 2-moderate growth* (2.4) *for $q = 2$ with a continuous, monotone decreasing function $\psi(r) > 0$ as in* (2.5). *Then $\omega \equiv 0$.*

PROOF. If $\psi(r) > 0$ is a continuous, monotone decreasing function, then $\frac{\psi(\rho)}{\rho^{\lambda-2}}$ is monotone decreasing. Hence, ψ satisfies (9.3) for λ as in (8.8). In view of Theorem 9.3, $\omega \equiv 0$. □

10. Constant Dirichelet problems for generalized harmonic 1-forms and harmonic maps

Let $F : [0, \infty) \to [0, \infty)$ be a strictly increasing C^2 function with $F(0) = 0$.

DEFINITION 10.1. $\omega \in A^k(\xi)$ $(k \ge 1)$ is said to satisfy an $F-$*conservation law* if for any vector field X on M,

$$\begin{aligned} &F'(\frac{|\omega|^2}{2})\langle i_X d^\nabla \omega + d^\nabla i_X\omega, \omega\rangle - \langle i_{\text{grad}(F'(\frac{|\omega|^2}{2}))}\omega, i_X\omega\rangle \\ &\quad + F'(\frac{|\omega|^2}{2})\langle \delta^\nabla \omega, i_X\omega\rangle - F'(\frac{|\omega|^2}{2})\langle d^\nabla i_X\omega, \omega\rangle = 0\,. \end{aligned} \tag{10.1}$$

We recall F-*lower degree* l_F is defined to be

$$l_F = \inf_{t\geq 0} \frac{tF'(t)}{F(t)} \tag{10.2}$$

A bounded domain $D \subset M$ with C^1 boundary is called *starlike* (relative to x_0) if there exists an inner point $x_0 \in D$ such that

$$\langle \frac{\partial}{\partial r_{x_0}}, \nu\rangle|_{\partial D} \geq 0 \tag{10.3}$$

where ν is the unit outer normal to ∂D, and for any $x \in D\backslash\{x_0\} \cup \partial D$, $\frac{\partial}{\partial r_{x_0}}(x)$ is the unit vector field tangent to the unique geodesic emanating from x_0 to x.

It is obvious that any disc or convex domain is starlike.

THEOREM 10.1. *Let D be a bounded starlike domain (relative to x_0) with C^1 boundary in a complete Riemannian n-manifold M. Assume that the radial curvature $K(r)$ of M satisfies one of the following five conditions:*

(10.4)

$(i) - \alpha^2 \leq K(r) \leq -\beta^2$ with $\alpha > 0, \beta > 0$ and $(n-1)\beta - 2d_F\alpha \geq 0$;

$(ii)\, K(r) = 0$ with $n - 2d_F > 0$;

$(iii) -\frac{A}{(1+r^2)^{1+\epsilon}} \leq K(r) \leq \frac{B}{(1+r^2)^{1+\epsilon}}$ with $\epsilon > 0, A \geq 0, 0 < B < 2\epsilon$ and

$$n - (n-1)\frac{B}{2\epsilon} - 2e^{\frac{A}{2\epsilon}} d_F > 0;$$

$(iv) -\frac{A(A-1)}{r^2} \leq K(r) \leq -\frac{A_1(A_1-1)}{r^2}$ with

$$1 \leq A_1 \leq A \quad \text{and} \quad 1 + (n-1)A_1 - 2d_F A > 0;$$

$(v) -\frac{A}{r^2} \leq K(r) \leq -\frac{A_1}{r^2}$ with

$$0 \leq A_1 \leq A \quad \text{and} \quad 1 + (n-1)\frac{1+\sqrt{1+4A_1}}{2} - d_F(1+\sqrt{1+4A}) > 0.$$

Assume that $l_F \geq \frac{1}{2}$. If $\omega \in A^1(\xi)$ satisfies F-conservation law and annihilates any tangent vector of ∂D, then ω vanishes on D.

PROOF. The cases $(i), (ii), (iii)$ follow from the proof of Theorem 10.1 in [**W3**], p.207-208. The curvature assumptions are slightly more general than those in Theorem 6.1 in [**DW**] in which p-form are discussed. The cases $(iv), (v)$ are precisely Theorem 10.1 in [**W3**], p.207. □

THEOREM 10.2. [**W3**, Theorem 10.2] *Let M, D, and ξ be as in Theorem 10.1. Assume that the radial curvature $K(r)$ of M satisfies one of the following conditions in (10.4). Let $u : \overline{D} \to N$ be an F-harmonic map with $l_F \geq \frac{1}{2}$ into an arbitrary Riemannian manifold N. If $u|_{\partial D}$ is constant, then $u|_D$ is constant.*

PROOF. Take $\omega = du$. Then $\omega|_{\partial D} = 0$. Hence ω satisfies an F-conservation law and annihilates any tangent vector η of ∂D. The assertion follows at once from Theorem 10.1. □

COROLLARY 10.1. *Suppose M and D satisfy the same assumptions of Theorem 10.2. Let $u : \overline{D} \to N$ be a harmonic map (resp. p-harmonic map, $p \geq 1$) into an arbitrary Riemannian manifold N. If $u|_{\partial D}$ is constant, then $u|_D$ is constant.*

PROOF. For a harmonic map (resp. p-harmonic map, $p \geq 1$), we have $F(t) = t$ (resp. $F(t) = \frac{1}{p}(2t)^{\frac{p}{2}}$). Obviously $d_F = l_F = 1$(resp. $d_F = l_F = \frac{p}{2}) \geq \frac{1}{2}$. Take $\omega = du$. This corollary follows immediately from Theorem 10.1 or Theorem 10.2. □

THEOREM 10.3. *Suppose the radial curvature $K(r)$ of M satisfies one of the five conditions in* (10.4) *in which $d_F = 1$. If $\omega \in A^1$ on M is a 2-balanced solution of $\langle \omega, \Delta\omega \rangle \geq 0$ or a harmonic 1-form, for $q = 2$, or for $1 < q(\neq 2) < 3$ with ω satisfying* Condition W (1.2) (cf. (1.3)), *and annihilating any tangent vector of ∂D, then ω vanishes on D.*

PROOF. By Theorem 1.2, ω is both closed and co-closed on M. Thus ω satisfies a conservation law (8.5), The assertion follows from Theorem 10.1 in which $F(t) = t$. □

References

[AR] D. Alexandru Rugina, *L_p-integrabilité des formes harmoniques K-finies sur les espaces hyperboliques réels et complexes* ([[L_p-integrability of K-finite harmonic forms on real and complex hyperbolic spaces]]), Rend. Sem. Mat. Univ. Politec. Torino **54** (1996), no. 1, 75–87. MR1490015

[AV] A. Andreotti and E. Vesentini, *Carleman estimates for the Laplace-Beltrami equation on complex manifolds*, Inst. Hautes Études Sci. Publ. Math. **25** (1965), 81–130. MR175148

[CW1] B.-Y. Chen and S. W. Wei, *Growth estimates for warping functions and their geometric applications*, Glasg. Math. J. **51** (2009), no. 3, 579–592, DOI 10.1017/S0017089509990012. MR2534009

[CW2] B.-Y. Chen and S. W. Wei, *Sharp growth estimates for warping functions in multiply warped product manifolds*, J. Geom. Symmetry Phys. **52** (2019), 27–46. MR3931269

[DW] Y. Dong and S. W. Wei, *On vanishing theorems for vector bundle valued p-forms and their applications*, Comm. Math. Phys. **304** (2011), no. 2, 329–368, DOI 10.1007/s00220-011-1227-8. MR2795324

[DLW] Y. Dong, H. Lin, and S. W. Wei, *L^2 curvature pinching theorems and vanishing theorems on complete Riemannian manifolds*, Tohoku Math. J. (2) **71** (2019), no. 4, 581–607, DOI 10.2748/tmj/1576724795. MR4043927

[GW] R. E. Greene and H. Wu, *Harmonic forms on noncompact Riemannian and Kähler manifolds*, Michigan Math. J. **28** (1981), no. 1, 63–81. MR600415

[HLRW] Y. Han, Y. Li, Y. Ren, and S. W. Wei, *New comparison theorems in Riemannian geometry*, Bull. Inst. Math. Acad. Sin. (N.S.) **9** (2014), no. 2, 163–186. MR3237065

[KW] H. Karcher and J. C. Wood, *Nonexistence results and growth properties for harmonic maps and forms*, J. Reine Angew. Math. **353** (1984), 165–180. MR765831

[K] L. Karp, *Subharmonic functions on real and complex manifolds*, Math. Z. **179** (1982), no. 4, 535–554, DOI 10.1007/BF01215065. MR652859

[PRS] S. Pigola, M. Rigoli, and A. G. Setti, *Vanishing and finiteness results in geometric analysis: A generalization of the Bochner technique*, Progress in Mathematics, vol. 266, Birkhäuser Verlag, Basel, 2008. MR2401291

[W1] S. W. Wei, *p-harmonic geometry and related topics*, Bull. Transilv. Univ. Braşov Ser. III **1(50)** (2008), 415–453. MR2478044

[W2] S. W. Wei, *On 1-harmonic functions*, SIGMA Symmetry Integrability Geom. Methods Appl. **3** (2007), Paper 127, 10, DOI 10.3842/SIGMA.2007.127. MR2366895

[W3] S. W. Wei, *Comparison theorems in Riemannian geometry with applications*, Recent advances in the geometry of submanifolds—dedicated to the memory of Franki Dillen (1963–2013), Contemp. Math., vol. 674, Amer. Math. Soc., Providence, RI, 2016, pp. 185–209. MR3565255

[W4] S. W. Wei, *Dualities in comparison theorems and bundle-valued generalized harmonic forms on noncompact manifolds*, preprint.

[WLW] S. W. Wei, J.-F. Li, and L. Wu, *Generalizations of the uniformization theorem and Bochner's method in p-harmonic geometry*, Commun. Math. Anal. **Conference 1** (2008), 46–68. MR2452403

[WW] S.W. Wei and B.Y. Wu *Generalized Hardy type and Caffarelli-Kohn-Nirenberg type inequalities on Finsler manifolds*, to appear in International Journal of Mathematics

[Wu] H.-H. Wu, *The Bochner technique in differential geometry*, CTM. Classical Topics in Mathematics, vol. 6, Higher Education Press, Beijing, 2017. New edition of the 1988 original [MR1079031]; Expanded version of [MR0714349]. MR3838345

[Y1] S. T. Yau, *Some function-theoretic properties of complete Riemannian manifold and their applications to geometry*, Indiana Univ. Math. J. **25** (1976), no. 7, 659–670, DOI 10.1512/iumj.1976.25.25051. MR417452

[Y2] S. T. Yau, *Some function-theoretic properties of complete Riemannian manifold and their applications to geometry*, Indiana Univ. Math. J. **25** (1976), no. 7, 659–670, DOI 10.1512/iumj.1976.25.25051. MR417452

Department of Mathematics, University of Oklahoma, Norman, Oklahoma 73019-0315

Email address: wwei@ou.edu

Selected Published Titles in This Series

756 **Joeri Van der Veken, Alfonso Carriazo, Ivko Dimitrić, Yun Myung Oh, Bogdan D. Suceavă, and Luc Vrancken, Editors,** Geometry of Submanifolds, 2020

755 **Marion Scheepers and Ondřej Zindulka, Editors,** Centenary of the Borel Conjecture, 2020

754 **Susanne C. Brenner, Igor Shparlinski, Chi-Wang Shu, and Daniel B. Szyld, Editors,** 75 Years of Mathematics of Computation, 2020

753 **Matthew Krauel, Michael Tuite, and Gaywalee Yamskulna, Editors,** Vertex Operator Algebras, Number Theory and Related Topics, 2020

752 **Samuel Coskey and Grigor Sargsyan, Editors,** Trends in Set Theory, 2020

751 **Ashish K. Srivastava, André Leroy, Ivo Herzog, and Pedro A. Guil Asensio, Editors,** Categorical, Homological and Combinatorial Methods in Algebra, 2020

750 **A. Bourhim, J. Mashreghi, L. Oubbi, and Z. Abdelali, Editors,** Linear and Multilinear Algebra and Function Spaces, 2020

749 **Guillermo Cortiñas and Charles A. Weibel, Editors,** K-theory in Algebra, Analysis and Topology, 2020

748 **Donatella Danielli and Irina Mitrea, Editors,** Advances in Harmonic Analysis and Partial Differential Equations, 2020

747 **Paul Bruillard, Carlos Ortiz Marrero, and Julia Plavnik, Editors,** Topological Phases of Matter and Quantum Computation, 2020

746 **Erica Flapan and Helen Wong, Editors,** Topology and Geometry of Biopolymers, 2020

745 **Federico Binda, Marc Levine, Manh Toan Nguyen, and Oliver Röndigs, Editors,** Motivic Homotopy Theory and Refined Enumerative Geometry, 2020

744 **Pieter Moree, Anke Pohl, L'ubomír Snoha, and Tom Ward, Editors,** Dynamics: Topology and Numbers, 2020

743 **H. Garth Dales, Dmitry Khavinson, and Javad Mashreghi, Editors,** Complex Analysis and Spectral Theory, 2020

742 **Francisco-Jesús Castro-Jiménez, David Bradley Massey, Bernard Teissier, and Meral Tosun, Editors,** A Panorama of Singularities, 2020

741 **Houssam Abdul-Rahman, Robert Sims, and Amanda Young, Editors,** Analytic Trends in Mathematical Physics, 2020

740 **Alina Bucur and David Zureick-Brown, Editors,** Analytic Methods in Arithmetic Geometry, 2019

739 **Yaiza Canzani, Linan Chen, and Dmitry Jakobson, Editors,** Probabilistic Methods in Geometry, Topology and Spectral Theory, 2019

738 **Shrikrishna G. Dani, Surender K. Jain, Jugal K. Verma, and Meenakshi P. Wasadikar, Editors,** Contributions in Algebra and Algebraic Geometry, 2019

737 **Fernanda Botelho, Editor,** Recent Trends in Operator Theory and Applications, 2019

736 **Jane Hawkins, Rachel L. Rossetti, and Jim Wiseman, Editors,** Dynamical Systems and Random Processes, 2019

735 **Yanir A. Rubinstein and Bernard Shiffman, Editors,** Advances in Complex Geometry, 2019

734 **Peter Kuchment and Evgeny Semenov, Editors,** Differential Equations, Mathematical Physics, and Applications, 2019

733 **Peter Kuchment and Evgeny Semenov, Editors,** Functional Analysis and Geometry, 2019

732 **Samuele Anni, Jay Jorgenson, Lejla Smajlović, and Lynne Walling, Editors,** Automorphic Forms and Related Topics, 2019

For a complete list of titles in this series, visit the
AMS Bookstore at **www.ams.org/bookstore/conmseries/**.